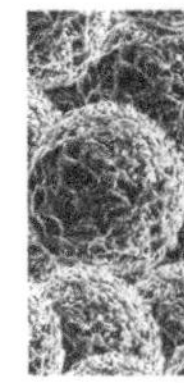

# RESPONSIBLE NANOBIOTECHNOLOGY

ARMIN GRUNWALD

# RESPONSIBLE NANOBIOTECHNOLOGY

## Philosophy and Ethics

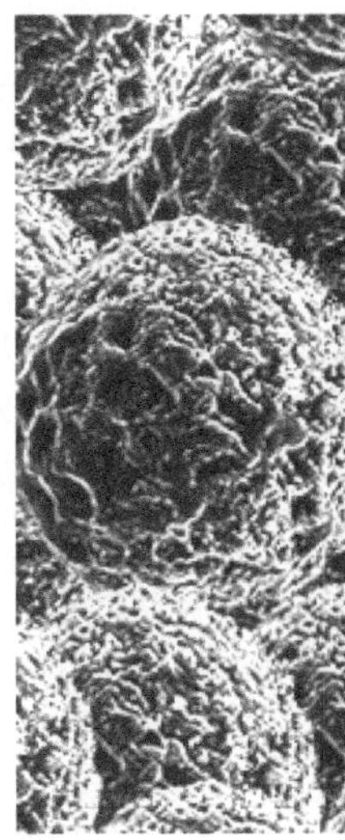

PAN STANFORD PUBLISHING

*Published by*

Pan Stanford Publishing Pte. Ltd.
Penthouse Level, Suntec Tower 3
8 Temasek Boulevard
Singapore 038988

Email: editorial@panstanford.com
Web: www.panstanford.com

**British Library Cataloguing-in-Publication Data**
A catalogue record for this book is available from the British Library.

**Responsible Nanobiotechnology: Philosophy and Ethics**



ISBN 978-981-4316-80-4 (Hardcover)
ISBN 978-981-4363-33-1 (eBook)

Printed in the USA

# Contents


*Preface* xiii

**1 Agenda and Overview** **1**

1.1 The Motivation 1

1.2 Objectives, Conceptual Framework, and Premises 5

1.3 Quick Guide Through the Book 8

1.3.1 Chapter 2: Nanotechnology in Context 8

1.3.2 Chapter 3: Ethics, Technology, and Risk 9

1.3.3 Chapter 4: Ethics of Nano(bio)technology: The Program 9

1.3.4 Chapter 5: Ethics of Nano(bio)technology: An Overview 10

1.3.5 Chapter 6: Synthetic Nanoparticles 10

1.3.6 Chapter 7: Toward Creating Artificial Life 11

1.3.7 Chapter 8: Animal Enhancement 11

1.3.8 Chapter 9: Human Enhancement 11

1.3.9 Chapter 10: From Applied Ethics to an Explorative Philosophy of Nanotechnology 12

1.3.10 Chapter 11: Conclusions and Perspectives 13

**2 Nanotechnology in Context** **15**

2.1 History of Nanotechnology 15

2.2 The World of Nanotechnology in a Nutshell 18

2.2.1 Nanometer-Scale Analysis and Manipulation 19

2.2.2 Characteristics of Nanomaterials 21

2.2.3 Areas of Activity and Applications 23

2.2.3.1 Synthetic Nanomaterials 23

2.2.3.2 Nanoelectronics 25

2.2.3.3 Nanobiotechnology 26

2.2.3.4 Nanomedicine 26

2.3 Defining Nanotechnology 27
2.4 The Interdisciplinary Nature of the Nanocommunity 33
2.5 Philosophical Interpretations 35
2.5.1 Triumph of Homo Faber 36
2.5.2 Huge Increase on Uncertainty 37
2.5.3 Nanotechnology as a Symbol of the Future 39
2.6 Public Perception 41
2.6.1 The "Grey Goo" Scenario 42
2.6.2 The "Prey" Scenario 42
2.6.3 The "Cyborg" Scenario 43

**3 Ethics, Technology, and Risk 49**
3.1 Problem-Oriented Ethics 49
3.1.1 Ethics for Resolving Moral Conflicts 50
3.1.2 Standard Situations in a Moral Respect 55
3.1.2.1 Pragmatic Completeness 56
3.1.2.2 Local Consistency 56
3.1.2.3 Sufficient Lack of Ambiguity 56
3.1.2.4 Acceptance 57
3.1.2.5 Compliance 57
3.1.3 Beyond Standard Situations in a Moral Respect 60
3.1.4 Ethical Expertise as Conditionally Normative Advice 63
3.2 Ethics of Technology 67
3.2.1 Normative Uncertainties Emerging from Technological Progress 67
3.2.2 Cross-Cutting Issues 70
3.2.2.1 Human Autonomy vs. Technicalization 71
3.2.2.2 Distributive Justice 71
3.2.2.3 Technology and the Environment 72
3.2.2.4 Technology and Life 73
3.2.2.5 Uncertainty of Our Knowledge of the Consequences 73
3.2.3 Ethics of Technology as Part of Technology Governance 74
3.2.3.1 Political Decisions 75

3.2.3.2 Entrepreneurial Decisions 76
3.2.3.3 Engineering 76
3.2.3.4 Consumer Behavior 77
3.2.3.5 Public Debate 77
3.2.4 Technology, Science, and Responsibility 78
3.3 Ethics and (Unclear) Risk 81
3.3.1 Classical Risk Management and Its Limitations 81
3.3.2 Ethical Issues in Dealing with Unclear Risk 84
3.3.2.1 Acceptability of Unclear Risk 85
3.3.2.2 Weighing Benefits against Unclear Risks 85
3.3.2.3 Normalizing the Situation under Consideration 86
3.3.2.4 Comparisons of Man-Made Situations of Unclear Risk with Natural Situations 87
3.3.2.5 Learning from Historic Cases 87

**4 Ethics of Nano(bio)technology: The Program 89**
4.1 Motivations of Nanoethics 89
4.1.1 Avoiding to Endanger Innovation 90
4.1.2 Taking Care of Unintended Side Effects as Early as Possible 92
4.1.3 Reacting to Apocalyptic Fears 93
4.2 Nanoethics as a New Field of Applied Ethics? 95
4.3 Problem-Oriented Ethics of Nanotechnology 102

**5 Ethics of Nano(bio)technology: An Overview 107**
5.1 Literature Overview 108
5.1.1 Interdisciplinary Expert Studies 108
5.1.2 Position Papers from Nongovernmental Organizations 111
5.1.3 Selected Edited Books 114
5.1.4 The Journal Nanoethics 118
5.2 Ethical Questions Related to Nano(bio)technology Applications 119
5.2.1 Nanomedicine: Risks and Benefits 120
5.2.2 Nanoelectronics: Surveillance and Privacy Issues 124

5.2.3 Using Processes of Life for Technological Purposes 126
5.2.4 Human Enhancement 128
5.2.5 Animal Enhancement 129
5.2.6 Military Applications 132
5.3 Cross-Cutting Ethical Issues 134
5.3.1 EHS: Environment, Health, and Safety 134
5.3.2 Distributive Justice: Nanotechnology and Developing Countries 137
5.3.3 Responsibility for Future Generations 140
5.4 Selection of Issues for In-Depth Studies 143

**6 Synthetic Nanoparticles 147**
6.1 Synthetic Nanoparticles: Fields of Application and Expectations 148
6.1.1 Surface Treatment 149
6.1.2 Food 150
6.1.3 Cosmetics 152
6.2 Possible Risks and Types of Risk 152
6.2.1 Health Risks 154
6.2.2 Environmental Risks 156
6.2.3 Nanoparticle Risks as "Unclear Risks" 157
6.3 Approaches to Dealing with Unclear Risk 159
6.3.1 Philosophical Approaches 159
6.3.1.1 The Consequentialist Approach 159
6.3.1.2 The Imperative of Responsibility 160
6.3.1.3 The Principle of Pragmatic Consistency 162
6.3.1.4 Deontological Advice 163
6.3.1.5 Projected Time 164
6.3.2 Operational Approaches 165
6.3.2.1 The Precautionary Principle 165
6.3.2.2 The Prudent Avoidance Approach 168
6.3.3 Interim Conclusions 170
6.4 Dealing Responsibly with Nanomaterials 171
6.4.1 Conditionally Normative Reflection 171
6.4.2 Informed Consent and Consumer Freedom 174

6.4.3 Regulation, Code of Conduct, and the Common Good 175
6.4.4 Operative Approach: Remarks on the Next Steps 181
6.4.5 Epilogue and Reflection: Risk Ethics and Nanoparticles 187

**7 Toward Creating Artificial Life 191**
7.1 Nanobiotechnology and Synthetic Biology 191
7.1.1 Nanobiotechnology 192
7.1.2 Synthetic Biology: Engineering Life 193
7.2 Chances and Risks 197
7.2.1 Chances 197
7.2.2 Risks 199
7.3 Ethical Issues 203
7.3.1 Dealing with Risks Responsibly 204
7.3.2 The Moral Status of Created Organisms 207
7.3.3 Quasi-ethical Concerns: Humans "Playing God"? 209
7.4 Hermeneutic Dimensions 213
7.4.1 Technicalization of the Natural or a More Natural Technology 213
7.4.2 The Relationship Between Technology and Life 217
7.5 Responsible Governance of Synthetic Biology 219

**8 Animal Enhancement 227**
8.1 (Nano)Technology for Intervening in Animals 228
8.2 The Semantics of Animal Enhancement 232
8.2.1 The Semantics of Enhancement 232
8.2.2 Animal Enhancement 234
8.3 Relevant Ethical Challenges and Normative Frameworks 237
8.3.1 Animal Experiments 238
8.3.2 Elimination of Animals' Capacity for Suffering 240
8.3.3 Transgressing the Boundary Between Humans and Animals 243
8.4 Changing Human–Animal Relationship 244
8.5 Summary and Conclusions 247

**9 Human Enhancement** **251**
9.1 Improving Human Performance of Converging Technologies 251
9.1.1 The Vision of Converging Technologies 252
9.1.2 Improving Human Performance: The Cultural Background 255
9.1.3 Enhancement Utopia 1: Neuroenhancement 258
9.1.4 Enhancement Utopia 2: Antiaging and Immortality 261
9.2 Semantics of Technical Enhancement 263
9.2.1 Enhancement Beyond Healing 263
9.2.2 Healing, Doping, Enhancement, and Alteration 265
9.2.3 Technical Enhancement 269
9.3 Human Enhancement: Ethical Analysis 272
9.3.1 Normative Uncertainties 272
9.3.2 Patterns of Ethical Argumentation 275
9.3.2.1 Ethical Consideration of the Consequences 275
9.3.2.2 The Naturalness of Man 278
9.3.2.3 The Question as to Ought 280
9.3.3 Assessment of the Current Status of the Ethical Debate 281
9.4 Changing Relations Between Humans and Technology 284
9.4.1 Neuroelectric Interfaces 284
9.4.2 Technicalization of Man by Nanotechnology? 290
9.5 Conclusions for Responsible Action 293
9.5.1 Need for Orientation on Human Enhancement 293
9.5.2 Responsible Action 297
9.5.3 Approaching an "Enhancement Society?" 300

**10 Explorative Nanophilosophy: More Than Applied Ethics** **303**
10.1 The Debate on "Speculative Nanoethics" 304
10.1.1 The Main Diagnosis: "Most Nanoethics Is Too Futuristic" 305
10.1.2 How Speculative Is "Speculative Nanoethics"? 306
10.1.3 The Anxiety that Unjustified and Artificial Concerns Might Emerge 308

10.1.4 The Opportunity–Costs Argument 310
10.1.5 Resume 311
10.2 Searching for Orientation by Investigating Futures 312
10.3 Futures as Social Constructs 314
10.4 Explorative Philosophy of Nanotechnology 317
10.4.1 Explorative Philosophy Beyond Applied Nanoethics 318
10.4.2 Elements of an Explorative Philosophy of Nanotechnology 321
10.4.2.1 Nano Epistemology 321
10.4.2.2 Nano Anthropology: The Relationship Between Humans and Technology 322
10.4.2.3 Nanotechnology Hermeneutics: Philosophical Interpretations of Nanotechnology 323
10.4.3 Epistemological Grounding 323

**11 Conclusions and Perspectives 327**
11.1 Ten Years of Nanoethics: What Has Been Achieved? 327
11.2 Moral Arguments Feeding a Broad Antinano Movement? 331
11.3 The Future of Nanoethics 335
11.3.1 Nanoethics as Concomitant Reflection on Nanotechnologies 335
11.3.2 Nanoethics as Interdisciplinary Research 337
11.3.3 Disentanglement of Nanoethics 339

*Bibiliography* 343
*Index* 369

# Preface

Nanotechnology is one of the most prominent emerging technologies. It has been heralded as a key technology for the twenty-first century that — according to the expectations of a broad alliance of policymakers, scientists, and industry representatives — will contribute to economic prosperity and sustainable development. Via its enabling role in nanobiotechnology and in concert with "converging technologies," nanotechnology could also influence the future of human nature and play a role in creating artificial life. The development of nanotechnology is thus also related to new debates about the *human condition* and the future of society as well as to man's relationship with nature and technology.

In view of the revolutionary potential frequently attributed to the nanosciences and nanotechnology with respect to nearly all fields of society and individual life, it is not surprising that research and reflection on their presumed societal consequences started early. Technology assessment (TA) and studies of the ethical, legal, and social implications (ELSI) began analyzing issues related to nanotechnology and society about 10 years ago. The analysis, deliberation, and assessment of the expected impact of nanotechnology on future society are now regarded as necessary contributions to the present and further development of nanotechnology and its enculturation into society. Ethical reflection on nanotechnology, in particular on its relationship with living systems in nanobiotechnology, has emerged quickly and led to the new term "nanoethics," which has been elaborated on and debated at workshops, conferences, and summer schools as well as in books and scientific journals.

In the present book, I review the considerations of nanotechnology elaborated in philosophy, ethics, and the social sciences and systematize and develop them further. The focus is on issues of ethical responsibility regarding chances and risks of nanotechnology and its possible applications. From this analysis of the normative challenges posed by nanotechnology, my goal is to derive orientation for further, responsible work in research and development. I thus put the book in the context of the keywords "responsible innovation" and "reflective sciences" which have been central concepts in the debates about the relationship between science and society for the last few years.

To a not unsubstantial extent, the analyses presented in this book are based on my own previous studies. Nearly 10 years ago I began to concern myself with the societal and ethical aspects of nanotechnology. This research has led to a number of publications that I have been able to use as a starting point for the present work. I would like, above all, to mention the book *Auf dem Weg in eine nanotechnologische Zukunft. Philosophisch-ethische Fragen* (The Path to a Nanotechnological Future: Philosophical and Ethical Issues), published in German by Verlag Karl Alber in 2008. The major developments versus that book are the introduction to the historical development of the field, the mapping of the ethical issues posed by nanotechnology, the addition of the field of animal enhancement, the programmatic focus on the explorative role of ethics and philosophy, as it has been formed in the debates on speculative nanoethics, and the concentration on aspects of responsibility.

My thanks go to my many colleagues in Germany, Europe, and around the world with whom I have had the opportunity in the last few years to discuss the philosophical and ethical aspects of nanotechnology. This has taken place both within these disciplines as well as within the framework of interdisciplinary dialogue, for example in the fields of technology assessment and of STS studies (i.e., on science, technology, and society). To represent these many people, I would here like to name just a few: Alfred Nordmann, Arie Rip, and Tsjalling Swierstra. My special thanks for long and ongoing cooperation on questions related to nanotechnology and for a number of valuable substantive suggestions go to my colleagues in Karlsruhe Christopher Coenen, Michael Decker, Torsten Fleischer, and Peter Hocke-Bergler as well as to Hans-Jürgen Link.

My thanks also go to Alison Hepper and Michael Wilson for ensuring the quality of the language and for their professional translation. I heartily thank Monika Zimmer for preparing the layout rapidly and carefully.

**Armin Grunwald**
Karlsruhe, January 2012

## Chapter 1

# Agenda and Overview

High expectations are placed on nanotechnology, for example, regarding health care, economic growth, and sustainable development. Parallel to them, however, there are also concerns about side effects and possible risks it poses. Commissions and expert groups have been dealing with the ethical, legal, and social implications (ELSI) of nanotechnology since an early stage of its development. The quest to achieve an ethical understanding in and for nanotechnology is an element of both public debate and scientific self-reflection. After having researched and debated the societal and ethical issues of nanotechnology for some ten years, the time is ripe for me to attempt to summarize, systematize, and assess the status of these issues as well as to think about the further perspectives of the field.

## 1.1 The Motivation

Since its beginning, nanotechnology has been viewed as something special, unlike any other technology, such as the technology of microsystems, propulsion technology, or catalytic technology. Starting with the controversial visions of Eric Drexler (1986), nanotechnology has contained a utopian and visionary element. After many in society were disillusioned with technical progress, especially as a result of the unintended consequences for the environment and society (Grunwald, 2009a), nanotechnology brought back hopes that technology might be a positively

*Responsible Nanobiotechnology: Philosophy and Ethics*
Armin Grunwald

ISBN 978-981-4316-80-4 (Hardcover), 978-981-4363-33-1 (eBook)
www.panstanford.com

redeeming force. Yet it also triggered anxieties about just the opposite, namely about dystopian developments. This coincidence of new technical opportunities and extremely far-reaching expectations — but also anxieties — provided from the beginning the central motivation for systematically studying the social and ethical issues of nanotechnology (for a comprehensive overview see Guston, 2010).

Viewed against this backdrop, the social discussion about nanotechnology does not revolve solely around the future of a specific line of technology or around the resulting consequences for society. It also revolves around such grand topics as the future of human nature, the future of the relationship between man and technology or between man and nature, and the sustainability of human development. Far-reaching questions arise precisely there where nanotechnology encounters living systems, such as in nanobiotechnology and in nanomedicine. This was the reason for me to speak of "nano(bio)technology" in this book. The boom in visions of the future and the clustering of reflective scientific studies in this field (e.g., Baird *et al.*, 2004; Nordmann, 2005; Dupuy, 2005; Selin, 2005a,b; Brown *et al.*, 2000; Grunwald, 2007; Fiedeler *et al.*, 2010) are an expression of this particular manifestation of the debate about nanotechnology, which makes it especially appealing to philosophical analyses. It should not be surprising when in the process methodological issues sometimes move to the focus of attention, in particular whether and how it is possible to make statements with regard to visions of the future or to derive orientation without slipping off into ideas that are purely speculative and arbitrary (Nordmann and Rip, 2009; Grunwald, 2010). In the meantime, the debate has differentiated into two branches, a practically oriented one and a futuristic one:

(a) Of interest among the practically oriented issues of nanotechnology are those that are already politically relevant, such as risks to our environment and health or issues of equity. Questions have been and are still being asked about the toxicity of nanoparticles, and debates have begun about the regulation of nanomaterials. The public debate about nanotechnology has been spurred on by position papers published by nongovernmental organizations (e.g., ETC, 2003), while ministries and other government bodies hold workshops and public discussions on the possible risks of nanomaterials. Researchers and nongovernmental organizations are dealing with issues of equity, in particular concerning the access of developing countries to the expected benefits of nanotechnology. The UNESCO has commissioned an expert group to prepare the field for action (ten Have, 2007).

(b) In a rather futuristic respect, the grand issues mentioned above — such as about the future of human nature and about man's relationship to nature and technology — continue to be discussed. This takes place above all in the field of "converging technologies" (Roco and Bainbridge, 2002), which builds on nanotechnology, especially on nanobiotechnology. Currently, the two great topics are human enhancement and synthetic biology.

The focus in both branches is on implementing the concepts of "responsible innovation," "responsible development," and "reflective science" (Siune *et al.*, 2009). Instead of there being a division of labor between science and innovation on the one hand and attempts to cope with their societal consequences on the other, as was frequently the case in earlier times, research and reflection on the possible consequences of science and innovation are now supposed to be pursued *as part of* research and development. Their results shall be taken into consideration in the further shaping of the scientific agenda and technical progress (Kaiser *et al.*, 2010).

In the last few years, ethical reflection on nanotechnology has developed quickly and identified many ethically relevant issues (e.g., Kushf, 2004; Grunwald, 2005). The early ethical studies on nanotechnology (about 2003) focused above all on the need to have ethics in and for nanotechnology. The ethically relevant aspects of nanotechnology that were named in those studies are an indication of a grouping and a very tentative (or positively phrased, of an open) approach to this relatively new field of reflection on science and technology. The concept "nanoethics" was quickly coined in this connection. Since then, a number of anthologies have been published about the social issues raised by nanotechnology in general and about ethical issues in particular (e.g., Schummer and Baird, 2006; Allhoff *et al.*, 2007; Banse *et al.*, 2008; Fiedeler *et al.*, 2010). The international journal *Nanoethics* was also founded during this period.

There are problems of definition, however, with the *object* of ethical reflection about nanotechnology, namely the field of nanotechnology itself (Schmid and Decker, 2003; Schmid *et al.*, 2006; Chapter 2). Nanotechnology has so far been less technology than science since it primarily takes the form of basic science conducted in a laboratory, which is the reason that people often speak of the nanosciences. The applications of nanotechnology that have already made it to market, such as sun creams, automobile tires, and specially treated surfaces, hardly appear sufficient to constitute a new challenge to ethics. They are, however, relevant with respect to risk and precautionary issues (Chapter 6). The features of nanotechnology that exert conceptional and methodological fascination on ethics lie elsewhere. The fascination — even an ethical challenge — stems from the expectations and fears, the hopes and concerns, the promises and the anxieties that accompany or are linked to nanotechnology. The objects of ethical reflection on nanotechnology include but go far beyond aspects of *current* nanotechnology and address in particular the *futures* that, metaphorically speaking, nanotechnology promises or threatens (Coenen, 2010; Grunwald, 2007a).

This situation has already led some to take the position that the time is premature for nanoethics (Keiper, 2007), and has created anxieties that nanoethics might now get lost in speculation (Nordmann, 2007a). One of the obvious tasks of this book is to examine the problematic nature of uncertain knowledge and the ethical arguments based on it. This in turn leads to another and a central motivation for the writing this book, which is referred to by the book's subtitle "Philosophical and Ethical Exploration of Nano Futures." In contrast to the widespread opinion that nanoethics is a subset of applied ethics, I proceed from the thesis that the core challenges associated with nanotechnology and the converging technologies (i.e., those futuristic aspects listed under "b" above) do not fall in the field of applied ethics, which customarily focuses on application-related issues and political action (DEEPEN, 2009). The involvement of the futures and issues named above — many of which go far afield — leads instead to my view that philosophical and ethical exploration of the possible futures must take place before applied ethics can play a role. Depending on the context, this exploration includes epistemology, anthropology, philosophy of technology, philosophy of the mind, or other philosophical disciplines. One of this book's central statements is that in this context philosophy plays a role on a substantially wider front than does "simply" ethics (Chapter 10).

## 1.2 Objectives, Conceptual Framework, and Premises

Many ethically relevant issues of nano(bio)technology have been uncovered and discussed in the last years. What has not been named, however, are well-justified criteria for determining why certain topics, such as nanoparticles crossing the border between technology and living systems, should be considered ethically relevant. No consensus has yet been reached on these criteria. In particular, it is often not clear what is *novel* about the ethical questions touched on by developments emerging from nanotechnology compared to those in well-known fields of technology. Furthermore, although ethical issues posed by nanotechnology have been identified, their analysis and the elaboration of proposals about how to deal with them in society are still in their beginnings. Despite its thematic broadness and the intense ongoing discussion, the ethics of nanotechnology is, therefore, still an emerging field.[1]

In view of this situation, the purpose of this volume consists primarily in studying the current and foreseeable developments in nano(bio)technology from the viewpoint of philosophical ethics and other subdisciplines of philosophy. Some of the questions I will look at are: Which developments are ethically relevant, and why are they ethically relevant? Are there ethical questions which have already been tackled by ongoing or recent discussions in the ethics of technology or in bioethics? Could analysis of the ethical aspects of nanotechnology benefit from other philosophical and ethical discussions? Are there developments which pose completely novel ethical questions? What can be said about the role that highly uncertain and contested nanotechnology "futures" play in ethical analysis and inquiry?

The object of this reflection are developments linked to nano(bio)technological research and developments, their applications, and their futures. The word "nano(bio)technological" is used because it permits associations in two directions. On the one hand, we are mainly dealing with nanotechnology and its consequences. On the other hand, it takes into consideration that some of the relevant ethical debates are now being conducted under different terminological designations, such as "bionanotechnology"

[1] Little interest was given to the criteria why certain issues of nanotechnology should be regarded ethically relevant (an exception is Lewenstein, 2006).

(Jotterand, 2008a), nanobiotechnology and synthetic biology, or human enhancement, although the roots of these fields and of these reflections are in the field of nanotechnology.

In this book I do not adopt with regard to either knowledge or reflection the specific perspective of any individual subdiscipline of applied ethics or of the ethics of an individual field, such as information ethics or bioethics (Nida-Rümelin, 1996a). The focus is instead on those philosophical and ethical issues that can be associated with the named objects of reflection. In this sense, the book is a cross-sectional analysis relative to the subdisciplines of applied ethics and philosophy in general in the context of nanotechnology. The analysis pursues the following objectives:

- Since the technical basis that the ethics of nanotechnology is supposed to refer to tenaciously defies all attempts to be defined (in contrast for example to other fields of technology, such as genetic engineering or nuclear fission), there must be reflection on what ethics is supposed to refer to in this connection and on what follows from the problem of giving its definition.
- The previous discussion of the ethical issues posed by nanotechnology is to be reconstructed in a systematic form and presented in a methodical manner. At issue is overcoming the partial arbitrariness that exists in identifying ethical issues in favor of a systematic approach. Following the explorative phase of nanoethics, the time is now ripe for a *systematic* overview based on clear criteria. This book thus will make the *criteria of the ethical relevance* of nanotechnological developments an explicit topic.
- The "newness" of ethical challenges continues to be a topic of discussion. Another task of this book consists in providing grounded information about the degree of novelty of the philosophical and ethical issues of nanotechnology.
- The analyses of ethical issues are to be deepened by case studies in certain, particularly relevant fields. The criteria for selecting these fields are also to be named.
- As already mentioned, ethics in nanotechnology consists to a large extent in the fact that the *futures* of nanotechnology are a subject of debate from an ethical perspective. This is an indication of the early point in time at which the (possible)

consequences of nanotechnology are being considered, which itself is associated with the problem of the uncertainty of the knowledge about the object of our reflection. This requires epistemological consideration. One of the goals of the book is thus to contribute to a better understanding of the relationships between the epistemological and ethical evaluations of technological futures.

- The frequently expressed expectation that nanoethics should offer political orientation on concrete issues in nanotechnology leads astray in some fields. Instead, an *explorative philosophy* of certain facets of nanotechnology, including hermeneutic, epistemological, and anthropological issues, is frequently more appropriate. One task of this book is to work out this program.

A transparent procedure is needed in order to reach these goals. As part of this procedure, it is programmatically important for me to disclose the following initial premises of this book:

- Considering the ethical issues of nanotechnology means mobilizing the instruments and knowledge of *philosophical ethics*. This takes place in contrast to the creeping change in usage in which "ethics" has frequently been employed simply as a placeholder for generally considering the social aspects of technology (e.g., in Allhoff *et al.*, 2007).
- According to the terminological order of this book, ethical questions are raised in the case of normative uncertainty, i.e., if the accepted moral orientation for action does not suffice for dealing with a specific issue but instead leads to conflicts, to uncertainties, or to problems in orientation. This concept (first in Grunwald, 2000a) makes it possible for the diagnosis of whether something in nanotechnology is ethically relevant to be put on a clear normative footing.
- Examining the ethical issues of nanotechnology also means entering into an *interdisciplinary* dialogue about the challenges arising from the nanosciences and nanotechnology instead of holding a monologue. This dialogue also includes issues of governance (Kaiser *et al.*, 2010).

Following this approach, our view is first directed at the existing field of ethical and philosophical consideration of nanotechnology, which can be accessed by studying the available literature. The larger objective can be summarized as overcoming the restriction of issues to those specific to applied ethics and making accessible the larger opportunities offered by an ethical — philosophical exploration of nanotechnology. This will demonstrate that such an explorative approach is also suitable with regard to other, future debates about new and emerging technology.

## 1.3 Quick Guide Through the Book

This book's contents are divided into three parts. The first part (Chapters 1–3) is dedicated to laying the necessary groundwork and includes an introduction into nanotechnology (Chapter 2) and a clarification of the ethical understanding of issues related to technology and risk (Chapter 3). The second and largest part (Chapters 4–9) covers substantive analyses of ethical issues in nanotechnology and its applications. It starts from considerations about the location of the ethics of nanotechnology in an systematic overview of the sciences (Chapter 4), and consists in an overview (Chapter 5) and in-depth examinations of the four primary fields, namely of using nanomaterials (Chapter 6), of acting responsibly in nanobiotechnology and synthetic biology (Chapter 7), and of the consequences of employing nanotechnology in animal enhancement (Chapter 8) and in human enhancement (Chapter 9). The third part focuses on conceptional issues that cross the individual areas of a philosophical–ethical exploration of nanotechnology (Chapter 10) and provides points of reference for further developing nanoethics as a field of research and reflection (Chapter 11).

### 1.3.1 Chapter 2: Nanotechnology in Context

It is of course appropriate at the beginning to describe the topic to be made the object of ethical reflection as much as necessary. Even though nanotechnology is still a young field, a history of its grounding has already become part of its self-understanding. The presentation of the subject matter itself initially covers the analytic and technical means that are employed to study nanostructures as well as to influence them in a dedicated manner. It also provides an insight into the world of nanoscale objects and an overview of the primary fields of application. The presentation of the problem of defining nanotechnology is an indispensible part of its self-understanding, just as the description of its essentially interdisciplinary nature is.

Gaining insight into the external perspectives that it has provoked also belongs to getting to know nanotechnology. Nanotechnology has in particular also provided reason for some to voice philosophical interpretations that go far beyond the technical and that have strongly molded the social debate. The picture is completed by a view of the social debate, both with reference to the high expectations as well as to the risk anxieties.

### 1.3.2 Chapter 3: Ethics, Technology, and Risk

Ethics shall provide orientation on controversial moral issues. The problems that ethics is supposed to take on are characterized by *normative uncertainty* in situations demanding decisions and/or action. Ethical reflection is not required as long as the normative orientations that are accepted in practice permit decisions that are clear and uncontroversial. Situations of normative uncertainty arise, however, for example, if there is a moral conflict or if questions are raised with regard to nanotechnological products that cannot be answered with the usual moral means. Criteria are deduced from this consideration as to when such a situation makes ethical consultation necessary. The description of the moral contents of technology serves in general to structure the detailed analyses of nanotechnology in which the issue is to search for normative uncertainty (whether real or only imagined) with regard to nanotechnology and its applications. Comments about the relationship between ethics, technology, and risk identify the terminological and conceptual framework of the analysis.

### 1.3.3 Chapter 4: Ethics of Nano(bio)technology: The Program

First the arguments are reconstructed that are raised in favor of the need for ethical consideration of nanotechnology. It is an open question whether "nanoethics" can or should be a new subdiscipline of applied ethics and thus a field of its own, or whether it is rather a subfield in the existing traditions of ethics. The hypothesis is formulated in this chapter that nanotechnology is precisely *not* an object of ethical reflection that can legitimize the establishment of a new and independent area of ethics for an individual field of technology. Since ethical reflection on the nanosciences and nanotechnology consists in part in a conflation of different traditions of ethical reflection, "nanoethics" should not be understood as a new subdiscipline but rather as a platform on which philosophical exploration, ethical deliberation, interdisciplinary dialogue, and democratic debate on these issues can take place.

### 1.3.4 Chapter 5: Ethics of Nano(bio)technology: An Overview

Based on the existing literature, I attempt to systematically map the ethical issues posed by nanotechnology. Among the ethical issues to be named in this context are those related to risks and the necessity of precautionary steps, to distributive justice, to threats to one's private sphere, to anxieties of being controlled by being accessible technologically, to opportunities for medical use, to military applications, to transcending the interface between technology and life, and to the technological enhancement of man. The issue in all of these areas is to find the answer to the double question: Where are the ethical issues in the sense of normative uncertainty? In which sense are these ethical issues new? On the basis of this overview, arguments are also given for the selection of the case studies to be presented in the following chapters.

### 1.3.5 Chapter 6: Synthetic Nanoparticles

Intense toxicological research is being conducted on nanoparticles' mechanisms of action in the human body and in the environment as well as on their long-term retention. Our current knowledge is however limited and can by no means provide answer to all the questions regarding responsible handling of such particles. This lack of knowledge about the possible consequences for our health and the environment raises ethical questions about how we handle uncertainty and about precautionary steps that might become necessary. Some negative experiences, especially those associated with chemicals and asbestos, led to early demands for a moratorium on the commercial utilization of nanoparticles. Others opt for soft measures such as self-commitments. In this chapter I identify the controversial normative issues that are hidden behind these different positions and suggest a strategy for rational conflict management and responsible development.

### 1.3.6 Chapter 7: Toward Creating Artificial Life

Fundamental processes of life take place at the nanoscale since important components (e.g., proteins) are precisely this size. In nanobiotechnology, classical machine language — for example, terms such as factories, rotors, pumps, and reactors — is employed to speak about elements of life. At this level, nanotechnology could make it possible to create technical replicas, dedicated modifications, or engineering of cells. It is increasingly becoming possible to transcend the classical border between the technical and the living. In as much as it is possible to a previously unknown degree to technically reconstruct the living and to make it controllable, synthetic biology that could create artificial life out of abiotic source material comes within our grasp. In this chapter, I analyze the far-reaching questions regarding responsibility that are raised in this connection and answer them as far as possible in the current situation.

### 1.3.7 Chapter 8: Animal Enhancement

The use of animals for human purposes, such as for food, as working animals, or for sports, has a long tradition in human civilization. Although genetic engineering and molecular biology posed profound steps toward even further reaching interventions in the nature of farm animals, it appears that nanotechnology will enable man to take further steps to increase his/her influence over animals. This chapter first describes some of the directions of research that are currently in progress or are anticipated in this field. It then asks about the semantics of animal enhancement, in part in preparation for considerations of human enhancement (Chapter 9). Against the backdrop of the existing normative frameworks, the ethical challenges are identified and lead over the level of applied ethics to issues from hermeneutics and the philosophy of science that are related to the relationship between man, animal, and technology.

### 1.3.8 Chapter 9: Human Enhancement

The technical enhancement of human beings is the topic that determines a large portion of the recent philosophical–ethical debate on nanotechnology. It requires interfaces between the nervous system or the brain and technical systems that nanotechnology can make possible. The convergence of biotechnology, nanotechnology, the neurosciences, and information technology opens, according to this vision (Roco and Bainbridge, 2002), new dimensions of technical opportunities. Most important of these is that the human body and its mental capacities appear — at least in a scientific utopia — to be open to technical modification. Among the ensuing ethical issues are the argumentative role of the naturalness of the human body, the question as to the criteria and limits of enhancement, the consequences for a future society, and anthropological questions as to the future of man's nature. The analysis of these issues reveals that hermeneutic work is necessary for a precise discussion of the ethical aspects: to understand what is meant here by technical enhancement, to determine how this differs from healing and doping, and to identify the criteria according to which an enhancement is considered *enhancement*.

### 1.3.9 Chapter 10: From Applied Ethics to an Explorative Philosophy of Nanotechnology

Pursuing ethics for nanotechnology means — as demonstrated especially by the four in-depth analyses — that one must be able to deal with some far-reaching and speculative statements or visions of the future and with the uncertainty that is necessarily linked to them. A set of tools is needed in order to consider the relative merits of multiple controversial or diverging futures argumentatively on the basis of standards of rationality. The necessary epistemological reflection that is part of the ethical concern with nanotechnology's futures leads to the question as to how to deal in an ethically responsible manner with what may be highly speculative futures, whether they be far-reaching promises or apprehensions. Examination of the criticism of speculative nanoethics makes it clear that the opportunities for ethics as a form of reflective support to overcome normative uncertainties decrease with the increasing speculative nature of its object. In turn, other, in particular hermeneutic functions of this reflection move to the forefront, taking the form as it were of pre-ethical consideration to provide enlightenment about emerging normative uncertainties. I use the term "explorative philosophy" to refer to this expansion of the application-oriented approaches of applied ethics and discuss it.

### 1.3.10 Chapter 11: Conclusions and Perspectives

The debate about the ethics of nanotechnology is still young. The first object of this chapter is to provide a diagnosis of the previous achievements and consequences of the ethical debates on nanotechnology as measured against public expectations. Building on the results of the previous chapters, I draw conclusions and conceive perspectives for this field. These are initially related to the question of whether the previous ethical studies on nanotechnology exhibit signs of a front that rejects nanotechnology on moral grounds. The chapter closes with considerations of the future development of ethics for nanotechnology.

# Chapter 2

# Nanotechnology in Context

This chapter describes some of the fundamentals of nanotechnology (for further information see Schmid *et al.*, 2006; Schmid, 2008; Sattler, 2010) and its societal contexts to the extent necessary for understanding the debates on societal and ethical issues.[1] The topics covered are the history of nanotechnology (Section 2.1), the basic principles of physics and metrology and important areas of application of nanotechnology (Section 2.2), the definitions of nanotechnology and the difficulties surrounding them (Section 2.3), the interdisciplinary nature of the scientific community working on nanotechnologies (Section 2.4), the philosophical interpretations of nanotechnology (Section 2.5), and its social perception (Section 2.6).

## 2.1 History of Nanotechnology

In terms of its basic idea, the history of nanotechnology extends far back. Its guiding idea — namely, the intentional manipulation of matter at the level of atoms or molecules — was formulated back in 1959 by the American physicist and Nobel Laureate Richard P. Feynman in a lecture that is often quoted as representing the founding document of nanotechnology:[2]

[1]The title of this chapter follows that of the book edited by Nordmann *et al.* (2006, in German).

[2]It should be noted, however, that the relationship between Feynman's speech and nanotechnology is a construction or discovery *ex post facto*. At that time nobody took the thoughts expressed by Feynman as a recommendation to invent nanotechnologies because it was simply so far beyond the technical and scientific capabilities of the time.

*Responsible Nanobiotechnology: Philosophy and Ethics*
Armin Grunwald

ISBN 978-981-4316-80-4 (Hardcover), 978-981-4363-33-1 (eBook)
www.panstanford.com

> But I am not afraid to consider the final question as to whether, ultimately — in the great future — we can arrange the atoms the way we want; the very atoms, all the way down! What would happen if we could arrange the atoms one by one the way we want them [...]. What would the properties of materials be if we could really arrange the atoms the way we want them? [...] Atoms on a small scale behave like nothing on a large scale, for they satisfy the laws of quantum mechanics. [...] At the atomic level, we have new kinds of forces and new kinds of possibilities, new kinds of effects. The problems of manufacture and reproduction of materials will be quite different. [...] The principles of physics, as far as I can see, do not speak against the possibility of maneuvering things atom by atom. It is not an attempt to violate any laws; it is something, in principle, that can be done; but in practice, it has not been done because we are too big. (Feynman, 1959)

These visionary words — visionary compared to the technical possibilities of the time — express the essential characteristics of nanotechnology: the extension of man's technical competence into the dimension of atoms, which is the guiding motif of nanotechnology (NNI, 1999), the use of new effects at the border between classical physics and quantum physics, the "atomic reductionism"[3] of all effects to the planned technical manipulation and arrangement of atoms (Heckl, 2004), as well as the reference to a possible obstacle, namely that we humans are too large to be able to operate at the nanometer level (Smalley, 2001).

The term "nanotechnology" was first used by the Japanese physicist Norio Taniguchi (1974). His basic idea was an ever smaller miniaturization, according to which nanotechnology would ultimately be the logical further development of fine mechanics and microsystems technology: the manipulation of matter at the level of atoms or molecules.

The foundation for the technical abilities to at least even begin to carry out planned operations at the atomic level was laid by discoveries and inventions made in the 1980s. Scanning tunnel microscopy (STM, invented in 1981), atomic force microscopy (AFM), and scanning probe techniques make it possible on the one

[3]The concept of reductionism is here not being employed in the terminological sense of the philosophy of science (see on this Oppenheim and Putnam, 1958), but rather as a description of positions of scientists (which in various details may be very different) who view the macroscopic world as being determined by processes at the atomic level.

hand to probe surfaces by employing the distribution of electrical potentials that are strongly dependent on distance, and on the other to carry out manipulations at the nano level (Section 2.2). This makes it possible, at least in theory, to carry out technical operations at the atomic level. Scientists from IBM achieved the spectacular success in 1990 of making this vision come true. They managed to use 35 individual atoms to "write" the IBM logo on the surface of an ice crystal (Eigler and Schweizer, 1990). This image is today one of the "icons" of nanotechnology.

The history of nanotechnology and of its public reception includes the visionary ideas of Eric Drexler (1986) though many scientists rejected them. Drexler carried Feynman's idea of intentionally arranging atoms to the extreme by envisaging the construction of a molecular assembler that reproduced and further developed itself. By reordering the atoms, this assembler was supposed to be able to use any arbitrary source material to produce something entirely different. Drexler linked his far-reaching expectations of solving practically all of man's problems to this vision. This proved to be the primary trigger for the boom in expectations surrounding nanotechnology in the following 20 years:

> In short, replicating assemblers will copy themselves by the ton, then make other products such as computers, rocket engines, chairs, and so forth. They will make disassemblers able to break down rock to supply raw material. [...] Teams of nanomachines in nature will build whales. [...] Assemblers will be able to make virtually anything from common materials without labour, replacing smoking factories with systems as clean as forests. (Drexler, 1986; Chapter 4)

As speculative as these expectations were and although distinguished scientists expressed well-founded and fundamental doubts about the possibility of building such an assembler (e.g., Smalley, 2001), Drexler's thoughts exerted a strong influence on nanotechnology's perception in society and its support among politicians, especially in the United States. The National Nanotechnology Initiative (NNI, 1999) established by Bill Clinton and Al Gore appropriately carried the distinctive title "Shaping the World Atom by Atom." This was the starting point of the boom in research in and financial support for nanotechnology that has continued until this day (Paschen *et al.*, 2004). The idea of arranging atoms or molecules and the atomic reductionism that was tied to it probably represent the guiding motif (leitmotif) for nanotechnology. "It can be said of this vision of nanotechnology that no prudent researcher believes in it officially but that its impact in the background is extremely important" (Nordmann, 2007b, p. 220).

The countermovement to this boom, which however had not (yet) jolted nanotechnology in any substantial way, was the social debate about the risks posed by nanotechnology from 2000 on (Section 2.6). Starting from the article *Why the Future Doesn't Need Us* (Joy, 2000) by Bill Joy, at the time the chief designer at Sun Microsystems, criticism focused initially on the visionary aspects of nanotechnology. The visions that were originally meant positively (e.g., the molecular assembler) were reinterpreted as leading to horror scenarios. From 2003 on, the debate shifted to the more practical issues of the possible environmental and health consequences of nanoparticles (Section 2.6).

At present it is possible to speak of a broad and quite "normal" field of research and development in nanotechnology. A large portion of nanotechnology takes place in the framework of the customary structures of science, such as in institutes and research programs, sometimes within the classical scientific disciplines, and sometimes more explicitly labeled as nanotechnology in an interdisciplinary mode. Mostly this research might be denoted as advanced materials research aiming at concrete applications (Chapter 6). The more visionary or futuristic elements of nanotechnology — Drexler's legacy — have migrated to the debates about "converging technologies" (Roco and Bainbridge, 2002; Banse *et al.*, 2008; Fiedeler *et al.*, 2010) and on synthetic biology (Boldt and Müller, 2008). Each of them is inconceivable without nanotechnology and has led to challenging debates on ethics. They are therefore prominently taken into consideration in this book (Chapters 7–9).

## 2.2 The World of Nanotechnology in a Nutshell

Nanotechnology originated from the convergence or mingling of developments from different disciplines (see Section 2.4). The sources are mesoscopic physics, especially solid-state physics, those portions of chemistry that work at the interface between molecular and macroscopic dimensions, the technology of microsystem (or microelectromechanical) and microprocess engineering inasmuch as their techniques can be further miniaturized (Heckl, 2004), and also molecular biology and biotechnology if living systems are involved.

In order to provide a concrete idea as to what research and development in nanotechnology is and with which means it takes place, this section includes a short introduction to the world of measurement and manipulation techniques (Section 2.2.1), an overview of the essential "objects" of the nanoworld (Section 2.2.2), and a brief presentation of those fields of application that are considered particularly important today (Section 2.2.3). Expert literature is available for more in-depth knowledge (e.g., Schmid *et al.*, 2006; Schmid, 2008; Sattler, 2010; ITA, 2010).

### 2.2.1 Nanometer-Scale Analysis and Manipulation

The nanometer dimension that both characterizes and provides the name for nanotechnology ranges, according to common understanding, from 1 to 100 nanometer (nm).[4] The unit of 1 nm is defined as one billionth of a meter, or in other words as one millionth of a millimeter or one thousandth of a micrometer. The small size of objects in this dimension is illustrated by the fact that the thickness of a human hair is approximately 20,000–180,000 nm and the size of bacteria 200–2000 nm. Viruses are some 10–400 nm large. The average diameter of the DNA double helix is about 2 nm. The small size of such objects means that the nanometer dimension is inaccessible to man's customary forms of perception. Highly developed and complex technology is required to develop "eyes" for our perception and "fingers" for technical manipulation.

*Scanning probe microscopes* play a central role in such analysis. Since the electrical exchange between a surface's structure and the tip of a microprobe that is placed very near to it (in atomic dimensions) depends strongly on the distance between them, it is possible to employ a probe tip to scan a surface structure very precisely and to analyze it. The scanning tunneling microscope, which was developed in 1981, takes advantage of the quantum mechanical effect that

[4]Although this characterization is found in many definitions, it is nevertheless arbitrary. Under normal circumstances there is not necessarily any difference in principle between 98 nm and 102 nm that would support such a size-dependent definition (on the problem of defining nanotechnology, cf. Sec. 2.3). The relevance of a size definition for ethical questions is even harder to identify (Chapter 4).

electrons can jump (tunnel) between a very small tip of a probe and the surface of a sample without the tip touching the sample. Since this tunneling flow between the sample and the tip of the probe can be measured and is strongly dependent on distance, it can be used as a measure for the distance between the tip and the sample. The measured tunneling flow acquired when the tip is scanned line wise over the surface determines the topographic image of the surface, whose resolution (ideally) is at the atomic level.

In *atomic* (or scanning) *force microscopy*, the attracting or repelling forces of the interaction between the surface and the tip are used to determine the surface's structure. The electric interaction — repulsion or attraction — is transformed into the mechanical deflection of a cantilever. The degree of deflection versus normal indicates the distribution of electric potential across the surface being studied. This makes it possible, for example, to directly image organic layers on insulating substrates and to analyze biological specimen. An entire arsenal of scanning probe techniques has in the meantime been developed from these basic forms for different applications.

Scanning probe techniques can not only be used to analyze nanostructures but also to produce them. When a scanning probe microscope is used as a finger to manipulate a sample, it is possible in principle to rearrange atoms intentionally on the surface. This can, for example, in AFM, be done purely mechanically by increasing the tip's pressure on the surface, thus removing atoms from the surface (nanoengraving). It is also possible by electrical and optical means. Another possibility for employing an atomic force microscope as a production tool in nanotechnology is what is referred to as *dip-pen lithography*. In this procedure, a nanoscale pattern is applied to a specimen by employing the tip of an atomic force probe coated with a liquid film. The fluid contains molecules that align themselves on the specimen, forming a structure 10–20 nm wide. These clusters of molecules can be used to attribute a function to the surface, for example, to apply detector molecules to a biochip with nanometer precision.

The production of nanoparticles is, on the one hand, mechanical; powder containing smaller and smaller particles (e.g., graphite or sand) is produced by grinding. On the other hand, chemical procedures such as the precipitation of substances from solutions are used in which the precipitation process is triggered and influenced by the addition of suitable substances. Nanopores produced by lithography, etching, or self-organization can be used as a template for the production and spatial arrangement of nanowires, nanosticks, and nanotubes. The material for the nanowires is secreted by the pores electrochemically, chemically, or physically. The pore surfaces form the spatial limit and can also serve as an electrode or condensation core. Organic layers that are nanometer thick can be prepared by the accretion of long-chain molecules. Polymer particles, viruses, cells, and metal clusters can be used as a template for the synthesis of hollow nanocapsules that have multiple applications. They can serve as a nanotransporter, nanoreactor, or a model system for studying the processes in biological cell membranes.

### 2.2.2 Characteristics of Nanomaterials

A distinction is made in nanotechnology between dot-like structures, which are nanoscale in all three dimensions and are thus smaller than 100 nm according to common understanding, rod-like structures, which are nanoscale in two dimensions, and disk-like ones, which are nanoscale in only one dimension.

A *dot-like* structure can occur either as a crystalline or amorphous nanoparticle or as a cluster, nanoisland, or larger molecule. Compared to their volume, nanoparticles have a very large surface area and thus react more strongly. Their electrical, optical, and chemical properties vary strongly depending on their size. They may differ substantially from those of a solid-state body made of the same material; for example, as very small particles, a metal is an insulator. Many of the macroscopic properties of materials containing admixtures of nanoparticles are determined by the surface properties of the nanoparticles. This is true, for example, for the ability of sunscreen lotions containing titanium dioxide nanoparticles to block ultraviolet radiation. One category of materials that is becoming more and more important because of their unusual properties is that of *fullerenes*, which are pentagonal or hexagonal cage molecules of varying size that consist solely of carbon. Larger fullerenes provide, for example, the option of encapsulating foreign atoms. In addition to cages with single walls, some fullerenes are onion-like with multiple walls and conical or frustoconical in shape. Fullerenes can be manufactured in macroscopic quantities by vaporizing graphite. This procedure produces soot, some 10% of which is fullerenes. Because of their good biocompatibility, there is hope that they can be used in medical applications. For examples, radionuclides encapsulated in fullerenes could be used for targeted drug delivery in diagnostic and therapeutic applications (Sattler, 2010).

*Rod-like nanostructures* are produced primarily in the form of nanotubes, nanowires, and nanogrooves on surfaces. The rod-like structures that have been studied the most intensively are by far *nanotubes* made of carbon. They consist of cylinder-shaped layers of graphite, and their structure is closely related to that of fullerenes. Their diameter ranges from almost 1 to about 100 nm, and their length can currently reach several hundred micrometers. Carbon nanotubes can be produced by the kilograms or the ton. Numerous potential areas of application are currently being discussed. For example, these nanotubes are of interest as conducting connections for nanoelectronic modules, can function as transistors and logic modules, and are suitable for use as emitters in field emission displays (an ultra-flat alternative to cathode ray tubes). The storage of hydrogen in nanotubes is considered likely, even though it so far has not been possible to meet the high expectations.

*Disk-like structures* constitute an extremely important field of nanotechnology, the most advanced one in terms of applications at the industrial level. A disk's thickness is in the nanometer range, often just a few layers of atoms, which makes certain effects possible or is the precondition for the utilization of decisive properties of the disk material. For example, today's computer hard disks are inconceivable with nanolayers. The high storage density in them is based on the so-called *giant magnetoresistance*, which is present in a system of layers of iron that are separated from one another by a layers of chromium that are only a few nanometers thick (for the discovery of which Albert Fert and Peter Grünberg were awarded the Nobel Prize in 2007). Even the protective layer of carbon that protects the storage layer against oxidation and abrasion may not be more than a few nanometers thick; otherwise the read–write head would not be able to fly close enough over the magnetized layer (approximately 25 nm). Still in the implementation phase is the use of antibacterial, hydrophobic, and lipophobic textile coatings.

These developments lead to a multitude of new opportunities for combining known materials and the new nanostructures. The study of their properties and possible technological applications constitutes a large portion of today's nanoresearch (Schmid *et al.*, 2006). Much of this is still in the research or testing phases. Nanotechnology is frequently not a technology per se. The technical applications are mostly expected to be identified in the future. For this reason, people often prefer referring to the notion of "nanosciences."

## 2.2.3 Areas of Activity and Applications

Nanotechnology is tied to the hope for significant sales in almost all branches of the economy. According to most estimates, the areas with the greatest potential for innovation from nanotechnology are new materials, information and communication technology, and the life sciences and medicine.

### 2.2.3.1 *Synthetic Nanomaterials*

New materials are frequently decisive for achieving advances and breakthroughs in technology and its applications. The miniaturization of material structures in the nanorange frequently leads to the discovery of properties that are not present macroscopically for the same materials. Such technically useful properties include significantly higher levels of hardness, ultimate strength, and viscosity at lower temperatures, (greater) superplasticity at high temperatures, the formation of additional electron states, high chemical reactivity of the surface structures, and significantly increased surface energy. Let us look first at *surface treatments*. New and improved color effects result from the addition of nanoparticles to conventional lacquer. Other examples are self-cleaning surfaces that also exhibit hydrophobic and lipophobic properties.[5] Also already in use are optically functional surfaces for frontages, motor vehicles, solar cells, etc. (e.g., for antireflective coating, sun-protective glass, antireflexion coating of instrument panels). In the *cosmetics* industry, the mixture of nanoparticles with established products is being examined in order to reinforce specific desired characteristics. Sunscreen lotions with $TiO_2$ nanoparticles are probably the best-known products in the area of nanotechnology. A further area of application is in lightweight design. Crucial material factors (rigidity, wear-resistance, etc.) can be specifically improved by

[5]In this context, the now recognized lotus effect is often mentioned (Barthlott 1990), whereby the relevant characteristics of the lotus plant are exploited technically. However, this is not nanotechnology in the terminological sense.

introducing characteristic structural factors in the nanometer range. There are options for using polymers enriched with nanoparticles in particularly demanding cases of lightweight design or at high temperatures. In land and air vehicles, it has been possible to some extent to use stronger and lighter materials in place of traditional structural materials.

The world economy depends massively on a continuous flow of natural resources, which are returned to the ecosphere after a comparatively short phase of usage during which the initial resources have been modified (mostly degraded). The chemical industry is an important player in this process of the use and disposal of materials, which is very often accompanied by unintended side effects such as unwanted production emissions. Nanotechnology is expected to simultaneously help by considerably lowering the negative impact and by providing substantial additional economic value in the form of new and improved products. In particular, nanotechnology could help substituting specific hazardous chemicals by less dangerous ones (Fiedeler, 2008).

Catalysts were among the first manifestations of industrial nanotechnology. Nanoparticles and nanostructured materials offer new ways of designing and controlling catalytic functions, including providing for enhanced activity and selectivity for target reactions. Since the activity and selectivity of catalyst nanoparticles are strongly dependent on their size, shape, and surface structure, as well as on their bulk and surface composition, the development of the capability to synthesize particles at the nanoscale with defined physical and chemical properties is a crucial step toward controlling catalyst processes and increasing resource productivity. This is just one of the many places where nanotechnology can make new production processes possible in industry that could be superior both economically and environmentally.

One of the most flexible renewable energy technologies is the direct conversion of sunlight into electric power (Cheng and Fleming, 2009). Existing photovoltaic technologies, however, are limited by their low conversion efficiency and high cost. Nanostructured photovoltaic devices such as those using quantum dots have considerable potential for reducing costs. They allow light to be collected from a broader range of wavelengths of the sun's spectrum than conventional cells. Nanostructured absorber layers may further increase cell efficiency or permit the use of lower quality materials for the cells.

Another issue is the hydrogen economy. It is still facing a number of critical technological challenges, among them the procedures to produce and store hydrogen and to convert it into electricity. One strategy for producing hydrogen using renewable resources is the direct catalytic conversion of water into oxygen and hydrogen by using a nanostructured semiconductor catalyst or nanoscale additives. A prerequisite for building a hydrogen energy infrastructure, and especially for applications in transportation, is the creation of new lightweight, efficient, and safe hydrogen storage systems. Materials under investigation for hydrogen storage are nanostructured carbon-based materials, such as carbon nanotubes. Other energy technology options include new catalysts for fuel cells, high-efficiency devices for lighting or appliances, new materials for low-loss power transmission lines or new high-strength lightweight materials for transportation, construction, or electric power applications. Nanotechnology frequently provides the potential for using advanced materials in future applications.

#### 2.2.3.2 *Nanoelectronics*

The information and communication technologies are among the main driving forces in all advanced economies. Miniaturization is a key for further progress, as suggested by Moore's law postulating a fast decrease of the size of electronic devices such as microchips. Traditional technologies based on semiconductor physics and on the use of silicon are increasingly confronting physical limitations. Nanotechnology could lead to the next major step toward further miniaturization (Schmid *et al.*, 2006). In the field of nanoelectronics, new optic and optoelectronic devices for information storage and transport are being developed (Schmid, 2008; Sattler, 2011). The order of magnitude of the basic devices responsible for logic operations and data storage is approaching the nanometer dimension. In the future, information technology might rely on light-based technology, for example, those using purely optical circuits and photonic crystals. Quantum dots and carbon nanotubes are further possible elements of such new technology. Nanotechnology could also facilitate molecular electronics by using quantum mechanical effects to realize smaller and faster devices (quantum computing). Regarding the outstanding importance of information and communication technologies for the global economy, for approaching the knowledge society, for lifestyle and consumption, and for production patterns, it is easy to predict huge economic potentials in this field. Most of these developments are, however, still in the early phases of development and require a large amount of resources.

#### 2.2.3.3 *Nanobiotechnology*

*Nanobiotechnology* is currently a still little resolved, somewhat colorful term. There are also references to bionanotechnology (Jotterand, 2008a) and nanobionics (Chapter 7). Common to these directions of research is the attempt to build a bridge between inorganic and organic nature. The aim is to comprehend technically biological functional units from a molecular perspective and to produce functional building blocks on the nanoscale in a controlled way, including technical materials, interfaces, and boundary layers (VDI, 2002). The processes in a cell can be analyzed with nanotechnological procedures and made technically usable (Schmid *et al.*, 2006; Sattler, 2010). Nanobiotechnology is also interested in the mechanisms of energy production and transportation systems, as well as large-capacity data storage devices and data readers in which functional biomolecules work as components of light collection and conversion facilities, signal converters, catalysts, pumps, or engines. Synthetic biology has the aim of changing living systems technically in a targeted way or to reconstruct them (see Chapter 7). The *food sector* is often cited as an area of application for bioactive nanoparticles, on the one hand with regard to new packaging technologies, but also with a view to new techniques in preservation and for maintaining quality.

#### 2.2.3.4 *Nanomedicine*

There are great hopes set on the potential of nanotechnology toward providing improved *medical care* (Freitas, 1999). Miniaturization is an essential means of progress in many areas of medicine. Smaller samples for in vitro analysis allow less invasive and less traumatic methods of extraction. Better interfaces and biocompatible materials provide new opportunities for using implants and restoring damaged organic facilities. The field of medical application might develop into the largest field of future nanotechnological applications.

With the help of nanotechnology-based diagnostic instruments, it might be possible to discover diseases or predispositions for diseases earlier than at the present (see several chapters in Sattler, 2010). The development of "lab-on-a-chip" technology would further promote the emerging tendency toward personalized medicine (van Merkerk, 2007). In medicinal therapy, there is the prospect, with the help of nanotechnology, for targeted treatments to be developed that are free of side effects. The broad use of nanoparticle dosage systems could lead to progress in medicinal treatment. Nanotechnological methods can improve the biocompatibility of artificial implants. Drug-delivery systems could considerably enhance the efficiency of medication and minimize side effects (Bhaskar *et al.*, 2010; Sattler, 2010).

In the last few years, this debate has gone beyond the issues of healing diseases and re-establishing the lost functions of organs or other parts of the body to focus on the issue of "human enhancement" (Chapter 9). The idea of human enhancement is to replace, stimulate, and enhance functions of the body (e.g., of the human eye or the brain) by technical means, employing for example brain–machine interfaces, prostheses, or newly designed drugs. Nanobiotechnology and nanomedicine are essential research fields to this end.

Some of the applications have already reached general use. For example, carbon nanoparticles are used in automobile tires exhibiting both good grip and long life, titanium dioxide nanoparticles are contained in numerous sun creams, and biocidal silver nanoparticles protect against bacteria. Yet only some of these belong to the applications that are receiving attention in an ethical regard (Chapter 6). A large part of the debate about ethical issues in nanotechnology concerns the (*possible*) future applications and the (*possible*) future consequences of these applications. The issues involve an exponential of a potential event, which does not simplify the work of ethics in the least (Chapter 10).

## 2.3 Defining Nanotechnology

There have been many attempts to define "nanotechnology" as precisely as possible or at least to outline its contents. Although these attempts appear relatively similar at first glance — especially with regard to the dimensions implied by the use of the prefix "nano" — detailed analysis of the definitions reveals significant differences (Schmid and Decker, 2003; Decker, 2006), so much so that there has been talk of the uncertain — or even indeterminable — nature of the attempt to define nanotechnology (Schmidt, 2008). Here are three examples of these definitions:

(1) Nanotechnology is the understanding and control of matter at dimensions of roughly 1 to 100 nanometers, where unique phenomena enable novel applications. Encompassing nanoscale science, engineering and technology, nanotechnology involves imaging, measuring, modeling, and manipulating matter at this length scale. [...] At the nanoscale, the physical, chemical, and biological properties of materials differ in fundamental and valuable ways from the properties of individual atoms and molecules or bulk matter. Nanotechnology R&D is directed toward understanding and creating improved materials, devices, and systems that exploit these new properties. (NNI, 1999)

(2) Nanotechnology describes the creation and utilization of functional materials, devices and systems with novel functions and properties that are based either on geometrical size or on material-specific peculiarities of nanostructures. Purely geometrically the prefix "Nano" (Greek: dwarf) describes a scale 1000 times smaller than that of present elements of the micrometer-sphere (1 nm corresponds to the millionth part of a mm). This scale has become accessible both by application of new physical instruments and procedures and by further diminution of present microsystems. Also structures of animated and non-animated nature were used as models for self-organizing matter. (www.nanoforum.org)

(3) Nanotechnology describes the production, analysis and application of structures, molecular materials, internal interfaces and external surfaces with at least one critical dimension or with production tolerances of (typically) below 100 nm. Hereby it is decisive that solely on account of the nanoscaliness of the systems' components new functionalities and properties for the improvement of existing or the development of new products and application options result. These new effects and possibilities are predominantly founded on the relation of surface to volume atoms and on quantum mechanical behavior of building blocks of matter. [...] Accordingly, nanotechnology takes place in a transition range between individual atoms or molecules on the one hand and larger solids on the other. In this transition range, phenomena occur that are not observed at macroscopic items. (BMBF, 2004)

These definitions refer to the novelty of the functions and properties that appear at the nanometer scale. The physical, chemical, and biological principles at work at the nanoscale are mentioned as being characteristic. The reference in the definitions to applications results in a dimension being included that goes beyond that of pure natural science.

Probably the most ambitious definition in the sense of attempting to determine the specifically new aspect of nanotechnology is (Schmid and Decker, 2003; more precisely in Schmid *et al.*, 2006): "Nanotechnology is dealing with functional systems based on the use of subunits with specific size-dependent properties of the individual subunits or of a system of those." The following "specific size-dependent properties" were determined (Schmid and Decker, 2003, p. 24f.):

> Material properties cover magnetic, mechanic, electronic, optical, thermodynamic, and thermal features as well as the abilities for self-assembly and recognition. The specific-size dependence of these properties becomes evident when they
>
> 1. no longer follow classical physical laws but rather are described by quantum mechanical ones.
> 2. are dominated by particular interface effects.
> 3. exhibit properties due to a limited number of constituents, since the usual term "material" refers to an almost infinite number of constituents (e.g., atoms, molecules) displaying an averaged statistical behavior.

The basic idea is that nanotechnology does not consist solely in a miniaturization where simply effects of scale are manifested, but that something *qualitatively new* appears upon "entering nanocosmos."[6] Since this new quality cannot be tied to a clear indication of size such as 100 nm, this definition foregoes such arbitrary references. What precisely is to be understood by the new quality was specified by the authors in various tables that list the different scientific processes whose degree of newness was then discussed individually. The operationalization of the definition takes the form of a description of the cases to which it applies.

Yet these as well as other definitions have limitations. Attempts to apply them to concrete situations, such as assigning a concrete

[6] This metaphoric manner of speaking is suggested by expressions used in nanotechnology, which frequently use spatial terms to describe the nanometer dimension as uncharted territory (Nordmann, 2008, p. 226ff.).

researchprojecttonanotechnology,leadtoproblemsofdifferentiation. For example, there are no clear criteria as to what constitutes a *new* phenomenon or which operating principles are *new* (Decker, 2006). The reason for this lack of clarity is not, as is often claimed, that nanotechnology cannot be defined. It is, rather, that there is no consensus *as to the purpose* of the definition or that the question as to the purpose is not even posed. Based on a correspondence understanding of definition, attempts to define nanotechnology expect an "accurate" compilation of the objects without asking what "accurate" is supposed to mean in this connection and what the criteria and conditions of success of "accuracy" are — and will thus fail.

It is justified to ask whether and why a clear and unambiguous definition of nanotechnology is even necessary. Scientific progress takes place on the basis of very concrete procedures, experiments, theoretical hypotheses, and so on, which will take place independent of whether any of them are classified as nanotechnology. From an *internal perspective*, the question of a definition is irrelevant for nanotechnology itself (Decker *et al.*, 2004). The fact that the question of a definition has attracted such interest must have to do with other perspectives, which I will now examine briefly since the question of a definition has consequences for the issues studied in this book.

Any type of definition relates a specific notion to other notions. The term to be defined, the *definiendum*, is related to a subject area. The relation itself is called the *definiens*. Ideally, it should be possible to replace the *definiendum* by inserting the *definiens* without any loss or gain of information. There are some basic requirements for definitions, namely that a definition should

- only use terms that are already known or that have already been defined. If, in contrast, the *definiens* contains the *definiendum*, the definition is said to be circular
- lead to clear-cut decisions for each element of the subject area, whether it belongs to the subset given by the *definiens* or not. This principle is called *tertium non datur* — there is no third choice in between. In reality, however, there is frequently a gray area
- not contain exceptions
- be as simple as possible and unambiguous (nevertheless, the same expression can have different meanings in different

disciplines, as the term "crown" does in history, numismatics, or botany).

Definitions are supposed to perform certain functions and help someone achieve a goal. In general, such goals are to create a system, structure knowledge, make intersubjective understanding and cooperative action possible, or create a consistent basis for scientific disciplines. These general goals can be made more concrete for nanotechnology (Schmid and Decker, 2003; Decker, 2006):

- The definition is supposed to help make it possible to distinguish nanotechnology from already established fields of technology (such as microsystem technology or technical chemistry). In the process, the character of a developing field of research should become visible by reference to new knowledge, procedures or objects of research (clearest in the definition of Schmid and Decker, 2003 mentioned above). The purpose of this emphasis on the new is to position nanotechnology with regard to its perception by the public and political sphere and to legitimize its substantial financial support.
- A contribution is supposed to be made to constituting nanotechnology's scientific community and its identity. Discipline-related identities are represented in journals for the discipline, professorships, congresses, institutes, etc. Even if criteria for unambiguous definitions are not essential, the naming of nanotechnology's specifics is. In this regard, a definition serves to help the internal formation of the discipline by distinguishing it outwardly from other disciplines.
- An operative definition should make it possible to establish funding programs for supporting nanotechnology in ministries and other funding agencies and to guide their implementation.

Attempts to define nanotechnology as a whole have thus as a rule something to do with an *external* perspective of the field, i.e., in a very general sense one of science *policy*. At issue is the view of nanotechnology from the outside, as it is seen by politics, society, and the classical scientific disciplines. Even if attempts to define nanotechnology from the inside are made, i.e., by a researcher in nanotechnology, this takes place after an external perspective has been presented (as in the process that led to the definition in Schmid and Decker, 2003; Schmid *et al.*, 2006). The concept "nanotechnology" represents a term molded by research policy and research organization that correlates with external definitions of its goals and external perspectives of the discipline. These external goals and perspectives determine which requirements and criteria of success are applied to different proposed definitions.

Nanotechnology can also be regarded as a specific sociopolitical construct. Different actors with different perspectives and purposes confront one another in their attempts to determine nanotechnology conceptionally and in practice. The following quotation indicates the strange nature of the notion of "nanotechnology:"

> It could be said, based on Bruno Latour, the philosopher of science, that nanotechnology corresponds to a functioning coalition of molecules, probe microscopes, (ex-) chemists, visionaries, (nervous) investors, and even ethicists and philosophers of science. (Nordmann, 2007b, p. 216)

Far from what regulators would need in order to establish nanospecific rules or laws, and also far from all attempts at defining nanotechnology in a purely technological sense (cf. the quotations at the top of this section), this consideration makes it clear that there is also another dimension of "nanotechnology" which has to do with public perception, with the self-understanding of nanotechnology scientists, with the many images created by nanotechnology (Loesch, 2008), and with the fact that many borders can be crossed by nanotechnology, leading to the construction of new "boundary objects" (Kurath, 2010; Kurath and Maasen, 2006). This approach does not aim at defining nanotechnology but at understanding from an STS (science, technology, and society) perspective what "nanotechnology" means in the eyes of the actors and how it was constructed.

So which definition will be used in this book? Since the book will deal with ethical questions, it will apply the external perspective of the *question as to the normative uncertainties generated by nanotechnology* (see Chapter 3 for details). For these uncertainties, it is beside the point whether the physical limit of nanotechnology is considered to be 50, 100, or 200 nm. What is important is that relevant ethical issues arise, which we may not be able to determine if our definition is too narrow. For this reason, we will use a pragmatic characterization that is not clear cut (and avoid the word "definition"). Nanotechnology is taken to represent a collective term for technologies that are designed for or employ nanoscale systems (Fleischer, 2003) and that

- can be analyzed and manipulated in a purposeful and individual manner (as contrasted with statistically in the form of a large collective).
- make it possible to observe or create new effects and properties that are size specific, such as quantum mechanical effects.
- can or are supposed to demonstrate — at least in their intention — technical utility, as indicated by the word "technology" in the term.

The prefix "nano" is not qualified further; it is used solely as a guideline. It is irrelevant for ethical questions in this context whether the radius of the corresponding particles is less than 100 nm (which according to most definitions of nanotechnology would then be ethical questions of *nanotechnology*) or larger than 100 nm (Section 4.2). The scale of the materials and processes does, however, play a role, such as when the properties of a nanoparticle depend on its size or when a neuroelectric implant has to be miniaturized. The "technology" property of nanotechnology is central for this book since this property leads — via applications and their presumed consequences — to most of the ethical issues surrounding nanotechnology (Chapter 5). According to the situation that nanotechnology frequently is still more science than technology the envisaged "futures" of nanotechnology are also taken into account.

## 2.4 The Interdisciplinary Nature of the Nanocommunity

The disciplines of physics, chemistry, biology, and engineering and their subdisciplines such as solid-state physics and bioinformatics meet in the nanometer sphere, forming new interfaces to one another and with technical disciplines such as materials research and microsystem technology. In this sphere, the objects of research, the methods, and the results cannot simply be attributed to one of these classical disciplines or even to nanotechnology itself.

Nanotechnology can be viewed as the interface between these disciplines at the nanometer level:

> Nanoscience is the result of interdisciplinary cooperation between physics, chemistry, biotechnology, material sciences and engineering toward studying assemblies of atoms and molecules. (Renn and Roco, 2006)

Two different approaches are often distinguished (Bachmann, 1998). In the top-down approach, which is predominant especially in physics and physical technology, structures and components are continually being further miniaturized. In contrast, in the bottom-up approach, increasingly complex structures are being constructed, employing atomic and molecular components, and using the principle of self-organization. This approach has so far primarily been represented by chemistry and biology (for more on them see the corresponding chapters in Schmid *et al.*, 2006; Sattler, 2011).

The interdisciplinary interaction of physics, chemistry, biology, and the technical sciences that are involved is considered essential for the success of nanotechnology. There has in fact been talk of a paradigm change in these fields of science. Early on, the possibility or expectation was even mentioned that these classical fields of science might merge to form a new comprehensive discipline of nanoscience[7] and have a corresponding impact on the forms of university education (for a critical view of these euphoric expectations, see Schmid *et al.*, 2006; Chapter 7). Even chances for a newly unified science were seen:

> Most importantly, nanotechnology [...] reverses the trend of speciali-zation of scientific disciplines, providing unifying concepts for research and education, and leading to system integration in engineering and technology. (Renn and Roco, 2006)

Nanotechnology thus encompasses an unusually large number of different disciplines, including the natural sciences, medicine, and different engineering sciences. This multidisciplinarity is even larger than that in materials science and other fields of engineering. Education in a field based on so many different scientific disciplines

---

[7] The idea of converging technologies has contributed to a new optimism regarding the possibility of a unified science, including a vision of a "new renaissance" (Roco and Bainbridge 2002, p. 1ff).

— such as physics, chemistry, biology, engineering, materials science, and microtechnology — is a new challenge. Persuasive concepts for such education are still missing, especially with regard to the challenge as to how to transfer an education that at this time is a collection of different aspects of education in different disciplines to a new interdisciplinary field. Also, the relation of already existing university courses, in particular at the undergraduate level, on nanotechnology to the classical disciplinary education in physics, chemistry, and biology needs to be clarified. Currently, the situation in university education on nanotechnology issues might be described as more experimental than systematic.

In addition to this high degree of interdisciplinary work in the natural sciences and engineering, the social debates about nanotechnology have meant that the social and human sciences have — pursuing their nontechnological cognitive interests — engaged in the study of nanotechnology to an extraordinarily high degree. They include philosophy and ethics, political science, social science, economics, STS studies, technology assessment and ELSI (ethical, legal, and social implications) studies.

Taken together, it can be said that there is no well-defined nanocommunity in the sense of communities in scientific disciplines or subdisciplines which have their own journals, conferences, associations, etc. The nanocommunity has a patchwork character, and the cohesion of this patchwork is guaranteed in particular by the prefix "nano." The S.NET network (Society for the Study of Nanotechnology and Emerging Technologies), which has already organized two international conferences, could develop into a kernel of the interdisciplinary research around social and ethical issues of nanotechnology (Fiedeler *et al.*, 2010). The Encyclopedia of Nanotechnology & Society, which was just recently published (Guston, 2010), will also help develop such a community. Considering the difficulties in defining nanotechnology, however, the cohesion in this community will probably remain rather loose.

## 2.5 Philosophical Interpretations

Early on, nanotechnology stimulated interpretations that extended far beyond the horizons of science and technology. In particular, a number of the nanotechnological visions have led some to prepare diagnoses of the contemporary situation that are philosophical in nature, while leading others to use them to draft perspectives for the future of man and human civilization. These topics include (1) optimism about shaping the future that is based on atomic reductionism, (2) emphasis on the increasing uncertainty of the future caused by nanotechnology, and (3) the interpretation of nanotechnology as a symbol of the future.

### 2.5.1 Triumph of Homo Faber

Some of the commentaries on nanotechnology — primarily those from people involved in nanotechnology — clearly describe the return of optimism about shaping the future, even a return to a renewed claim to control and dominate nature:

> We are in the transition from being an amateur chess player to becoming a grand master, from observer to master of nature. [...] The age of discovery is coming to a close, and that of domination is beginning. (Kaku, 1998; cited after Schmidt, 2008)

These new (and obviously immodest) ideas regarding feasibility are fed by atomic reductionism, according to which all activity in the world can be traced back to causal processes at the level of atoms (see footnote 1). If nanotechnology were to offer the possibility of controlling these processes technically, then man would so to speak have control of the roots of all causal chains and could thus practically control everything. This interpretation envisages man's ultimate triumph, namely that armed with nanotechnology man could begin to manipulate the world atom by atom according to his own conceptions, which is ultimately the consummation of Bacon's thoughts (Schmidt, 2008). In deterministic thinking, control of the atomic dimension also means control over the spheres of living and of social phenomena. Drexler's molecular assembler is the technological side of atomic reductionism. Although this vision of "shaping the world atom by atom" (NNI, 1999) is profoundly incredible — "because it is exceptionally difficult and it would not be particularly efficient to create first molecules and then the entire world atom by atom" — it appears throughout nanotechnology as if it were a leitmotif (Nordmann, 2007b, pp. 220ff.). Ideas of being able to achieve anything arise inasmuch as man is viewed as the creator and controller of these developments:

> The aim of this metaphysical program is to turn man into a demiurge or, scarcely more modestly, the "engineer of evolutionary processes." [...] This puts him in the position of being the divine maker of the world. (Dupuy, 2005, p. 13)

These ideas are elevated to a theological level when man is described as a "created co-creator" and homo faber as a second god (Hefner, 2003). Representatives of this attitude presumably consider the recently expressed warnings about man's hubris and demands for "new modesty" to be an aberration of historic dimensions.

Objections to atomic reductionism come from epistemology (Schmid *et al.*, 2006). Nanoscientists frequently talk about atoms and molecules as if they were Lego blocks, which they can image, view, describe, and combine into larger complexes (after Janich, 2006; Schmid *et al.*, 2006). This view disregards, however, constructive elements that are essential for nanotechnological operations at the atomic level. One example of them is the notion of the "fat and sticky fingers" (Smalley, 2001) with which we and our equipment operate. These are said to prevent completely deterministic work with atoms, analogous to building with Lego blocks, because the interaction between manipulator and manipulated object cannot be disregarded. Even the visualization of atoms and molecules by imaging procedures such as scanning probe microscopy must answer questions such as to which premises affect the generation of its images and what are the constructive portions of the images. Imaging procedures reflect properties both of the apparatus performing the measurement and observation and of the probe surface being examined. According to epistemological arguments, atomic reductionism cannot be maintained because atoms are not (in contrast to Lego blocks) units independent of theory and cannot (in contrast to macroscopic building blocks) even be thought of independently of the roles of the producers and operator. This argument does not mean that technical operations at the level of atoms and molecules are impossible. It does, however, raise a warning sign against making far-reaching assertions on the basis of premature macroscopic analogies.

### 2.5.2 Huge Increase on Uncertainty

A second line of interpretation begins, initially quite innocuously, by viewing nanotechnology as an "enabling technology" (starting with Fleischer, 2003). This means that nanotechnology's core consequences are to be expected less from its direct inclusion in technical products, procedures, and systems, and that far-reaching advances from nanotechnology can be expected in many fields of technology in a more indirect manner (e.g., in electronics, energy technology, and medicine; see Section 2.2.3). As a cross-sectional technology, nanotechnology is available for use in any application and in many regards is *enabling*. This enabling property can appear at numerous levels: certainly at the level of new products using nanomaterials, particularly at the level of established key technologies such as information technology, communication technology, and biotechnology, and even more so by opening entirely new directions for technology, such as NBIC convergence (Roco and Bainbridge, 2002).

In this manner, nanotechnology contributes substantially to the multiple production of uncertainty, for example, with regard to conceptions, intentions, consequences, and ontology (Schmidt, 2008). Its openness toward being used in different applications means that we cannot know as much about nanotechnology's consequences for man and society as we would about those of "normal" technologies (i.e., those characterized by their applications). It is as if nanotechnology is becoming a "means in itself" or a means for which ends can be found (almost) arbitrarily. On the one hand, Rohbeck (1993), with reference to Dewey, referred to an "excess of means" to conceptualize technology's openness toward new applications as a result of the fact that it goes beyond the goals originally set for it. In contrast, in the enabling point of view the goals that nanotechnology is developed for are largely in the dark. Promotion of nanotechnology thus becomes an *end in itself*: people blindly promote the enabling property without knowing what precisely is to be enabled.

The extent of this uncertainty is illustrated, for example, by the extremely contrasting expectations toward nanotechnology, which range from catastrophe to paradise (Grunwald, 2007, 2009b). The paradisiacal expectations themselves range from Drexler's assumption that nanotechnology can be used to solve all of mankind's problems to the transhumanist hopes that "man as a deficient being" (Gehlen, 1986) can by means of technological improvement ultimately be made part of a perfect technological civilization. The anxieties about catastrophes concern, for example, man's complete loss of control (Joy, 2000). They appear, by the way, on both sides, both among the critics of nanotechnology (Joy, 2000; Dupuy, 2005) and among its adherents. In the latter, however, the meaning is just the opposite: "If we fail to chart the direction of change boldly, we may become the victims of unpredictable catastrophe" (Roco and Bainbridge, 2002, p. 3). The use of an anticipated catastrophe by both sides as a threat leads to an arbitrariness of conclusions and accurately illustrates the *maximum possible uncertainty*. In complete opposition to the assertion mentioned above about the return of thinking in terms of control and domination, the image here is of a development whose course cannot be predicted because of the enabling property of nanotechnology. Precisely because everything is possible, this development escapes our influence. Anticipations can range from paradise to apocalypse (Grunwald, 2009b; Laurent and Petit, 2006; see also Chapter 10).

### 2.5.3 Nanotechnology as a Symbol of the Future

A third perspective proceeds from the observation of the contrasting interpretations already mentioned and specifically addresses the opposition between them: "Yet while the statements of nanovisionaries express an optimistic technological Baconism, the nanofacts and their consequences suggest rather characteristics of uncertainty" (Schmidt, 2008, p. 2). The increased contingency of the human condition, which is a quality of nanotechnology (Grunwald, 2007), is thus seen as both an extension of uncertainty and the expansion of the opportunities for shaping the future — although making these opportunities for shaping the future become reality is becoming increasingly demanding. The transformation from passively accepting something given to considering it to be manipulable is a general feature of technological progress. To the same extent that the human ability to control is increased, new space for visions and shaping the future will appear, but also simultaneously — so to speak as a side effect — the same will be true for the challenges to identifying new forms of orientation in order to canalize and master the newly won spaces (which initially present themselves as increasing uncertainty). This in turn must become the topic of future debates — about the chances and risks — in which the process of reaching such social self-understanding can take place.

These debates often develop along specific symbols of the future (Grunwald, 2006a). Their function is to place our expectations of the future in the context of our present thinking, to reflect on them, to communicate the results of our reflections, and to deliberate on them in order ultimately to enable us to use the results for our current actions and decisions. The latter are not possible without notions of the future. Symbols of the future point to imagined *current futures* (Picht, 1971). The uncertainty of these symbols themselves is a precondition for their fulfilling their function of catalyzing social self-understanding.

Inasmuch as debates about nanotechnology transport and digest general social attitudes toward the future of man and society, they are placeholder events pointing beyond themselves to future developments — true "symbols." They are a placeholder for a discussion of society's future of considerable scope. In this sense, nanotechnology is one of today's symbols of the future. It is in good company. Some other prominent symbols of this kind are demographic change, climate change, the precautionary principle (von Schomberg, 2005), and sustainable development (Grunwald, 2007b). Each of them takes a look at different aspects of the future: for sustainable development, for example, issues of fairness and the limitedness of natural resources; for climate change, the vulnerability of man's way of doing business and lifestyle; and for demographic development, above all reproductive and migratory behavior. A common feature is their catalytic function in the context of the search for orientation in *today's* world via the (placeholder) route of deliberating about visions of the future (Chapter 10).

The specific feature of the symbol "nanotechnology" appears, according to the current situation, to be the future development of the relationship between humans and technology in a general sense (Grunwald, 2006a). This is, on the one hand, the case especially in view of the possibilities for "shaping the world atom by atom" (NNI, 1999) expected by some nanotechnologists (Section 2.1). For them, such shaping of atomic structures would not distinguish between animate and inanimate, between natural and artificial, or between human and nonhuman objects. Therefore, such far-ranging issues are involved, on the other, as "human enhancement" and fears of (or hopes for!) a coming strong technicalization of humans (Grunwald and Yulliard, 2007). Regarding nanotechnology as a "symbol of the future" seems to be adequate just because of these both possible developments and expectations (or fears) bringing with them a lot of grand challenges — including ethical ones.

These philosophical interpretations are, on the one hand, attempts at a better understanding of the changes nanotechnology will bring with itself. On the other, they are themselves part of traditional "grand narratives" accompanying human development and history (DEEPEN, 2009).

## 2.6 Public Perception

Polls show that nanotechnology's level of familiarity is rather low in the general population. Some 30% have heard of nanotechnology, but only 10–20% have a relatively concrete conception of it. The vast majority of the general population still shows only little interest in nanotechnology. If people take note of nanotechnology at all, this appears mostly as a blurred or nonspecific concept. Their views as to the possibilities and risks of nanotechnology correspond largely to their expectations regarding science and technology in general (Grobe, 2007).

For a long time, nanotechnology was perceived as a seemingly risk-free field. In the 1990s, public perception of it was low. A strong indication of its positive perception, however, was the use of the prefix "nano" in the media — not in the mass media but, for instance, in science magazines — as a synonym for "good" science and technology. In contrast to large power plants or big chemical factories — which frequently have a negative image in the public — nanotechnology promises a better, clean, and smart technological future based on the very positive appreciation of "the small." The nano hype in science and in the political communication about nanotechnology was also hyped with regard to public perception and media interest. It was presented as an outstanding example of a "positive story" with respect to scientific and technological progress.

This situation changed radically in 2000. The positive utopias of nanotechnology, based on technical access to "the small," were inverted into horror scenarios on the basis of precisely the same "small" technologies (Joy, 2000). The ambivalence of technology-based visions became obvious in a dramatic way (Grunwald, 2007a, 2009b). The public debate about the risk posed by nanotechnology emerged around issues concerning the visionary and more speculative developments. Within months, the topics of "grey goo," "nanobots," and "cyborgs" became well-known to many people (Schmid *et al.*, 2006).[8]

### 2.6.1 The "Grey Goo" Scenario

The media coverage of nanotechnology has invariably raised the specter of "grey goo," a doomsday scenario in which nanoscale robots self-replicate out of control, producing unlimited copies of themselves, consuming all available material, and ultimately laying waste to the planet: "Tough, omnivorous 'bacteria' could outcompete real bacteria: they could spread like blowing pollen, replicate swiftly, and reduce the biosphere to dust in a matter of days" (Drexler, 1986, p. 172). While most of the scientific community considers this to be science fiction, others have argued that it is a possible outcome of unregulated nanotechnology making use of concepts of self-organization (Section 7.3.1).

The public perception of risk was frequently influenced by arguments using the "grey goo" scenario to call for greater caution in scientific and technological progress. This scenario reminded many people of the large apocalyptic fears that were brought up in the 1970s and were based on technological innovations, and that motivated Hans Jonas (1984) to coin his famous "imperative of responsibility." Although apocalyptic illustrations of future developments have accompanied scientific and technological advances since the industrial revolution, their role and influence have varied considerably over time. The level of public and media interest in nanotechnology therefore justifies the question: "Is grey goo a real concern, or is it a distraction from the important issues?"

### 2.6.2 The "Prey" Scenario

The "prey" scenario is built on the same basic idea, i.e., of self-replicating nanorobots. This nanorobot idea was initially invented to describe a positive utopia expected from nanotechnology. Nanorobots could, according to this vision, invade the human body in order to act there as an improved health maintenance system

[8]Although these "dark sides" of the primarily positive visions were already present in Drexler's early book (1986), this did not become a public issue until following the publication of Joy's article.

(Chapter 9). Bill Joy (2000), however, drew a completely different picture. He argued that the "converging technologies" (in fact, he did not use this notion but pointed to developments that are currently being discussed under this term) nanotechnology, biotechnology, and information technology could indeed lead to the possibility of "intelligent" nanorobots. His concern is that humans might easily lose control over these armadas of nanorobots. Instead of being servants to human needs, nanobots could start acting independently of human orders and even gain control over the world. The completely technical civilization which could emerge from such a development would no longer need humans.

Losing control over technology is, just as the apocalyptic fears mentioned above, an old stimulus of social thought about technology. The concern that technology could invert the traditional relationship between humans and technology — technology serving human needs — is also a leitmotif of debates — both those philosophical and general in nature — on future prospects in a technicalized civilization. The more capabilities are given to technology, the greater the fears in this respect (Grunwald and Julliard, 2007). Accordingly, it is not at all surprising that such scenarios as "prey" have emerged and will also accompany future debates on more and more "autonomous" technology.

### 2.6.3 The "Cyborg" Scenario

Nanotechnological visions repeatedly contain elements that blur the boundary between what human beings are and what they create as a result of technical achievements and utilizing technical applications. The Foresight Institute and the "extropian movement" have developed futuristic visions which include the possibility of the technological storage of human consciousness. Furthermore, some researchers who participate in the U.S. National Nanotechnology Initiative assume that artificial brains might be created within the next 25 years (Cauller and Penz, 2002). Such visions pose the question of the extent to which technical or partly technical-partly biologically constructed man-machine chimeras can lay claim to the status of a person. An entire spectrum of anthropological and ethical questions follows from this. Precisely this aspect of nanotechnology is also a central theme in some American churches, merely from just the opposite point of view from that in the transhumanism discussion: "When God fashioned man and woman he called his creation very good. Transhumanists say that, by manipulating our bodies with microscopic tools, we can do better. Are we ready for the great debate?" (Hook, 2004). Nanotechnology currently acts as a playing field on which various philosophies of life compete.

The idea of cyborgs is not a new one. Starting with the science fiction literature of earlier periods (cf., e.g., the figures and stories of Golem and Frankenstein), there has been an ongoing debate on the relationship between humans and technology and on the border between them. The technicalization of humans on the one hand and the growing capabilities of autonomous robots on the other are apparently posing challenges to traditional convictions. Cyborgs lead to issues of control between people and machines, which are relevant to medical devices, especially active or neural implants.

Beyond those futuristic elements of public debate a second and more "down to earth" branch of the risk debate on nanotechnology emerged focusing on the question of possible harm to health and the environment caused by synthetic nanomaterials (see Chapter 6 for an in-depth analysis). Concerned groups began to think about the analogies and parallels between nanotechnology and the lines of technology that have a specific history in the public debate about risk: nuclear technology and biotechnology (ETC, 2003). Newspapers put nanotechnology in the category of risky technologies. Reinsurance companies quickly became aware of the possible risks posed by nanomaterials and of related governance questions (Munich Re, 2002; Swiss Re, 2004). We can thus witness, in the mirror of public perception and the mass media, the fall of nanotechnology from a synonym for "good" scientific and technological progress to a technology line that is expected to pose many risks, many of which are still unknown (Myhr and Dalmo, 2007).

One specific item made a major contribution to the "fall" of nanotechnology in public perception. This was the question of a possible analogy to asbestos. Asbestos provided a case study that showed what could result from the intensive use of materials if a careful impact analysis had not been performed. Some voices pointed to possible analogies of synthetic nanoparticles to asbestos: "Some people have asked whether the ultra-small particles and fibers that nanotechnology produces, such as carbon nanotubes, might become the new asbestos" (Ball, 2003). In fact, there are almost no analogies in a physical or chemical regard or with respect to the size or shape of asbestos fibers and today's synthetic nanoparticles. The relevance of asbestos to the nanoparticle risk debate originated above all in the dramatic case of what could happen if no or few precautions were taken. The miraculous engineering properties of asbestos had led to a rapid growth in its exploitation and use. Although adverse health effects (asbestosis) had been observed rather early and led to some workplace regulations in the 1930s, other relevant pieces of knowledge (concerning lung cancer and mesothelioma caused by asbestos fibers) had, however, been ignored or even suppressed. There was no comprehensive collection and assessment of data prior to the 1960s (Gee and Greenberg, 2002). The story of asbestos, especially the severe health and economic disasters associated with it, served as strong motivation to ask for a more precautionary approach in the field of nanoparticles.

The emergence of the risk issue in combination with the fact of there was practically no knowledge available about the side effects of nanotechnology led to severe irritations and to a kind of helplessness at the early stage of that debate. Statements from that time waver between an optimistic "wait-and-see" strategy (Gannon, 2003) on the one hand, and strict precautionary and sometimes "alarmist" approaches on the other:

> The new element with this kind of loss scenario is that, up to now, losses involving dangerous products were on a relatively manageable scale whereas, taken to extremes, nanotechnology products can even cause ecological damage which is permanent and difficult to contain. What is therefore required for the transportation of nanotechnology products and processes is an organizational and technical loss prevention program on a scale appropriate to the hazardous nature of the products. (Munich Re, 2002, p. 13)

The position on nanoparticle regulation that is probably still the most famous is the postulate of the ETC group for a moratorium:

> At this stage, we know practically nothing about the possible cumulative impact of human-made nanoscale particles on human health and the environment. Given the concerns raised over nanoparticle contamination in living organisms, the ETC group proposes that governments declare an immediate moratorium on commercial production of new nanomaterials and launch a transparent global process for evaluating the socio-economic, health and environmental implications of the technology. (ETC, 2003, p. 72)

The ETC work gave a significant push to debates about nanotechnology regulations in many countries including broader governance issues (Kaiser *et al.*, 2010). They also increased the fears of nanotechnology researchers of a broad public front of rejection and protest analogous to that in the histories of nuclear power technology and of biotechnology. A completely different but also far-reaching recommendation aims at "containing" nanotechnology research. This strategy would imply a secret and strictly controlled nanotechnology development that seems to be unrealistic and unsafe as well as undemocratic. Furthermore, this recommendation is irritating regarding the ideal of an open scientific community:

> CRN has identified several sources of risk from MNT (molecular nanotechnology), including arms races, gray goo, societal upheaval, independent development, and program of nanotech prohibition that would require violation of human rights. It appears that the safest option is the creation of one — and only one — molecular nanotechnology program and the widespread but restricted use of the resulting manufacturing capability. (Phoenix and Treder, 2003, p. 4)

All of these different proposals have enriched (and fueled) the public and scientific debate. Seen from today's perspective, these proposals are documents of a very specific situation. Nanotechnology, itself still in an embryonic state, experienced itself, more or less suddenly, as the subject of a public debate about risk. The situation was characterized by severe challenges: while high expectations of benefits were still dominant, no reliable knowledge about the possible side effects of nanotechnology was available.

The first years of the nanotechnology risk debate were thus characterized more by suspicions and uncertainties than by knowledge-based and rational deliberation. In the meantime, however, things have changed considerably. An intense debate on nanotechnology issues took place early in many countries. In 2003, the German Parliament held a debate about nanoparticles that was informed by a technology assessment study (Paschen *et al.*, 2004). A study by the Royal Society and the Royal Academy of Engineering (2004) resulted in numerous statements and recommendations aimed at closing the knowledge gaps and at minimizing the risks that might result from the production and use of nanoparticles. Following that report the UK Government promised to promote more research on the risks posed by nanotechnology, to conduct reviews of existing regulations with respect to their applicability to nanotechnology, and to support a public debate (UK Government, 2005). Ambitious research programs on the possible side effects of nanoparticles have also been started in many other countries in the meantime (see ENRHES, 2010, for a recent review).

In spite of these activities, public uneasiness concerning nanoparticles has not disappeared (Marcinkowski *et al.*, 2008). New postulates for a moratorium have been brought forward, in particular in the fields of cosmetics and food (Friends of the Earth, 2006; BUND, 2008). Empirical research on the public perception of nanotechnology shows persistent considerable concern, albeit in a rather small part of the population. Up to now there is no broad and radical opposition or rejection but rather a more or less "normal" societal debate on chances and risks (Section 11.1; see also Grunwald and Hocke-Bergler, 2010).

Recently, the public debate on nanotechnology, in particular on its more futuristic aspects, was analyzed in the DEEPEN project funded by the European Commission. One of the results was that cultural narratives form the background of many of the public debates and concerns. These narratives are (Macnaghten *et al.*, 2010; see also DEEPEN, 2009):

- Be careful what you wish for
- Opening Pandora's box
- Messing with nature
- Kept in the dark
- The rich get richer and the poor get poorer

This result shows that public perception of new and emerging technology not only depends on the specific parameters and impacts of that technology but also on underlying cultural attitudes and traditions. It also makes clear that the public debate on nanotechnology is not an isolated one but is rather part of an ongoing societal debate on new science and technology (von Schomberg and Davies, 2010).

# Chapter 3

# Ethics, Technology, and Risk

The ethics of technology is a relatively new field of applied ethics (Mitcham and Briggle, 2010). Few standards have crystallized in it, and there is still no consensus, not even on fundamental issues, such as on the direction and goals of the ethics of technology and on the relationship of philosophical–ethical expertise to the moral positions of affected individuals and stakeholders on the one hand and to policy decisions on the other. For this reason it seems advisable for me to provide a few preparatory clarifications regarding my understanding of ethics before beginning the discussion of the ethics of nanotechnology. This is important to ensure that both the basis of the later results on the ethical issues of nanotechnology and their claim to validity are transparent.

## 3.1 Problem-Oriented Ethics

Applied ethics is caught in the tension between normative, top-down approaches in which philosophical principles, for example, the categorical imperative, are supposed to be broken down into positions on concrete problems and bottom-up approaches that take into consideration the moral notions present in the affected lifeworld and search for rather pragmatic solutions to conflicts. In the tension between these two approaches, which has something to do with the controversies between continental European and Anglo-

*Responsible Nanobiotechnology: Philosophy and Ethics*
Armin Grunwald

ISBN 978-981-4316-80-4 (Hardcover), 978-981-4363-33-1 (eBook)
www.panstanford.com

Saxon conceptions, I will first examine and expound the motives behind the problem-oriented procedure, which will determine the further course of matters.

### 3.1.1 Ethics for Resolving Moral Conflicts

The distinction between factual morals and ethics, understood as the reflective discipline for considering moral conflicts or ambiguities, has been widely accepted in the modern discussion. This distinction takes into account the plurality of morals in modern society. It restricts ethics to those cases where pluralism as the basis for choosing the good life of an individual has to be transcended in order to achieve and justify the rules and procedures of collective life. Individual ideas of the good life, resulting from individual moral conceptions and preferences, have to be assessed, weighed, and compared to determine if there are conflicts between particular morals. The task of ethics is, thus, to analyze these ideas with respect to the nature of their justification, their presuppositions and premises, their normative content, and their justifiability. Ethical reflection can thus be seen as a methodological endeavor dealing with the set of methods, procedures, instruments, and tools to discursively manage and resolve conflicts arising from different moral assumptions in the absence of consensually accepted customs or moral systems. Ethics in this sense — which does not consider questions of an individual's "good life" — consequently always deals either with questions of collective actions or of individual actions with a collective impact and is, therefore, closely related to political philosophy.

This emphasis on moral conflicts being the *pragmatic sites* of ethical reflection and, therefore, creating the precondition of ethics' practical relevance leads to the conclusion that moral pluralism does not, as often asserted, *oppose* opportunities for ethical reflection but, on the contrary, is actually its necessary precondition. If, namely, *one* moral conception were binding and if its validity were not called into question, there would be no need for reflection on decision-making problems resulting from different and often conflicting moral conceptions. The pragmatic site of ethics is the location for discursively handling and resolving moral conflicts. Ethics is relevant to scientific and technological practice precisely when *participants* in technological development reflect on such situations of conflict (Grunwald, 2003).

In this context, the term "conflict" may not be considered in an overly narrow and restricted sense. As used here, it includes moral ambiguities, uncertainties, and indifferences, as well as new challenges for which moral customs have yet to be established or where there are doubts as to whether established moral traditions apply. In this sense, it serves as a framework for all situations where, for a given action or decision-making problem, there is no consensually accepted moral background from which orientation for decision making can be gained. In the following, I will refer to such situations as situations of *normative uncertainty*.

The implication of the fact that moral conflicts in this general sense are the pragmatic sites of ethical reflection is simply that ethical reflection is not required in their absence. This conclusion affects all areas of human life: ethical reflection is not required in many, probably even in most, of the cases when we act or face the need to make a decision. Indeed, looking at our lifeworlds, for example, it is clear that we do not reflect on all of our actions and decisions from an ethical perspective; on the contrary, ethical reflection is a type of activity which is undertaken rather rarely. This view is compatible with the approach of Habermas (1973), who distinguishes between the level of action (*Handlungsebene*) and that of discourse (*Diskursebene*). The latter is only reached in situations when there are severe problems which cannot be solved at the level of action. Applying this distinction to practical questions regarding how to decide or to act means that proposed actions and decisions can be assessed within the level of action *if* there is an accepted moral background. In these cases there is no need for ethical reflection, which — as a theoretical endeavor — belongs to the level of discourse. Ethical reflection is asked for if those assessments require the questioning of and reflection upon the background itself. This leads back to exactly the types of situations designated by the keyword "moral conflicts" as used above, namely, as the pragmatic sites of ethical discourse.

Viewing ethics as a reflective discipline for *de facto* moral issues means, therefore, that ethics can achieve practical relevance for the development of technology only if the decisions related to this technology involve moral conflicts. The moral implications of conflicts over technology form the thematic crux of ethics in technological development. Such conflicts are often about more than technical means and instruments. They are about visions of the future, about images of humanity, and future plans for society. Since technology is, in principle, not only a means but must always be analyzed with regard to its means–ends relationships, the level of ends — i.e., evaluative rationality (Rescher, 1984) — always plays an important role in discussions about technological means. In this way scientific, political, and public debates about new technologies often find themselves approaching questions such as what kind of society we *want* to live in, what self-image of man we implicitly assume, and whether or under what conditions future developments are acceptable or desirable. Such conflicts over technology, therefore, are ethically relevant and as such require that ethical reflection is explicitly undertaken.

Problems and needs stemming from social practice are at the methodological beginning of ethics, as it has been introduced here. This means that ethics is fundamentally problem-oriented in the sense that it is related to action outside of it. Pursuing ethics is thus a means to ends that lie beyond it.[1] This fundamental orientation of ethics around problems permits it to be considered a "needs science" (*Notwissenschaft* after Lorenzen, 1987). The "need" that ethics is supposed to contribute to overcoming can be an *already existing* problem in the form of normative uncertainty (see Section 3.1.2 for clarification). The consequence of limiting ethics to already existing problems would, however, with regard to science and technology be the known criticism that ethical reflection inherently comes too late and is thus not in a position to contribute to preventing undesirable consequences (Chapter 10). It is therefore necessary and for this book a constituent component that problems that are *anticipated or are envisaged in mental experiments* are a reason for employing precautionary ethical analysis.

The frequently discussed issue of the manner in which applied ethics is "applied" appears artificial in this perspective. The point is not for universalist principles such as the categorical imperative to be applied to praxis, but — just the opposite — that the universalist principles of ethics owe their origin and legitimation to the demands of praxis. Universalism is not an abstract construction that societal reality has to align itself with; it is rather a suggestion for solving certain types of moral problems.

---

[1] This view of ethics raises questions regarding its success and the criteria for its success. Inasmuch as ethical activity is not an end in itself but is supposed to serve ends outside itself, obvious tasks are to examine if and to what extent ethics contributes to the realization of these ends and how it could do this better.

In this sense the multitude of ethical conceptions in no way pose a difficulty to defining the concept of "applied ethics." On the contrary, they are different offers for satisfying the needs that develop out of praxis, and different (and certainly sometimes also competing) means for the purpose of coping nonviolently with normative uncertainties. The different schools of ethics can also be considered to be competing means in this means–end schema of ethical counseling. The situation that several means are in competition for one end is well known in action theory and from our lifeworld and is everything else but a problem, except of course that this situation leads to the problem of selecting and deciding and thus to the question as to the appropriate criteria.

The means–end view of ethics does entail problems and challenges. Initially this is above all the need to recognize the problems developing out of praxis, to reconstruct them in their own terms, and to assess their normative substance. Furthermore, since the goal is to strive to contribute to solving a problem, we must make sure that the ethical offer for solving a problem *is heard* in the respective area of praxis. Problem-oriented ethics thus needs a model of the field of society that it is related to. It needs both knowledge of the system as well as knowledge as to where it can contribute its part to solving a problem. This system can be a specific setting, such as how the respective "area" works (Nida-Rümelin, 1996a), for example, in the case of medical ethics empirical knowledge of the healthcare system, new medical technology, or the doctor–patient relationship. Yet it can also comprise issues of social theory, such as questions of interpretation in the context of reflexive modernity and the debate about side effects (Böschen *et al.*, 2006), and frequently has to take empirical knowledge from social science into consideration, such as how regulations and governance can be envisaged in differentiated societies in general and in the specific area in particular. Thus in problem-oriented ethics, normative uncertainties developing out of praxis confront problems of normative evaluation, issues of social theory, and specific items of knowledge from the fields of social praxis (*Lebenswelt*).

When ethics is considered in this manner as a means to certain ends, then its relationship to praxis is made up of four components.

1. Social praxis is the source of the problems that ethics must deal with. The objects of ethical reflection come from social praxis, either directly via existing problem situations or as problems envisaged by mental experiments.

2. It determines the requirements and criteria of success that will be applied to the proposed ethical solutions. Whether ethical reflection is successful and satisfies it own aspirations is determined by the relevant social praxis, not in the world of academia.
3. It provides the "source material" for predeliberative agreements (see Section 3.1.4 for clarification) which are necessary as the argumentational foundation for ethical proposals for solving moral conflicts and which form the basis of the acceptability of ethical orientation.
4. It is the addressee for the results of ethical analysis deliberation, which should then be better able to proceed with action and decision making.

This list makes the multiple relationships between ethics and social praxis or the lifeworld obvious. The relationships named above, however, are themselves as a rule not uncontroversial: problems are not simply supplied, criteria of success are not simply determined, the elements of predeliberative agreements are not simply provided, and results will not be simply implementable by addressees. On the contrary, each of these relationships first has to be reconstructed and interpreted. These achievements of interpretation and reconstruction belong to ethics' spectrum of tasks, and they themselves frequently constitute a good measure of the enlightenment that ethics can achieve.

The self-understanding of ethical reflection, just like that of other forms of social and political counseling, is that it is pragmatic, in contrast to the models of a decisionistic or a technocratic relationship between science, the public, and politics, all of which are equally deficient according to democratic theory (Habermas, 1970). According to this, scientific, social, and political consultation only plays a preparatory role in the advisory process that the "citizenry" must complete (in the sense of self-counseling):

> An element of anticipation is always contained in the integration of technical knowledge and the hermeneutical process of arriving at self-understanding. For it is set in motion by discussion among scientists isolated from the citizenry. The enlightenment of a scientifically instrumented political will according to standards of rationally binding discussion can proceed only from the horizon of communicating citizens themselves and must lead back to it. (Habermas, 1970, p. 74, German original, 1968, p. 137)

Problem-oriented ethics of nanotechnology must therefore put itself in the service of a more general social dialogue about issues posed by nanotechnology and provide ethical orientation to support the democratic debate, formation of opinion, and decision making (Grunwald, 2008a). Next, we will have to look closer at the issue of *normative uncertainty* by differentiating between standard and nonstandard situations in moral respect.

### 3.1.2 Standard Situations in a Moral Respect

In lifeworld matters, explicit ethical reflection hardly ever forms our initial observation. Most decisions take the form of goal–means deliberations at the action level (Habermas, 1973) without any particular reflection on their normative background. The normative background of these decisions is not analyzed, but is taken to be valid. The discourse level, at which this background could be called into question, is the exception. My hypothesis is that technology-relevant decisions are usually made analogously (Grunwald, 2000a). The initial observation that led me to formulate this thesis is that the great majority of technology-relevant decisions can be classified as "business as usual" in the following sense: the normative aspects of the basis for the decision are not made the object of special reflection, but accepted *as given* in the respective situation, thereby also accepting the normative framework this entails. No explicit ethical reflection is as a rule necessary when an engineer in a laboratory thinks over the question of whether she or he should use iron or aluminum for a certain component, when a manager has to decide whether she or he should order production unit A or B for a particular area, or when a licensing authority has to decide on an application for building a chemical manufacturing plant. This is true in all of these cases, even if normative elements obviously play an important role in these decisions.

The reason is that actors *can assume*, in making these decisions, a normative framework — the basis on which the decision can be made — to be given. A conflict in these cases does not carry moral connotations. Parts of this normative framework are (national and international) legal regulations, the standard procedures of the relevant institutions (e.g., corporate guidelines), possibly the code of ethical guidelines of the profession concerned, as well as general, uncodified societal usage. The demands on the normative framework which define a business-as-usual situation are formulated more precisely by the following criteria (expanding on Grunwald, 2000a):

#### 3.1.2.1 *Pragmatic Completeness*

The normative framework must pragmatically provide sufficient coverage of the pending decision from a normative point of view. None of the essential aspects may be left unanswered or considered to be indifferent from a normative point of view. As a rule, this is the case in all routine situations of action. This criterion cannot be satisfied for those situations for which normative routine attitudes are not yet present for new forms of action. Well-known examples are frequently tied to advances in science and technology, such as dealing with atomic weapons, issues of intensive medicine or reproductive medicine, genetic diagnostics, or reproductive cloning.

#### 3.1.2.2 *Local Consistency*

There must be a sufficient degree of consistency between the elements of the locally relevant normative framework, i.e., the framework in the respective field of action. In this context, "sufficient" means that the conclusions for the pending decision or opinion converge even if they follow from different morals (e.g., different religious attitudes). Otherwise there would be a conflict for which ethical reflection would be necessary. In the case of stem-cell research, for example, the different attitudes as to whether an embryo is to be attributed the status of a person lead to certain conflicting judgments regarding the circumstances under which such research can be carried out.

#### 3.1.2.3 *Sufficient Lack of Ambiguity*

Among the actors there must be sufficient consensus regarding the interpretation of the normative framework in the context of the pending decision. It is useless, for example, if all of the involved persons or positions proceed from the dignity of man but understand it differently. Problems currently arise with this demand for a lack of ambiguity particularly in the context of the concept of sustainability. This term, which is being employed in an inflationary manner, is frequently used with different meanings, which is especially true of its normative aspect.

#### 3.1.2.4 *Acceptance*

The normative framework has to be accepted as the basis of a decision by the actors and those affected. Inasmuch as this is not the case, there is obviously a conflict in a normative sense. It would thus be impossible to speak of a moral standard situation. This would even be the case if there were no agreement about the selection of the criteria to be employed to choose among different options for acting. An example is whether cost–utility considerations or ethical criteria based on principles should be employed; another is how the burden of proof should be distributed. In particular, such questions are relevant in connection with dealing with "unclear risk" (Chapter 6).

#### 3.1.2.5 *Compliance*

The normative framework not only has to be accepted in the respective field, but it must be *de facto* adhered to by the majority. This criterion refers to the necessity that announcements and deeds must correspond. Inasmuch as this is not the case, it can be a case for the states attorney, for example, if environmental protection laws are not observed. It can also compromise the dignity of negotiating parties, for example, if voluntary industrial commitments are repeatedly disregarded, jeopardizing trust. Or it results in such a far-reaching rejection of the consequences resulting from an existing normative framework that the entire framework itself has to be questioned.[2] Such a situation can hardly be called a moral standard situation, even though ethical reflection will hardly be able to correct the consequences.

These criteria can be interpreted as *conditions of coherence*. For a framework to be used as a means of orientation for action and decision making without further discussion, there must be a minimum amount of *pragmatic coherence* between the currently valid normative framework and the need for action and decision making.

If these conditions of coherence are satisfied in a specific context, then neither moral conflicts nor ambiguities exist. There is, consequently, no need for ethical reflection. Participants and others

[2] An example is a decision regarding the permanent disposal of radioactive waste. Even democratically legitimated decisions have largely been rejected in Germany and some other countries.

affected by a decision can take information about the normative framework into consideration as axiological information without having to analyze it. *Quasi-descriptive* technology assessment based on empirical value research is then possible. Although normativity must be taken into consideration in such a technology assessment, it is not expressly reflected in the assessment, but is inferred from the existing social circumstances. This is possible because the currently valid normative framework — according to the criteria listed above — is coherent enough to be able to provide undisputed orientation for acting and decision making. In such "business-as-usual" situations, the *criteria* for making decisions are *a priori* obvious and not questioned (e.g., a cost–benefit analysis in situations in which this is considered an appropriate method according to the accepted normative framework). This discussion raises the following questions: (i) Do such normative frameworks exist in modern society? (ii) How do they relate to society as a whole? (iii) How are they developed? and (iv) How can they be made explicit?[3]

1. The observation that numerous social practices, sectors, and subareas function as a rule without ethical reflection (e.g., in a company's internal industrial activity, a school's daily organization, or a licensing authority) can be explained by the assumption that a normative frame of reference does exist that is acknowledged as valid for decision making and action without further scrutiny. It is practically neither possible nor sensible to subject the entire arsenal of normative contributions in society to continuous ethical reflection, which would immediately result in a standstill of all social activities (Mitcham, 1987). Descriptivistic conclusions (from *is* to *ought*) are, therefore, in a certain sense not only indispensable but also pragmatically reasonable, and are actually pragmatically accepted in broad areas of society. In this context, "pragmatically" means as long as the respective normative framework works well enough.
2. With regard to society as a whole, there is no known normative framework that satisfies the conditions of coherence listed above. Modern society is characterized by moral heterogeneity, by a pluralism of values, and by differentiations in quite varied respects (Giddens, 1984; Beck, 1986). Decision making is,

[3] The question, what could be done if the criteria are not fulfilled will be dealt with in the following section (Sec. 3.1.3).

therefore, delegated to established decision-making procedures that provide legitimacy. This provides a means by which decisions can be made, in particular, when a generally accepted level of normativity does not exist (Luhmann, 1983). The basic normative conditions which satisfy the coherence criteria are always only valid contextually and "locally" (in time and space). There are numerous normative frameworks in a multitude of various contexts for which there are no coherent general and overarching frames of reference.

3. Basic normative conditions originate in the pressure of social practice and can therefore change over time: a once-valid normative framework can be put in doubt by even minor changes in certain parameters. Basic normative conditions are also extremely fragile. Since transgression of the limits of an acknowledged normative framework is problematic for ethics, the greater challenge consists in being able to recognize these limits at all. This is especially the case considering that such a border is not laid down *a priori*. On the contrary, it is subject to historical change and is socially constructed. It can also be challenged and altered by the introduction of new technologies and by new scientific results and abilities.
4. If the primacy of practice applies in this sense, then normative frameworks can only be made explicit by reconstruction. They are not codified (if they were, they would be accessible to a great extent in an unambiguous manner), but are the object of interpretation and explication. Recognizing the limits of the validity of normative frameworks can therefore not be attained by means of deductive, "top-down" reasoning. This itself requires interpretation and reconstruction, with all of the methodological problems this entails.

The following simile is proposed as a metaphor for this conception of society's normative structure: the scope of a local normative framework forms an "island of coherence" in the heterogeneous ocean of society. It is not an ontological unit, but develops out of the interpretation of social practice (i.e., a controversy that cannot be decided can be conducted over the existence and scope of certain islands). Islands of coherence are — over time — dynamic: their size and shores change, as does their relationship to one another. Such islands can vanish and reappear. On the islands, value research is sufficient to provide orientation for action and decision making. Value research cannot, however, provide adequate orientation as soon as the limits of these islands are transgressed.

### 3.1.3 Beyond Standard Situations in a Moral Respect

Social coherence is incomplete and unstable, a fragile mesh, which is continually being challenged and threatened by social developments or by innovations and by the conflicts resulting from them. Technical innovations (but not only these) can lead to various forms of *incoherence*. They can "disturb" ethical coherence and make ethical reflection necessary as part of a modification or supplementation of the normative framework. New scientific knowledge and technological innovation may transform earlier standard situations in a moral respect into *nonstandard situations* where one or more of the criteria given in Section 3.1.2 are no longer fulfilled (see Fig. 3.1).

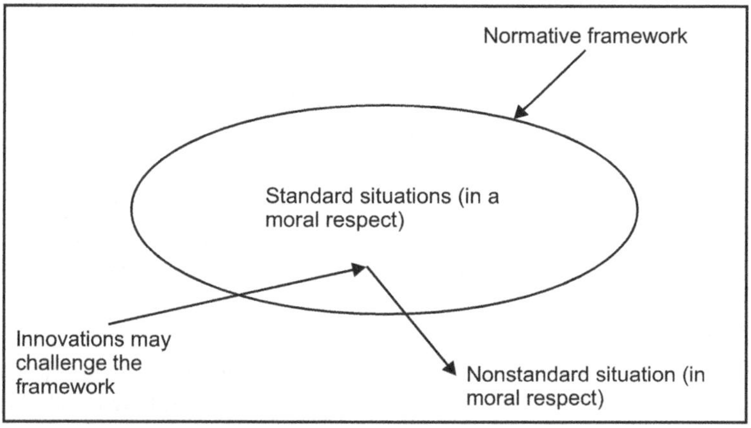


**Figure 3.1** The basic model of innovations challenging standard situations.

In this modified situation, there are three options to choose from:

1. Reject the innovation causing moral trouble — renounce its possible benefits and maintain the initial normative framework.
2. Try to modify the properties of the innovation responsible for causing moral trouble (maybe circumstances of its

production involving animal experiments or the location of a nuclear waste disposal site in a sacred region of indigenous people) in order to be able to harvest the expected benefits without causing moral trouble (see Fig. 3.2).

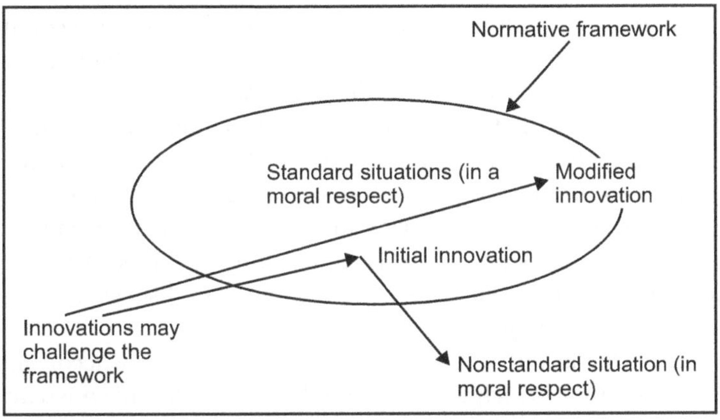


**Figure 3.2** Modify innovation.

3. Modify the normative framework, so that the new technology could be accepted (and the benefits harvested) in a way that would not lead to izncoherencies regarding the modified framework (see Fig. 3.3).

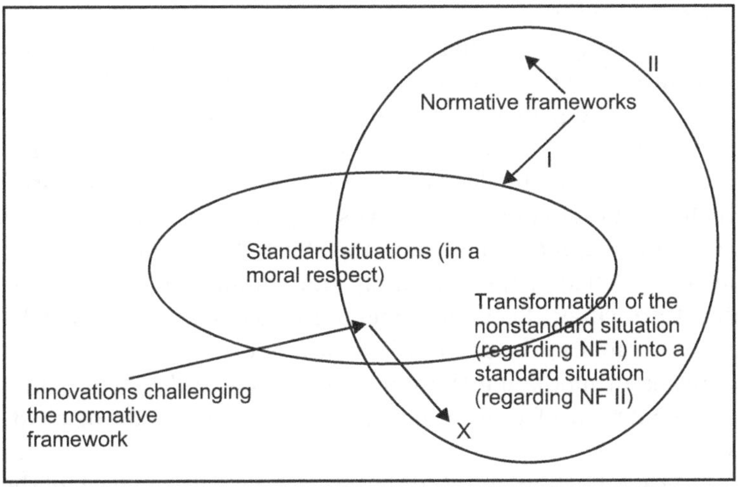


**Figure 3.3** Modify normative framework.

Ethical reflection plays a decisive role for anticipated actions in determining the choice between such alternatives and — in cases 2 and 3 — between different versions and for the concrete consequences. In these cases the reflection is an act of balancing the expected advantages of the innovation or the new technology against the moral or other costs if — as is probably the most common situation — there are no *categorical* ethical arguments for or against. The following can be said about the options:

*Option 1*: As a rule, this option is chosen if there are strong, i.e., categorical, ethical arguments against the new technology. An example is reproductive cloning. Such cloning or research on cloning is prohibited in many countries for ethical reasons. Although the argument of Habermas (2001) against interventions in the germ line also claims to be such a strong argument, in discussions it is an object of controversy.

*Option 2*: The option of shaping technology specifically according to ethical values or principles is behind the approaches of constructive technology assessment (CTA; see Rip *et al.*, 1995), of the social shaping of technology (Yoshinaka *et al.*, 2003), and of value sensitive design (van de Poel, 2010, pp. 1001ff.). The focus is on directing the shaping of technical products or systems along the relevant factors of the normative framework so that the products or systems fit the framework. This would so to speak in itself prevent normative uncertainty from arising (on critical questions to this approach, see Grunwald, 2000b).

*Option 3*: Frequently there are even more complex necessities to balance factors, such as when the (highly promising) use of a new technology or even research on it is not possible except by producing normative uncertainty. Examples are animal experiments undertaken for nonmedical purposes (see Chapter 8) or research in which the moral status of embryos plays a role. The issue is then to examine if and to what extent the affected normative framework can be modified without coming into conflict with the essential ethical principles. Even the handling of technical risks that have to be tolerated in order to utilize an innovation often takes place by means of modifying the normative framework, such as in the implementation of precautionary measures or a code of conduct (see Chapter 6).

Ethical reflection plays a different role, however, in each of these options. The results of the reflection have to be introduced to the different fields of action (e.g., politics, economics, law, etc.). In order for them to be accepted and taken into account, it is necessary for them to be accepted as being an expertise and not to be considered to be a party or subjective statement on moral issues. In the following I will consequently present a few comments on the structure of an ethical expertise (after Grunwald, 2003).

### 3.1.4 Ethical Expertise as Conditionally Normative Advice

The question "What does ethical expertise consist in?" is not a trivial one, but would probably be answered differently by various systems of ethics. Diffuse and obscure allusions predominate, such as "providing orientation," "clarification of the normative premises and presuppositions," "contribution to the basis for acting and deciding," "recommendations for action," "answers to the question, 'What now?'." Even the question of whether ethical advice consists in (more or less strong) normative recommendations or in descriptive analysis has not been answered unanimously. Ethical counseling can, according to the analysis given above, consist in answers to the question of the limits on business-as-usual situations and in the reconstructive analysis and evaluation of possible measures for cases of transgression:

1. Which topics are identified socially as ethically relevant (in the sense of conflicts between systems of morals), and what are the criteria for this classification?
2. Which subjects are treated as standard questions, and why are they not discussed as ethical questions (automobility is not addressed as an ethical problem, but xenotransplantation is)?
3. Which expedients are acceptable in cases of "transgressions" in view of the validity and acknowledgement of normatively coherent frameworks (prohibition/permission, moratoria, and modifications of society's normative framework)?
4. Which implications do the various possible strategies have with regard to aspects of justice, to intentions for universalization, to human rights, etc.?

Expertise in answering these questions requires knowledge which fulfills the demand for *transsubjectivity*: this knowledge and advice has to be able to lay claim to validity beyond the sphere of subjective opinions and beliefs. Expert knowledge generally means knowledge that can both

(a) lay claim to validity above and beyond the sphere of subjective opinion and belief, and
(b) be taken as advice for a problem which has to be solved (i.e., action-guiding knowledge).

Assuming that it would be possible to construct something like transsubjective validity in the field of moral postulates, the question would be: What is methodologically necessary in order to make this possible? To move toward an answer, I will recount the discursive constellation of an orderly debate between proponent and opponent (Habermas, 1988). The question to be decided in this type of *discourse* is whether the assertion can be defended argumentatively. "Ethical expertise has to be submitted to a procedure of being scrutinized if it is to be acknowledged as expertise."

The claim raised by ethical advice to be transsubjectively valid — and only this type of advice should interest us in the following — is therefore to be developed or approved *deliberatively by argumentation*. Linguistic analysis of chains of argumentation leads to the conclusion that every transsubjective form of consultation, arrangement, discourse, etc. has to be founded on a commonly acknowledged basis. It is necessary for certain *predeliberative agreements* to be generally accepted if any deliberation should come to a result (Grunwald, 2003).[4] Lifeworld or scientific deliberations actually often succeed, often even unproblematically, whenever there is such a common predeliberative basis. Otherwise, the argumentation could end in the abyss of the well-known Münchhausen trilemma (Gethmann, 1979):

> Common deliberation presupposes some common ground; without shared values and understandings, discussion quickly degenerates in unending dispute. (Burns and Ueberhorst, 1989, p. 91)

An ethical expertise, too, is not absolutely objective in the sense of being free of underlying, imputed assumptions, but is valid with reference to the preceding, predeliberative agreements.

[4] Whenever the preconditions for discursive communication are concerned, these agreements are designated by Gethmann as "pre-discursive agreements" (Gethmann 1979). These are the agreements which require a minimum of presuppositions. In practical contexts, such agreements are as a rule not sufficient but include further cultural or social agreements.

An argumentative if–then chain of reasoning culminates neither in an ultimate justification ("objective" ethics), nor in the mere subjectivity of the decision maker (decisionism), but in a generally recognized *predeliberative agreement*, which itself is anchored in practice (Grunwald, 2003). This serves as a foundation on which argumentation can be built that will lead to results that should be accepted by all of those *who share the same initial basis.*

What could a predeliberative agreement for ethical expertise possibly be? Predeliberative agreements make reference to the following:

1. The disposition among all involved actors that moral conflicts *should be resolved argumentatively.* There must be a general willingness among participants to acknowledge and to accept a better argument.
2. (Some) confidence among the participants that an argumentative approach has a certain chance of success.
3. A common terminological basis and to fundamental differentiations.
4. Criteria for the quality and types of admissible arguments. The most prominent question which has to be decided predeliberatively in this context is whether or not descriptivistic inferences can be permitted. Is the inference from *is* to *ought*, from *de facto* acknowledgement of normative postulates to their ethical justification admissible or not (in the latter case, would it be designated as a descriptivistic or naturalistic fallacy)?
5. In essence, the *de facto* acceptance of *rules* which lay down the criteria and methods for the actual deliberation without substantially prejudicing the results. Standards of rationality belong to these procedural elements: for cognitive rationality, the recognized standards of scientific rationality (Rescher, 1988); for instrumental rationality, the methods of reproducible verification or the principle of methodological order; and for evaluative rationality, the rules of generalization and their relationship to society's normative frame of reference. These rules for deliberation are supposed to guarantee procedural justice. Given the acceptance of the procedures of argumentation and of the rationality criteria, it is to be expected that the results will be accepted as long as the

rules of procedure are observed. "The acceptance of the rules for the verification procedure thus leads to the acceptability of the results from the application of the rules."

Ethical expertise must be grounded in cultural foundations *over which no one can dispose;* otherwise it would become arbitrary and ideological. The concept of predeliberative agreements refers to that "given" basis in social lifeworlds. Since members of societies live according to common practices and are socialized into them, there is no free choice (in the sense of a choice of just any criteria for judging the normative basis for investigations of coherence). Incremental changes in society's normative framework can be made (e.g., by passing a law) which then themselves have to be reflected carefully with regard to the former state. The question, then, is not why people should adhere to such a normative framework — this is already implied by their membership in society — but the task is to reconstruct the implicit elements of this societal rationality, in order to make them accessible to criticism and further development, and to employ them for the purposes of technology ethics.

As a consequence, ethical expertise as a result of deliberation is only valid relative to the acknowledged predeliberative agreements. The scope of the agreements' validity, i.e., the extent of their general acceptability, is then also the range of validity of the ethical expertise itself. Therefore, anyone who argues on the basis of a different agreement could arrive at different results. This reasoning implies that ethical expertise can only be *conditionally normative* rather than categorically normative.

Ethical expertise therefore does not consist in action-guiding recommendations that are categorical in nature, but is a specific type of *knowledge* presentable in form of if–then chains (Grunwald, 2003). In particular, ethical expertise cannot answer the questions concerning what should be done and how one should act. Society is left to itself as far as planning for the future and setting the course are concerned. Collectively binding normativity will be produced only in democratically legitimized procedures. These procedures, however, are not independent of arguments (Grunwald, 2008a), which is the opportunity for ethical expertise as well. This analysis results into requirements for ethicists to express their arguments in public debates and political decision-making processes and thus contribute to the process of forming opinions by argumentation.

In summary, for methodological considerations, ethical expertise cannot consist in top-down judgments that are categorical in nature. Instead, ethical expertise consists in an examination of the coherence between elements from different categories within the normative sphere of society (Section 3.1.2). Ethics as coherence analysis in situations that do not lend themselves to being classified as business as usual in the moral sense is therefore not normative but — because of its dependence on predeliberative agreements — *conditionally normative*. Ethical expertise does not consist in categorical action-guiding recommendations but is *knowledge* of the premises and presuppositions of moral standpoints, of consistency and coherence problems in society's normative framework. This is knowledge that should contribute to the societal processes of debate, forming opinions, and decision making.

## 3.2 Ethics of Technology

Although the application of ethical reflection to issues of technology is only a few decades old (Mitcham and Briggle, 2010), its necessity is now taken for granted. This brief introduction shall serve as a bridge between the more general picture of ethics that has been drawn above, and the following chapters on nanotechnology.

### 3.2.1 Normative Uncertainties Emerging from Technological Progress

The normative aspects of science and technology inevitably lead, in a morally pluralistic society, at the very least to social debates and often also result in conflicts over scientific progress and innovative technologies. As a rule, opinions as to what is desirable, tolerable, or acceptable in society are a matter of controversy. Such unanswered questions and conflicts in the context of science and technology are the starting point for the ethics of technology. Conflicts over technology are, as a rule, not only conflicts over the means of technology (e.g., in questions of efficiency), but also conflicts over visions of the future, over concepts of humanity, and over views of society.[5] The role of the ethics of technology consists,

[5] This is particularly the case for nanotechnology. See Chapters 7–9 for illustrative case studies of far-ranging moral questions emerging from technical possibilities enabled by nanotechnology.

generally speaking and with respect to Section 3.1, in analyzing the normative structure of technology conflicts, and in searching for rational, argumentative, and discursive methods of resolving them.

It has, however, long been a matter of controversy whether science and engineering have any morally relevant content at all (and could therefore be a subject of ethical reflection at all). Until into the 1990s, technology was frequently held to be *value neutral*. Numerous case studies have, however, since recognized the normative background of decisions on technology (even of those made in the laboratory, see van de Poel, 2001) and made it a subject of reflection (van de Poel, 2010). The basic assumption in this transition is that technology should not be viewed solely as a sum of abstract objects or processes, but to take the fact seriously that it is embedded in societal processes (Rip *et al.*, 1995). Technology is not something natural that originates on its own and develops according to inherent regularities, but is consciously produced to reach certain ends and purposes, namely to bring something about that would not happen of itself. Technology is therefore always embedded in societal goals, problem diagnoses, and action strategies. There is no "pure" technology in the sense of a technology completely independent of this societal dimension (Radder, 2010). Technology is, therefore, morally relevant, particularly concerning its purposes and goals, the measures and instruments used, and the evolving side effects. It is currently recognized as being an appropriate subject for moral responsibility (Jonas, 1984; Durbin and Lenk, 1987) and ethical reflection (Grunwald, 1999; van de Poel, 2010).

This is also true of science. The value neutrality of science was postulated in the era of positivism. Since then, there have been many developments that lead one to think about the ethical aspects of science and about science as being subject to human responsibility. Furthermore, theoretical analyzes have shown that science methodologically draws on normative fundaments (such as theories of measurement, procedures of proof and argumentation, and the quality criteria to be applied). Science — analogously to technology — does not operate by contemplating about how nature works; it is rather involved in societal purposes and strategies. Scientific knowledge not only explains nature but also delivers knowledge for action, manipulation, and intervention, a statement that obviously also applies to nanotechnology (Ball, 2003). Science and technology consequently exhibit morally relevant aspects, namely concerning

(1) the purposes they pursue, (2) the instruments they employ, and (3) the consequences and side effects they produce.

1. In order to shape technology and provide orientation to science, there must be imagination concerning the desired future developments, about the goals and visions of the future of society, or ideas regarding what science and technology should contribute to meeting future challenges. In many cases, the aims and goals of science and technology are not problematic. To develop therapies for illnesses such as Alzheimer's disease, to provide new facilities to support handicapped people, or to protect society against natural hazards — visions of this type can be sure to gain high social acceptance and ethical support. There are — at the normative level — no rational reasons to object to them (this might be different at the level of the instruments needed or the side effects to be expected). In other areas, however, there are social conflicts even at the normative level. The visions related to manned spaceflight, for example, are controversial in nature. In the field of nanotechnology, the discussions surrounding the "converging technologies" (Roco and Bainbridge, 2002), especially those concerning the vision of improving human performance, are the most likely to become subjects of controversy in a moral respect. These questions lead to the challenge posed by knowledge policy (Stehr, 2004): What knowledge we want to have in some future and what do we not want to have? Obviously, such questions will be subject to ethical analysis and deliberation.
2. Instruments, measures, and practices in research may lead to moral conflicts. Examples are the moral legitimacy of experiments with animals or of practices making use of human persons, embryos, or stem cells as subjects of research, but also experiments with genetically modified organisms or plants, especially their release outside of laboratories or, in earlier times, the experimental testing of nuclear weapons. Questions regarding the moral status of the human embryo and of animals as well as questions regarding the acceptability of the risks involved have led to severe moral conflicts. A further dimension in this field concerns the responsibility of scientists and engineers as professionals for safeguarding the quality of the scientific and technological output. In this

respect, nanotechnology raises questions, for example, as to how to deal responsibly with nanoparticles — regarding for example, their partially unknown behavior and consequences (cf. Section 7.2.1) — at one's place of work or in conducting experiments. The ethics of science and of engineering are the respective fields of reflection, for example, concerning a code of conduct or rules of good scientific and engineering practice (VDI, 2000) but going far beyond them.

3. Since the 1960s the unintended and adverse effects of scientific and technical innovations have been considerable, and some of them were of dramatic proportions: accidents in technical facilities (Chernobyl, Bhopal, Fukushima), threats to the natural environment (air and water pollution, ozone holes, climate change), negative health effects as in the case of asbestos, the social and cultural side effects (e.g., labor market problems caused by productivity gains), and the intentional abuse of technology (the attacks on the World Trade Center). The increasing *complexity of technical systems,* their diverse interlacing, and their connectivity with many areas of society increase the difficulties of being able to predict and consider the consequences of actions or decisions. This applies in particular to new cross-sectional technologies such as nanotechnology. This situation leads to societal and ethical challenges: How can a society that places its hopes and trust in innovation and progress protect itself from undesirable, possibly disastrous side effects, and how can it preventatively stockpile knowledge to cope with possible future adverse effects? What extent of risk or ignorance is morally acceptable? How is responsible action possible in view of the high uncertainties involved possible?

### 3.2.2 Cross-Cutting Issues

Issues for the ethics of technology manifest themselves in normative uncertainties such as they result from the development, use, and disposal of technology. These general issues include, for example, human rights, the ban on human instrumentalization, the principle of self-determination regarding personal information, and informed consent. Many issues for the ethics of technology thus have substantive interfaces to the ethics of science, of the environment, and of medicine, and also structural contacts to technology assessment, political ethics, and economic and business ethics.

Normative uncertainties regarding technology appear above all in several cross-cutting issues (discussed in the following sections) that appear in probably all ethical debates on technology. These topics are (1) the safeguarding of human autonomy, (2) the problem of distributive justice, (3) the relationship between technology and the environment, (4) the relation between technology and life and (5) the necessity of dealing with the uncertainties of our knowledge of the consequences.

### 3.2.2.1 *Human Autonomy vs. Technicalization*

Advances in science and technology lead, according to many interpretations, to an *expansion* of man's opportunities for acting. This increases human autonomy by creating the possibility to select among several different options. The flip side is that technological progress can also remove previous options or make them less accessible. For example, in a technologically perfected world the life of a handicapped person might get less appreciation. It is also possible that technological systems might force adjustments, limiting our freedom of choice and thus an individual's autonomy. Self-determination regarding information, a part of an individual's autonomy, can be limited by technology, just as technological means can be used to achieve external controls over humans. In particular, the word "technicalization" is frequently associated with a subordination of man under technology, a loss of control, a discontent caused by man's dependence on technology, and a loss of individuality, emotionality, and spontaneity. Finally, the relationship between technology and human autonomy also concerns the question of how far technology can be formed or whether it follows its own dynamics, forcing society to adapt. The relationship between human autonomy and technical progress is ambivalent.

### 3.2.2.2 *Distributive Justice*

The distribution of new opportunities provided by technology, such as regarding affluence and health, as well as of the possible but unintended consequences for persons and groups, does not correspond per se to the accepted standards of fairness. This leads to debates and normative uncertainties. The persons who benefit from a technology are, for example, often different from those potentially affected by the possible risks. This leads to *asymmetry* in the distribution of opportunities and risks. The *de facto* distribution of risks and benefits in society that stem from technology — in a spatial and temporal sense and with regard to the groups affected — frequently displays deficits in fairness caused above all by differences in affluence or power but also by decisions about locating something such as an atomic power plant, a garbage incineration plant, the route of a highway, or factories. Some of these problems of distribution arise locally as with the issues just mentioned, some regionally or nationally, and others globally with regard to problems of development (such as when hazardous waste is exported to third world countries). One particular aspect is that of fair *access* to new technology, such as in relation to the supply of energy or of information, which is also a relevant issue for nanotechnology (Section 5.3).

One aspect that is particularly important in this context is the problem *of assuming responsibility for future generations.* The consequences of technology often reach unforeseeably far in the future, as with the problem of establishing a final repository for radioactive waste or in the relevance to tomorrow's climate of the technology used today. The degree to which we today can compel future generations to tolerate the consequences of technological activity is a matter of controversial discussion in the ethics of technology (Jonas, 1984; Brown-Weiss, 1989).

### 3.2.2.3 *Technology and the Environment*

Our natural environment is being changed to a large degree by technical innovations. Technology needs natural resources, which are thus taken away from the environment. Technology's use of these resources leads to emissions and waste, which in turn has consequences for ecosystems. Technology and innovation are among the factors that substantially determine the degree of sustainable development on the part of the way the human economy functions. The relationship between technology and sustainable development is ambivalent as a matter of principle. On the one hand, technology is viewed as a *problem* for sustainable development and as the cause of many environmental problems; on the other hand, it is seen as the *solution* for problems of sustainability or at least as one component of the solution. In view of the further growing global population and the legitimate need for development in the poorer and emerging countries to help them catch up, it is impossible to imagine steps in the direction of sustainable development without innovative and more sustainable technology. The question is not whether technical progress works for or against sustainability, but rather how the scientific-technical progress would have to be organized for its consequences to be positive contributions to sustainable development. Issues from the ethics of technology in this context are closely linked with questions of environmental ethics (Rolston, 1988) and sustainable development (UNCED, 1987).

#### 3.2.2.4 *Technology and Life*

Classically, the ethics of technology has been concerned with artifacts of engineering technology (e.g., Ropohl, 1996). The scientific-technical development of the past decades has however begun to make the traditional border between inanimate technology and the realm of the living more permeable. One aspect of this is that technical interventions in the sphere of molecular biology have led to genetic engineering, which cannot be understood as a classical (natural) science but as technology or a technoscience (Latour, 1995). The recent debate referred to this under the keyword "biofacts" (Karafyllis, 2006). Another aspect is that in nanotechnology, or more precisely in nanobiotechnology, the approaches to transcend the classical borders go substantially further, extending even to the visions of a synthetic biology (Chapter 7). The elimination of these borders is a challenge to established ideas of morality, which, for example, speak of living beings as being "God's creatures," since artificially produced living beings would be man's creatures. It raises new questions as to the paths of development and possible risks since artificially produced or technologically modified life would develop further according to the principles of self-organization and could possibly reproduce and get out of control.

#### 3.2.2.5 *Uncertainty of Our Knowledge of the Consequences*

Large technical systems such as energy plants or infrastructure systems, new cross-cutting technologies such as nanotechnology but also control measures or support for research are marked by high complexity, poor prognosticability of the intended consequences, and the frequent presence of unintended consequences. The development and introduction of new technology and also attempts to influence them thus pose an act in a state of *uncertainty and risk*. With regard to the risks caused by decisions on technology policy or the direct use of technology, the ethics of technology is concerned with the normative basis of the questions as to how, to what degree, for which reasons, and under which conditions society can be expected to tolerate them. This is all the more true since precisely conflicts over technology and normative uncertainties are triggered by the question as to which risk is acceptable under which conditions (cf. nuclear energy, genetic technology, electrosmog, the hole in the ozone layer, etc.). The epistemological uncertainty regarding the consequences contains a moral dimension, which Hans Jonas formulated very pointedly in forming the question of whether the entirety can be made "a bet as part of a wager" (1984). Although the link between normative uncertainty and epistemological uncertainty may be a structural feature of every debate on technology, its consequences are particularly dramatic for questions of nanotechnology (Chapter 10).

### 3.2.3 Ethics of Technology as Part of Technology Governance

The objective of ethical reflection is to provide orientation for forming an opinion, making a judgment, and making a decision *ex ante*: "Ethical problems of the sort we are concerned with here presuppose that we are faced with a choice between different possible alternatives of action — where naturally to do nothing is one of these alternatives" (Hermeren, 2004, p. 95). In this way, ethics takes the perspective of a *participant* in ongoing communication in and over science and is thus part of the overall technology governance (for the case of nanotechnology, see Kaiser *et al.*, 2010). Ethical inquiry into technology, in particular in nanotechnologies, aims to influence the further course of development or, in short, *to make an impact* rather than being a mere ornament. The early stage of many developments in nanotechnology provides an advantageous opportunity: the chance and also the time for concomitant reflection, as well as the opportunity to integrate the results of reflection into the process of technology design, and thereby contribute to the further development of nanotechnology (Moor and Weckert, 2003).

At any rate, to be clarified is therefore not only how ethical deliberation can take place and upon what grounds it could be founded (see above) but also how its results can enter into decision-making processes. This is the meaning of this section's title: ethical reflection must occupy a verifiable spot in the framework of comprehensive technology governance. It would otherwise only be able to formulate moral appeals from an external perspective.

The concept of conditionally normative advice provided by ethics for ongoing deliberation suggests a modest understanding of the role of ethical reflection. Ethics is neither a censorship agency nor a decision-making entity. It does not have any legions and thus has to seek a spot in technology governance solely on the basis of the force of its arguments. Otherwise, it would be worthless and senseless. One approach is to identify the possible pragmatic locations of ethical consultation in technologically relevant decision-making processes. Pragmatic locations in the ethics of technology are those situations in social praxis in which there is normative uncertainty on the basis of technological issues. This determines who the partners are in an interdisciplinary and transdisciplinary dialogue about ethical issues of technology. These partners in a dialogue represent social praxis in the four relationships between reflection and praxis mentioned in Section 3.1.1: they formulate the respective problem, they provide information about the relevant contexts, they are the ones with whom consideration of a predeliberative agreement must take place, and they are the addressees of ethical advice.

A complete picture of these pragmatic locations would require a complete theory of technology in society. Such a theory would have to include theories of the origin of technology, the route technology takes through society during the phase of its utilization, and the manner in which society deals with a technology after its use is discontinued — this would be no less than a comprehensive theory of society. The following sections will briefly describe the most important pragmatic locations of the ethics of technology as can be found in the relevant literature.

#### 3.2.3.1 *Political Decisions*

Since governmental technology policy creates obligations for everyone, the influence it exerts on technology in a morally pluralistic society is always a stage on which normative uncertainties will probably manifest themselves. It is thus a pragmatic location for ethics in shaping technology. Policy consultation by problem-oriented ethics can, for example, take place in the preparatory phase of legislation relevant to technology. In the run-up to policy decisions it is possible for ethical reflection to carry out enlightenment with regard to the relevant normativity. This positioning of conditionally normative consultation affects all constellations in which state action influences technology. These include:

(a) Direct state-run or at least state-dominated technology development, for example, in the fields of space travel, military technology, and transportation infrastructure
(b) Indirect state control of technology by means of programs promoting research and technology, for example, in materials science, on regenerative sources of energy, or in stem-cell research
(c) Indirect state control of technology by setting boundary conditions such as environmental and safety standards or laws stipulating recycling
(d) The role of the state as a *user* of technology, for example, with regard to the observance of sustainability standards

Ethical advice that helps to overcome normative uncertainties in these areas can on the one hand be handled in the context of professional consultation on policy, such as in technology assessment (Grunwald, 2009). On the other, they can also be dealt with in ethically enlightened public debates and participatory processes (e.g., Berlosznik, 2007).

#### 3.2.3.2 *Entrepreneurial Decisions*

The development of technology takes place primarily in the economy at market conditions. The shaping of technology by and in enterprises is operationalized by means of requirement specifications, project plans, and strategic entrepreneurial decisions. These in turn take place on the prescriptive basis of an enterprise's headline goals, general principles, plan goals, and self-understanding but also including assumptions about later consumers of the technology and future market conditions. They are based on images of man and the estimates of the goals and purposes set for the technology, issues which are accessible to ethical reflection. Inasmuch as normative uncertainties play a role in these processes, ethical reflection on the moral foundations of entrepreneurial decisions is also a pragmatic location for ethics in shaping technology.

#### 3.2.3.3 *Engineering*

Engineers and engineering scientists are confronted in a special way with attributions of having responsibility because of their close links with the processes of the development, production, utilization, and disposal of technology (Durbin and Lenk, 1987; van Gorp, 2005). Reflection on the moral foundations of engineering activity also are a pragmatic location of the ethics of technology inasmuch as normative uncertainty arises during this activity, such as in cases of conflicts over the evaluation of safety and environmental issues between an engineer as an employee and the enterprise as the employer (see Kastenhofer *et al.*, 2010 for the case of sustainability).

#### 3.2.3.4 *Consumer Behavior*

The individual preferences of users and consumers of technical systems and products help determine the success of technology developments in two ways: first, by means of their purchasing and consumer behavior, and second (and less noted), by means of their comments in market research. Their preferences also contain moral backgrounds and values. In the purchase of an automobile, for example, criteria such as sportiness, cost, status, and environmental compatibility play a large role that varies from person to person. The influence on technological development resulting from consumer behavior arises however almost naturally from the concurrence of the actual purchasing behavior of many individual persons. This can be referred to as normative uncertainty when the insight into certain relationships, although present, does not lead to a change in purchasing behavior. A well-known problem is, for example, that awareness of a problem with regard to the deficient environmental compatibility of certain forms of behavior — though definitely present — may not lead to a change in behavior. Such normative uncertainty — a divergence between knowledge and action[6] — can provide the reason for psychological concern just as it can for ethical reflection. This is also a pragmatic location for the ethics of technology, the goal being to make certain normative aspects of consumer behavior obvious by sensibilizing consumers and thus to contribute, according to certain criteria, to attaining ethically better justified consumer behavior.

#### 3.2.3.5 *Public Debate*

The course of technical development is also decided by public debates, above all by those in the mass media. Public discussion in Germany influenced, for example, political opinion on atomic

[6] Compare this with the criterion of compliance as the necessary condition for the presence of a standard situation in a moral sense (Section 3.1.2).

energy, thus providing much of the basis for the decision in 2000 to phase out atomic energy in that country. Similarly, the public discussion about genetically modified organisms has influenced the regulatory attitude of the European Union and the official acceptance of the precautionary principle. This can also be recognized by the fact that different regulations were established in those countries in which the public debates were very different, such as in the United States. Most of the public debates conducted in the media have also influenced the shaping of the policy framework, with its indirect influence on technology.[7]

These considerations make it possible to name the pragmatic locations for the ethics of nanotechnology. These locations are the points and constellations in social development and communication in which the normative uncertainties elicited by nanotechnology and its embedding in society are reflected. To them belong, without any claim to completeness, debates in research policy on priorities in funding research projects and programs; debates about regulations of nanotechnology; debates in the media and public, frequently triggered by nongovernmental organizations and social movements; discussions about the responsibility of nanoscientists and nanotechnologists or other actors in the context of nanotechnology with regard to newly emerging moral issues; controversial issues in dealing with nanomaterials and the related processes in industrial manufacturing and production; debates between different groups of the population as well as between differently developed regions of the world about the distribution of the advantages expected from nanotechnology; discussions in nanomedicine about new diagnostic and therapeutic procedures with consequences even for the doctor–patient relationship or for the cost structure in the healthcare system; debates about societal futures in view of the far-reaching promises made for nanotechnology.

### 3.2.4 Technology, Science, and Responsibility

The concept of responsibility has been used repeatedly in connection with technical progress (Durbin and Lenk, 1987). It associates

[7] The incipient debate on nanotechnology in food is currently this type of debate. It affects the public and media reporting, but not consumer behavior. The fact that an identification requirement has not been implemented means that the consumer does not have the opportunity to consciously choose in favor or against such food (see Chapter 6).

ethical questions regarding the justifiability of decisions with the possible actions of concrete persons and groups (i.e., governance) and with the challenges posed by uncertain knowledge of the consequences (discussed below). The term is also used in the context of nanotechnology, namely in many central spots, especially political ones. Here is a prominent example:

> Responsible development of nanotechnology can be characterized as the balancing of efforts to maximize the technology's positive contributions and minimize its negative consequences. Thus, responsible development involves an examination both of applications and of potential implications. It implies a commitment to develop and use technology to help meet the most pressing human and societal needs, while making every reasonable effort to anticipate and mitigate adverse implications or unintended consequences. (National Research Council, 2006, p. 73)

Responsibility is neither a quasi-ontological predicate nor a natural object, but the *result of an act of attribution* (Grunwald, 1999), either if actors attribute the quality to themselves, thus making a statement about their judgment of their own actions or of their handling of the consequences — whether ex ante or ex post — or if the attribution of responsibility is made by others. The attribution of responsibility is itself an act that takes place relative to *rules of attribution*. These rules must also be justified, for example, by limiting the circle of individuals capable of being responsible and indicating the criteria — i.e., the preconditions that individuals must fulfill — for being able to hold someone responsible.

A four-place reconstruction is suitable for discussing issues of responsibility in technical progress. *Someone* (a responsible subject) assumes responsibility for *something* (considering the results of acts as the object of responsibility) relative to *a body of rules* and relative to *a level of knowledge*. The moral dimension is reached when the question is posed as to the *body of rules according to which* responsibility *should* be assumed. These rules form the normative context for judging acts to be responsible. Insofar as normative uncertainties arise (see Section 3.1.2), for example, because of moral conflicts, ethical reflection on these rules and their justifiability is needed. The relation to the status of the available knowledge is indispensible in issues of prospective responsibility. The comment that nothing else comes from "mere possibility arguments"

(Hansson, 2006) is an indication that in debates over responsibility it is essential that the status of the available knowledge about the futures to be accounted for is determined and is critically reflected from epistemological points of view. Consequently, questions arise in three dimensions in prospective debates over responsibility:

(a) *The epistemic dimension*: What is known? What can be known? Which uncertainties continue? How can they be qualified? And what is at stake if worse comes to worst?
(b) *The ethical dimension*: Can the acts in question (e.g., scientific developments) be justified? Under which conditions? Which ethical patterns of reflection and argumentation are pertinent?
(c) *The political dimension*: How is responsibility supposed to be distributed? Which social groups are affected and should help decide about the distribution of responsibility? Are these questions that concern the "polis"? Or can they be delegated to groups or subsystems?

Debates over responsibility are frequently limited to level (b) and treated exclusively in the context of the *ethics* of responsibility. The results are the familiar allegations of being simply appellative, of epistemological blindness, and of being politically naïve. The analysis shows, however, that issues of responsibility are inevitably interdisciplinary, which is precisely the means by which it acquires its specific quality — both specific problems as well as its specific appeal. This is precisely the reason that the concept of responsibility can be found at a key spot in the title of this book. The issue is not one of abstract ethical judgments but of responsible innovation, which entails the observance of concrete contexts and governance factors. This is also shown by the following quotation from a report by the European Commission (Siune *et al.*, 2009, p. 31):

> European Commission documents, particularly those on nanotechnology, often refer to responsible innovation. Recently, a further step was taken by preparing and publishing a code of conduct for nanoscience and nanotechnology (N&N) research (EC, 2008a). The restriction of the code to "research" was necessary, because of the limited remit of the European Commission in this respect, but the code is broader, and refers also to public understanding and the importance of precaution. There are explicit links to governance: the guidelines "are meant to give guidance on how to achieve good

governance," and when this is further specified, there is also this interesting item: "Good governance of N&N research should take into account the need and desire of all stakeholders to be aware of the specific challenges and opportunities raised by N&N. A general culture of responsibility should be created in view of challenges and opportunities that may be raised in the future and that we cannot at present foresee." (EC, 2008a)

## 3.3 Ethics and (Unclear) Risk

From the beginning, unintended and undesired consequences of technology have been at the center of reflection in the ethics of technology, namely the technology-dependent risks, the problems of the uncertainty of our knowledge, and the question of the ability to bear responsibility (Hansson, 2010). All of these topics also have a special significance in nanotechnology.

### 3.3.1 Classical Risk Management and Its Limitations

Risk assessment and risk management strategies and approaches have been developed and established in many areas of science, medicine, and technology (Renn *et al.*, 2008), for example, in dealing with new chemicals or pharmaceuticals. Such "classical" risk regulation is adequate if the level of protection is defined and if the risk can be quantified. Such risk quantification can, for example, be the product of the probability of the occurrence of the adverse effects and the assumed extent of possible damage. Another procedure is the use of other quantifying schemes, such as one that defines risk as "the probability of an adverse effect in an organism, system, or (sub)population caused under specified circumstances by exposure to an agent" (IPCS, 2004, p. 13). In such situations, thresholds can be set by law, by self-commitments, or by applying participatory procedures, and measures can be taken to keep particular effects well below predefined thresholds (von Schomberg, 2005).

In many cases, however, the conditions for such classical risk regulation are not (or not sufficiently) fulfilled. There may be situations of an "inconclusive risk assessment" or of "unclear risk," i.e., it might not be clear whether the situation under consideration could involve risks at all or that the extent of possible damage might not be known (Wiedemann and Schütz, 2008).[8] Reliable knowledge concerning possible adverse effects might not be available yet or might be persistently controversial and hypothetical. Empirical evidence could still be missing, and data or their interpretation might be conflicting. In this situation it is difficult to make legitimate and rational decisions about precautionary measures without either running into the possible high risks of a wait-and-see strategy or putting too much stress on precautionary argumentation. The consequence of the latter might be that one is no longer able to act at all or that one might run into problems in implementing measures involving huge economic costs and consuming public resources, keeping them from being spent on other urgent public tasks.

Ethical reflection and guidance could contribute to dealing with such situations. To this end it is necessary first to identify the entry points for ethical reflection that are or might be involved in situations of unclear risk or of high uncertainty. In order to explain the notion of "unclear risk" I will examine a decision-making situation or the respective preparatory or preceding debate where knowledge about possible harmful effects has to be taken into account. The specific difference distinguishing the notion of unclear risk from "standard" risk will be the availability and validity of the information about possible harmful effects (following von Schomberg, 2005).

"Standard" risk is a *quantified magnitude* denoting the probability of the occurrence of the adverse effects multiplied by their impact. In situations where such quantitative risk measures are available, policy makers can respond with a classical risk management approach whereby, for example, thresholds can be set and risks can be either minimized or kept below a certain level according to the respective level of protection. Because there is *scientifically sound and consolidated* knowledge concerning the adverse effects in question — with respect to the probability of their occurrence as well as concerning the expectable damage — society can act by applying the familiar preventive measures (for example, by building higher dams if higher floods are expected). Laws such as the Toxic Substance Control Act in the United States (Wardak, 2003) determine what has to be done in such standard risk situations. Consequently, ethical reflection is not required, and the problem under consideration

[8] A perfect illustration of the problems and open questions related to such a situation of unclear risk is provided by the present debate on the possible harmful effects of nanoparticles (Chapter 6).

may be approached directly using the established risk regulation measures.

While "standard" risk is considered to be a quantifiable parameter, "unclear" risk denotes a nonstandard situation in a moral sense (Section 3.1.3) since such risk is characterized by limits to its quantifiability, by a lack of knowledge and data, by conflicting data or conflicts over their interpretation, by epistemic uncertainties, and/or by unsolved scientific controversies (von Schomberg, 2005). The following types of cases can be distinguished analytically:

1. The first and most relevant case in this respect is simply *lack of knowledge* about cause–effect relations, which makes it impossible to answer the question whether there will be any adverse effects at all. Concerning the far-reaching difference between *possible* and *probable* effects, this case covers the situation of debate over the *merely possible* adverse effects. In this case, neither the probability of the occurrence of adverse effects nor the magnitude of possible damage can be calculated. This would be a typical situation of unclear risk. The current debate about the possible harmful effects of nanoparticles on human health or the environment may again be taken as an example.
2. Secondly, classical risk approaches are limited with respect to their applicability when particular cause–effect relationships have not yet been scientifically established while at the same time the adverse effects can already be observed. The BSE (Bovine Spongiforme Encephalopathy) case was of this nature until scientific evidence about the cause–effect chain leading to the "mad cow" phenomenon was established. In such cases the harmful effects are visible but it is not possible to identify the main origins of the effects or to calculate the probability of their occurrence.
3. It may also be difficult to apply the "classical" risk approach if the scientific information that serves as the basis for estimating possible adverse effects is not fully reliable. This is notably the case when an epistemic debate is ongoing in science, as when different disciplines use competing models, analogies, or basic assumptions in order to acquire new knowledge about the subject matter under investigation. Fundamental contradictions in the basic scientific conceptualizations create an unclear situation concerning the data available and

their interpretation (cf. von Schomberg, 2005, for the case of genetically modified organisms).

Decision making in these cases can obviously not be grounded on classical risk assessment. We will denote these types of situations as decision making with unclear risk (Wiedemann and Schütz, 2008).[9]

Yet is it possible to distinguish unclear risk in this sense from mere speculation? If, in the first case mentioned above, there is no knowledge at all about possible harmful effects, then suspicions concerning unclear risk would be purely speculative, with all the imaginable consequences for decision making. Part of the questions to be addressed should therefore be whether arguments (even weak ones) anticipating adverse effects should be considered on the basis of some uncertain "evidence" or the plausibility of the assumptions but without any real scientific evidence. There will be a need not only to distinguish between quantified risk and unclear risk but also between unclear risk and mere speculation — which is, however, a challenge to epistemology rather than to ethics.

### 3.3.2 Ethical Issues in Dealing with Unclear Risk

A first approach to dealing with unclear risk would be to attempt to transform the "unclear" into a "classical" risk by means of empirical research, for example, in toxicology or environmental medicine. Ethics cannot support this approach because ethical reflection does not help to fill in gaps in our knowledge about the possible toxic effects of new chemicals. But ethical reflection is needed during the phase where the knowledge gaps still exist and decisions must be made regarding the existence of the gaps in case of uncertainty and unclear risk.

Decision-making situations involving unclear risk are directly subject to ethical inquiry according to the understanding of ethics presented above:

> Many scientific disputes do not refer to differences in methodology, measurements, or dose–response functions, but to the question of whether the observed or assumed impacts violate or meet predefined

[9] Frequently, the notion of uncertainty is used for these cases, and "risk" and "uncertainty" form the basic distinction. However, uncertainty highlights the fact that quantifiable risk is also, in a certain sense, uncertain because the information is preliminary in nature and incomplete. The terms risk and unclear risk used here are both uncertain but contain a different degree of uncertainty.

> values. Often it is also contested which values are (will be) actually of issue or are (will be) subjected to discussion and how essential these values are for which groups. (Renn and Roco, 2006)

Such situations are nonstandard in a moral respect because the normative framework — in this context, the system of risk regulation in place — governing the situation does not allow either a conclusive statement or a clear indication for corresponding action. In this indifferent or ambiguous situation we have to look for other types of orientation. There are some typical questions in this respect (see also Hansson, 2010).

#### 3.3.2.1 *Acceptability of Unclear Risk*

The most general questions that are ethically relevant are whether and under which preconditions we should tolerate unclear risk, and what criteria should govern the question of acceptability. A strong precautionary approach could motivate the postulate that unclear risks should be completely avoided. In questions of technology, for example, this would imply that any new material, product, or system would not be introduced without full knowledge of possible harmful consequences and their impact (cf. Rippe, 2002). Jonas restricted this strong precautionary approach to situations with potentially apocalyptic character (Section 6.3). On the other hand, there are more or less hidden wait-and-see strategies that take a lack of knowledge about adverse effects to be permission. It is easy to see that there are different moral positions behind the differing approaches to the problem of unclear risk. Moral conflicts over unclear risk may therefore be expected (and this was the case, for example, in the case of genetically modified organisms; von Schomberg, 2005), and ethical inquiry will be required.

#### 3.3.2.2 *Weighing Benefits against Unclear Risks*

There are often clear und well-founded assumptions about benefits expected from the further development of specific technologies. For example, a huge market potential is usually ascribed to nanoparticles and nanomaterials. This can trigger the ethical question as to the role that such *opportunities* play in decision-making situations involving unclear risk. Is it ethically legitimate to continue proceeding with the development and implementation of a technology given the existence of large gaps in our knowledge or of scientific controversies about risk? Two problems become visible at this point. First, the question of principle is whether it is justified at all as a decision-making approach to weigh benefits against unclear risks. There might be ethical knockout arguments against an unlimited ability to weigh pros and cons, for example, in potential cases extremely high damages, including a catastrophe for humankind (see Jonas, 1984). Second, there is a problem at the methodological level: how can we perform such balancing in the absence of quantitative measures, at least for possible risk? At the heart of ethical inquiry is the question — which frequently leads to conflicts — as to the criteria upon which (often well-known) benefits and (in the situation of uncertainty, unclear) risks could be weighed against each other as the basis for a rational decision-making process.

#### 3.3.2.3 *Normalizing the Situation under Consideration*

A familiar approach to providing orientation is to propose analogies between the new situation and situations we are familiar with. A new (and perhaps irritating) decision-making situation might be "normalized" by referring to a well-known situation. This approach, however, depends on the acceptance of some premises. Forming an analogy is not a value-neutral step because it always presupposes the similarity of the new to older situations. Assessing the similarity, however, depends on the criteria employed. The determination of the criteria is a normative step in building analogies. For example, the promoters of genetically modified organisms constructed an analogy between the genetic modification of organisms in laboratories and traditional methods in agriculture. The aim of both approaches is to change the genome of the affected organisms in a desired manner. The opponents rejected the analogy, however, pointing to differences in other respects, for example, concerning the time span of the modifications. The choice of criteria for an analogy may thus cause conflicts and is therefore a candidate for ethical reflection in cases of uncertainty. The question is whether the comparison of one type of unclear risk with another type would be justified. For example, a question with the potential for moral conflict might be whether the criteria gathered from experience in the fields of new chemicals or pharmaceuticals could be transferred to nanoparticles to assess the risks they pose.

#### 3.3.2.4 *Comparisons of Man-Made Situations of Unclear Risk with Natural Situations*

The attempt is sometimes made to normalize new and irritating situations by pointing to natural situations. A well-known example from the field of "clear" risk is the attempt to normalize the potential danger from exposure to radioactive radiation by pointing to the fact that we are exposed to natural radioactivity, in some regions even to a considerable dose. A problem similar to that in the previous point arises at this point. Such comparisons are valid only relative to the criteria employed. At least one criterion is not considered in the example mentioned here: the fact that exposure to natural radiation is truly a "natural" hazard and therefore beyond human influence while there can be no doubt that radiation from a technical installation is man-made. This led many concerned people not to accept the intended normative implication of that comparison, namely that they should accept exposure to artificial radiation because it was less than that to natural radiation. The case shows the potential for conflict at the level of normative criteria with the consequence that ethical reflection might be required to argue for or against the comparability of exposure to man-made and to natural radiation.

#### 3.3.2.5 *Learning from Historic Cases*

Some voices point to an analogy between artificial nanoparticles created in industry and asbestos (Section 2.6). Can we learn something from the asbestos story (Gee and Greenberg, 2002) or from other stories (Harremoes *et al.*, 2002)? And under which conditions could we transfer that knowledge to newly emerging cases such as that of nanotechnology and nanoparticles? In this case, ethics would again look for the normative criteria for transferring knowledge and forming analogies.

The contribution of ethics to this subject consists of a value judgment of the situation (the values involved, the possible moral conflicts, the applicability of the precautionary principle), of a clarification of whether a new unclear risk is comparable to other types of risk, and of the disclosure of the normative presuppositions and implications involved, as well as of the investigation of the normative basis for practical consequences. The questions of the acceptability and comparability of risks, the advisability of weighing up risks against opportunities, and the rationality of action under uncertainty are at the heart of ethical reflection in cases of unclear risk (Chapter 6).

Ethical advice provided on these issues must then be brought into political or societal debates in order to ascertain the impact in terms of science and technology governance, for example, under the notion of reflective science and responsible development (Section 3.2.4; see more generally also Siune *et al.*, 2009).

# Chapter 4

# Ethics of Nano(bio)technology: The Program

Following the preparatory introductions to nanotechnology (Chapter 2) and the ethics of technology (Chapter 3), the issue now is to formulate a program for philosophical and ethical reflection on nanotechnology that can be worked through in the following chapters. To this end, I will first describe the motivations, expectations, and definition of goals given in the available literature for an ethics of nanotechnology (Section 4.1), then cope with the question of whether nanoethics might claim the status of a separate subfield in applied ethics (Section 4.2), and finally mark out the field of a problem-oriented ethics of nanotechnology (Section 4.3).

## 4.1 Motivations of Nanoethics

The presumed depth of penetration of nanotechnology and its potential for social transformation are often given as reasons for the necessity of studying it from an ethical perspective. In this vein, Allhoff *et al.* (2007, see the Preface) assert that "nanotechnology will eventually impact every area of our world." In view of the revolutionary potential that many attribute to nanotechnology, the ethical, legal, and social implications (ELSI) have been examined by commissions and study groups for years (e.g., Coffrin and MacDonald, 2004; NanoForum, 2004) and demands have been raised, especially by politicians, that nanotechnology be developed in a responsible manner, postulating new forms of ethics:

*Responsible Nanobiotechnology: Philosophy and Ethics*
Armin Grunwald

ISBN 978-981-4316-80-4 (Hardcover), 978-981-4363-33-1 (eBook)
www.panstanford.com

> In this regard, the advocate could maintain that issues are transformative or revolutionary in some particular way and that, whatever other ethical frameworks we have already developed, those frameworks will be ill-equipped to deal with the force that nanotechnology represents. (Allhoff, 2007, p. 198)

The debate on converging technologies (Roco and Bainbridge, 2002), based on the convergence of different technologies on the nanoscale, has strengthened the motivation for nanoethics. It widened the area of expected transformations to the very nature of humans by proposing "human enhancement." Anthropological questions about the future of humankind, about changing relations or disappearing borders between humans and technology, about a technicalization of humans (Grunwald and Julliard, 2007) and a humanization of technology illustrate the large-scale nature of transformations (possibly) related to nanotechnology (Chapter 9). A greater in-depth view shows that there are different colors and accents in the debate which will be distinguished in the following.

### 4.1.1 Avoiding to Endanger Innovation

The first comments on the necessity of addressing the social issues, including ethical ones, surrounding nanotechnology were made in the years following the announcement of the American National Nanotechnology Initiative (NNI, 1999). In summary, the predominant patterns of argumentation that were raised regarding the urgent need for ethical reflection were:

> Without an attention to ethics, it would not be possible to ensure efficient and harmonious development, to cooperate between people and organizations, to make the best investment choices, to prevent harm to other people, and to diminish undesirable economic implications. (Roco, 2007, p. xi)

When the debate on the risks posed by nanotechnology began, it was noted that nanotechnology and ethical reflection on it developed at two very different speeds (Mnyusiwalla *et al.*, 2003). One train of thought ran as follows. While nanotechnology developed rapidly, there was hardly any interest in ethical issues. For example, not a single project of this nature received any financial support in the United States in 2003, while work was being conducted in hundreds of purely technical projects:

> As a result, the ethics of nanotechnology have not really been addressed even when funds have been specifically allocated for doing so. In 2001 the US National Nanotechnology Initiative allocated between $16 and $28 million for the study of its societal implications but only about half that budget was used. (Ball, 2003)

This concern, which we refer to as one of a "policy of innovation," is that the serious and growing gap between the rapid advances made in nanotechnology and its inadequate ethical appraisal could lead to a development in which the expected advantages and opportunities promised by nanotechnology might not be achieved, for example, because of public rejection: "We believe that there is danger of derailing NT [nanotechnology] if serious study of NT's ethical, environmental, economic, legal and social implications [...] does not reach the speed of progress in the science" (Mnyusiwalla *et al.*, 2003, p. R9). Study of the ethical issues and of the consequences of nanotechnology is thus necessary to be able to introduce innovation in modern societies. Its absence might otherwise result in the threat of public resistance, which would hinder both progress and the social utilization of the expected advantages and benefits of science and technology:

> The only way to avoid such a moratorium [following ETC Group 2003, A.G.] is to immediately close the gap between the science and ethics of NT. [...] either the ethics of NT will catch up or the science will slow down. (Mnyusiwalla *et al.*, 2003, p. R12)

These authors suggest measures for helping ethics in this chase, based on the experience gained from the ethical research that accompanied the human genome project. This experience includes the appropriate support of activities regarding the ethical, legal, and social implications, the motivation of the next generation of scientists in these fields (building capacity), an interdisciplinary and transdisciplinary networking instead of ethical reflection in an ivory tower, and the participation of developing countries and the public. In view of the possibility that ethical reflection might have a negative impact on innovation, the demand for us to "mind the gap" appears to be an astute one for a policy of innovation: the earlier that possible ethical problems from nanotechnology innovations are recognized, the easier it might be to avoid them constructively.

Recently, the debate on "speculative nanoethics" (Nordmann and Rip, 2009; Grunwald, 2010) came back to this type of argument.

Nordmann and Rip (2009) fear that the arbitrariness of speculative nanoethics could lead to completely unfounded and unjustified ethical problems, which could then become real issues for nanotechnology and the public debate. In this way, there could be serious damage to the further development of nanotechnology, even if the ethical concerns were completely artificial: "worries about the most futuristic visions of nanotechnology can cast a shadow on all ongoing work in nanoscience and technology" (Nordmann and Rip, 2009, p. 274). Artificial problems might be created, and their communication to the public might really damage innovation policy in the field of nanotechnology and prevent us from exploiting the opportunities it offers (see Chapter 10 for my reaction to this argument).

### 4.1.2 Taking Care of Unintended Side Effects as Early as Possible

The fact that science and technology often have unintended consequences is one of the most frequent causes of ethical reflection. The *unintended consequences* of technology include the risks from developments in science and technology to society and the environment that take the form of catastrophic accidents or creeping changes. This situation illustrates why many optimistic expectations relating to future technological progress have currently been abandoned. The rising range of negative effects in time and space, reaching even a "global" technological level, emphasizes the relevance of all of this and led to the implementation of institutions and projects for technology assessment (Rip *et al.*, 1995; Grunwald, 2009a) and other types of reflective research. Classic questions are: Which risks can or should be tolerated in view of the positive consequences hoped for? How should cost–benefit assessments, risk–opportunity analyses, and comparative risk evaluations be carried out? When should the precautionary principle be applied in view of our insufficient knowledge (Harremoes *et al.*, 2002)? And how are we to deal with situations in which it is unclear whether there is any risk at all (Section 3.3)?

Even before the ETC Group issued its call for a moratorium (ETC, 2003), toxicologists had drawn attention to unanswered questions in connection with nanomaterials, especially nanoparticles (Colvin, 2003). The background for this was previous experience, namely, the use of new materials, especially chemicals, can have negative consequences on the environment and our health. The history of asbestos, in particular, is a warning (Gee and Greenberg, 2002). Initially this led to demands for empirical research into the consequences, often referred to as environment-safety-health (or ESH) studies, as a form of responsible science. The normative background of this argumentation consists primarily in the perception of the responsibility of science as it has developed in view of the unintended consequences in previous cases. Ethics can contribute to these issues in a particular way (Chapter 6).

### 4.1.3 Reacting to Apocalyptic Fears

One of the factors present when demands for an ethics council were first raised were fears of a possible apocalyptic side to nanotechnology (Joy, 2000). Moor and Weckert (2004), for example, tied an ethics of nanotechnology to the expectations that nanotechnology "would include [...] how to minimize the risk of runaway robots." Jean-Pierre Dupuy put the apocalyptic dimension of Joy's fears at the center of his conceptualization of an ethics of nanotechnology (Dupuy, 2005; Dupuy and Grinbaum, 2004). He argues that, inasmuch as nanotechnological dreams make use of the principles of self-organization and thus could release self-organizing nanomachines into the world, they would ultimately and inevitably lead to an absolute catastrophe (Dupuy and Grinbaum, 2004). A focus on apocalyptic scenarios has been an acknowledged pattern of argumentation in studies that consider the ethics of technology following Hans Jonas' "imperative of responsibility" (Jonas, 1984) and is certainly part of the spectrum of ethical considerations of nanotechnology.

Apocalyptic fears related to nanotechnology are more or less speculative in nature. They include the gray goo scenario, the cyborg scenario, and the loss of control scenario (cf. Section 2.6). There are, on the other hand, bright visions attributed to nanotechnology, related, for example, to the molecular assemblers and enhancement technologies. Such visions can have real influence:

> [...] that men dream science before doing it and that these dreams, which can take the form of science fiction, have a causal effect on the world and transform human condition. (Dupuy, 2005, p. 5)

However, because of their speculative nature, the visionary stories can be told in different ways. They might even be reversed, resulting in dystopian views on future societies built on nanotechnology. The "loss of control" scenario, for example, is nothing more than the "dark side" of some positive visions given by Drexler (1986) in the context of the molecular assemblers. In this way, a deep ambivalence is shown regarding far-ranging futures: visions meant positively, including utopian ideas, might in the eyes of other actors or the public lead to apocalyptic fears (Grunwald, 2007a,b): "Tremendous transformative potential comes with tremendous anxieties" (Nordmann, 2004, p. 4). The task of ethics in this respect would primarily be to test and evaluate the apocalyptic visions with regard to ethical aspects.

However, this motif of ethics for nanotechnology overlooks those features of nanotechnology that might serve ethics as the possible starting points for ethical considerations that are below the threshold of apocalyptic threats and, thus, all developments that are already at a technical or nearly technical stage of development (such as the potential risks from nanoparticles). Moreover, another risk in restricting oneself to this manner of argumentation is that ethical reflections will not be perceived or taken seriously by those who reject as nonsense, or at least as pure speculation, visions of the future in which these apocalyptic fears are based on technology. On the other hand, the fact that apocalyptic dangers are made a topic signals the presence of normative uncertainty and thus the necessity of ethical reflection.

In these various lines of argument developed by early authors as to why nanoethics should exist, no convergent line can be recognized that can explain the need for an ethics of nanotechnology transparently and in a way consistent with a complete formulation of ethical understanding. There is only agreement on the fact that an ethics of nanotechnology is necessary because of the presumed huge impacts on future development, but neither on the reasons given for this nor with respect to its design. On the one hand, the reason given for needing philosophical and ethical study of the expected transformative potential appears logical heuristically: if the transformative potential of nanotechnology is great and if likewise the uncertainties about the forms it might take and its consequences are also great, then it is very probable that normative uncertainty and conflicts would manifest themselves in a pluralistic society. On the other hand, however, this argument does not provide any specific reasons as to where the ethical challenges lie or how we can search for them. It only states that if we were to look, then it is quite probable that we would find them. The approach of a problem-oriented ethics based on the diagnosis of normative uncertainty (Chapter 3) shall provide more insight and legitimization about the need for and design of ethics of nanotechnology.

In their very diversity, the approaches mentioned beg the question of why, then, the need for ethics in nanotechnology is expressed with such vehemence. A comparison with other areas of science in two directions makes justification of this question even clearer:

1. For other areas of technology assumed to exert a considerable influence on future society, the demand for a separate ethics cannot be so clearly discerned, for instance with regard to the revolution brought about by the introduction of synthetic materials several decades ago, microsystem technology, or new energy technologies.
2. In other research areas such as stem cell research, genetic technology (e.g., reproductive cloning), or reproductive medicine, there was no necessity to substantiate the relevance of ethics because it was part of the debate from the very beginning: normative uncertainties and conflicts are openly visible. In those places, however, where it was evident that normative uncertainties and thus ethical questions are involved, the necessity of ethical reflection must not be specifically stressed. The fact that ethical reflection in nanotechnology is not simply practiced but that its necessity is justified over and over again and by different arguments seems an interesting phenomenon in itself.

## 4.2 Nanoethics as a New Field of Applied Ethics?

The concept of "nanoethics" has risen rapidly to fame in the past few years. The concept seems to stem from about the year 2002. From 2003 on, initially a few individual works on nanoethics can be found in the literature, which then strongly increase in number. Currently, the concept of nanoethics is used by several philosophers but above all by researchers into nanotechnology from the political and social sciences as well as the science, technology and society

(STS) community. This rapid rise of this concept and some (actually just a few) statements from relevant actors have led to the following diagnosis:

> Alongside the social scientists there are the many other academics — mostly humanities scholars — who have begun to adopt the neologism "nanoethics" to describe their work. Here "ethics" is generally not meant to describe a professional code of conduct but a specialty of applied ethics, in the sense of all the other "ethics of" sub-disciplines in which rarified academic orientations coexist only very uneasily with the "real world" — from computer ethics to environmental ethics to engineering ethics to bioethics. (Keiper, 2007, p. 63)

There are two sides to this interpretation. On the one hand, nanoethics is understood as a subdiscipline of *applied* ethics. Applied ethics concerns itself with practical questions relating to pending actions and decisions, which thus have some bearing on policy making (DEEPEN, 2009). However, as will be seen in the following chapters, this definition is too narrow. It is far more the case that philosophical reflections on nanotechnology relate not only to questions of applied ethics but also to questions of anthropology, technological philosophy, or hermeneutics, which can in no way all be understood as "applied"(cf. Chapter 10).

On the other hand, there is the suggestion that ethical questions of nanotechnology should be handled in the context of *a separate subdiscipline* in applied ethics. It is said that the established subdisciplines in applied ethics such as bioethics, medical ethics, and environmental ethics are not prepared or suitable for dealing with questions of nanotechnology.

A whole range of developments can be now observed which might lead us to believe that the process of establishing a separate discipline of nanoethics is well underway. These include the warnings mentioned at the outset of the type "Mind the Gap!" (Mnysuiwalla *et al.*, 2003) and the rapid spread of the word nanoethics (even as a book title, cf. Allhoff *et al.*, 2007). In addition, we might mention the founding of the specialist journal *Nanoethics*, although it does not regard itself as the vanguard of a movement, but rather leaves the issue open as to whether it is to be viewed as the central organ of a newly developing subdiscipline of nanoethics or as an open platform for ethical questions regarding "technologies which converge at the nanoscale" (Weckert, 2007), and which itself provides a forum for conducting this discussion (Section 5.1). Workshops and sessions devoted to ethical–social issues at large conferences and congresses on nanotechnology, as well as increasingly also calls for project proposals by the large research funding organizations and an increasing number of workshops for young scientists, summer schools, and academy congresses could equally be interpreted as a sign of a quasidisciplinary institutionalization of nanoethics.

In a comprehensive anthology on these issues (Allhoff *et al.*, 2007) four reasons are given for creating a subdiscipline of nanoethics (see also Allhoff and Lin, 2006):

> First, nanoethics also commands a significant amount of attention and money, though far less than the amount poured into nanotechnology. In the U.S., the NNI currently sets aside approximately $43 million for the "identification and quantification of the broad implications of nanotechnology for society, including social, economic, workforce, educational, ethical, and legal implications"[17]. So it would certainly be strange that there would be so much invested by various government agencies, universities, publishers and other organizations globally, if nanoethics were not a distinct or intelligible field. [...]
>
> Second, it is unclear why we should accept the litmus test that, to be a true discipline, nanoethics must either raise new ethical issues or larger ethical issues than other technologies. Looking again at chemistry, for example, whether or not we can properly categorize it as a subset of physics, there is no existential dilemma about its status as a legitimate category; no one is proposing to do away with the name or reorganize the university chemistry lab under the physics department. Therefore, it is unclear why such a dilemma would exist with nanoethics, even if nanoethics can be wholly contained within another field or set of fields.
>
> Third, to the extent that nanotechnology is a convergence of many disciplines in the first place, it should be no surprise that nanoethics is a convergence of many ethical areas as well. So even if a new area of ethics requires raising new or larger issues, that standard may no longer apply with the discovery or creation of nanotechnology, which uniquely draws from other disciplines like no other discipline before it. [...]
>
> Fourth, nanoethics does seem to raise new ethical issues insofar as it adds a new dimension or "flavor" to current ethical debates. For instance, though privacy may be a relatively old debate, the

> possibility of creating near-invisible and undetectable devices did not meaningfully exist prior to nanotechnology; so nanotechnology brings a new urgency and reality to the issue of privacy. (Lin and Allhoff, 2007, p. 9f.)

Ultimately, the force of the discussion is defused by the statement that the choice of framework within which the ethical questions of nanotechnology are discussed is not significant, the important issue being that these questions *are* discussed at all. Even if the latter point can be initially granted, the question still remains then, given such a pragmatic defusing statement as this, as to why then so much energy is spent on gathering arguments for nanoethics as a new discipline or at least a subdiscipline. Perhaps the authors themselves felt that their arguments are weak:

- Argument 1 is a false conclusion on the basis of a faulty version of the concept of ethics with regard to the logic of the terms used. When the "identification and quantification of the broad implications of nanotechnology for society, including social, economic, workforce, educational, ethical and legal implications" (the authors are implicitly following NNI, 1999, here) are here understood to constitute nanoethics, "ethical" encompasses a whole range of aspects which themselves include "ethical" — this is a circular and thus logically deficient definition.
- Argument 2 is based on creating an analogy between nanoethics and chemistry. This analogy does not hold, however, for two reasons. On the one hand, the analogy is distinctly weak and can at best be understood metaphorically. No arguments are given as to why the analogy is a meaningful one. On the other hand, even if the analogy were justified, the example of chemistry that is used is not acceptable. Science theory publications on the basic principles of chemistry have proven that chemistry does indeed possess its own knowledge perspective and cannot be reduced to physics. Physicists do not answer chemistry questions (cf. Bensaude-Vincent, 2009; Janich and Rüchhardt, 1996).
- Argument 3 is more a hope than an argument. Although it is widely considered that nanotechnology mainly consists in the convergence of physics, biology, chemistry, and technological thought, the opinion is just as widespread that this very convergence leads to the emergence of something

demonstrably new and discrete (e.g., Schmid *et al.*, 2006). If one draws an analogy to nanoethics here, then that must directly imply that the convergence of various subdisciplines leads to genuinely new challenges in nanoethics in order to be able to speak of nanoethics. This, however, is only an expectation or hope. There is no logically convincing argument beyond a poor analogy.

- Argument 4 is not contested by anyone, not even by the many opponents of a discrete form of nanoethics and thus provides no contribution to the argumentation for creating a discipline of nanoethics.

The argumentation in favor of nanoethics constituting a discipline of its own thus fails completely.[1] This failure is not surprising, as a glance at the established subdisciplines of ethics shows. Applied ethics consists of subdisciplines such as bioethics, medical ethics, information ethics, or economic ethics that are tied not to *technologies* but to *ethical questions* such as dealing with information (information ethics, cf. Capurro, 2004), dealing with life (bioethics), or dealing with patients (medical ethics). Technology often plays a role here by creating new normative uncertainties, for example, with regard to informational self-determination or the relationship between physician and patient. The dominant perspective of examination here is oriented, however, on the normative uncertainties occurring in the relevant social practice and not to the technology itself. Although terms such as nuclear ethics, computer ethics, or genetic ethics are occasionally used, they do not designate a particular subdiscipline of ethics but function as a shorthand summary for the ethical handling of the relevant field of technology. The examples of stem cell research, nuclear energy, genetic diagnosis, or the internet show that ethical reflection on these technologies is pursued precisely not as a subtype of ethics in these areas of technology but in the context of the subdisciplines of ethics oriented to ethical issues. Considering the customary division of ethics into subdisciplines oriented on types of ethical questions, the formation of a subdiscipline of ethics which connects its specific *definiens* to an area of technology such as a nanoethics would thus not be plausible. In ethics, the subdisciplines are not arranged according to fields of technology, but according to the moral issues raised in various "areas" (Nida-Rümelin, 1996a):

[1] We should mention that one of the authors relativized this stance shortly after the book was published (Allhoff 2007).

> It is difficult to specify exactly what could make an area of technology so special that it needs its own ethics, but a minimal requirement must be that it either raises ethical issues that are not raised by other kinds of technologies, or that it raises ethical issues of a different (i.e., larger) magnitude than other technologies. (Holm, 2005, p. 1)

Many of the ethical questions that are raised by nanotechnology are already being handled in other contexts of ethical reflection (Ach and Siep, 2006; MacDonald, 2004). The ethics of technology, bioethics, medicine ethics, anthropology, the philosophy of artificial intelligence, and even the theoretical philosophy of technology concern themselves with issues of sustainability, risk evaluation, the interface between humans and technology or between living things and technology. In the same way, a European statement comments:

> The enabling character of nanotechnology generates familiar biomedical ethics like the gap between diagnostics and therapy or sensitivity of genetic information. This means we build on a familiar pool of ethical and social discussions. (EU, 2006)

Even so, with a view to the chapters that follow, we must regard two problem areas as relatively new: the emerging option of producing artificial life (Chapter 7) and the options of human enhancement on the basis of converging technologies (Chapter 9). But even in this regard, there are traditions in the discussions in philosophy and anthropology that can be drawn upon. Since the degree of novelty in fundamental ethical questions is low, there is no incentive for founding a new subdiscipline.

Furthermore, one argument against the demand that nanoethics be considered a separate subdiscipline of ethics is that ethical questions of nanotechnology can only be evaluated as normative uncertainties in the (real or imagined) *context* of technical applications or innovations. To this end, ethical questions are examined which a product or process based on nanotechnology would entail in its area of application. Individual atomic or molecular manipulations per se have no moral dimension and would therefore not be placed in the focus of a technology assessment or technology ethics. In the same sense, nanotechnology is not an object of ethical reflection per se, but there is a need for reflecting (at least envisaged) *applications* and their aims, means, and consequences.

Furthermore, we must remember that the question of whether a product contains nanotechnology does not per se play a role for ethical reflection. It is only important if there are specific limitations or enhancements on its use or consequences that are bound up with the "nano" property. Products that are new, modified, or made possible by nanotechnology are generally evaluated in their various domains of use. These areas of application, specifically their normative structures and the possibly newly developing normative uncertainties — and here things have come full circle with regard to the considerations above — are already the object of consideration in the context of established subdisciplines of ethics. This thought also leads to the conclusion that the relevant existing subdisciplines constitute the right place to conduct ethical reflection on nanotechnology. If nanotechnology is utilized in medicine, that means medical ethics; if it is an issue concerning surveillance technologies, the right place would be information ethics; if it is a matter of environmental risks, environmental ethics is relevant, and so forth.

It is not the "nano" property that is crucial for ethics but the properties of a new product in the context of an application. If, in the course of ethical reflection, it should turn out that it is not the nanotechnology but the microsystem technology that has provided the crucial contribution to technical facilitation, this then would be entirely insignificant for ethical reflection. This can be made clear by an intellectual exercise: supposing a new technological product causes normative uncertainty and thus requires ethical advice. If one considers nanoethics as the place to handle these issues for products emanating from nanotechnology, then to be consistent a "micro" ethics must be responsible for technical innovations from microsystem technology and, if the technological components are larger than this, perhaps a "milli" ethics would come into play. This exercise makes it clear how inadequate a conceptualization oriented on the physical size of technical manipulation is for ethical reflection.

Overall, there is no point of reference for founding nanoethics as a discrete subdiscipline of ethics (see in the same vein Swierstra and Rip, 2007). Ethical reflection should be organized according to aspects of purposive rationality and practicability. If one takes the fact into consideration that nanotechnology raises normative uncertainties in such diverse areas as distributive justice, securing the private sphere, the border between technology and life, and the technical enhancement of humans (Chapters 5 ff.), then the idea of gaining something from a concentration of the ethical reflection on nanotechnology in its own field of nanoethics does not seem plausible. The orientation of the subdisciplines of ethics according to ethical issues suggests that these subdisciplines are the places to treat such ethical questions as are raised by nanotechnology (in this regard, see also Ebbesen *et al.*, 2006). Problem-oriented ethics (cf. Chapter 3) in general should not be attached to fields of technology but to ethical questions or challenges which take their origin in a social practice (such as risk management or information handling) in which normative uncertainties occur. The right place for ethical reflection in nanotechnology is where there is a tradition of dealing with pertinent normative uncertainties — such as in the ethics of technology, information, or anthropology, or in medical ethics. Nanoethics is, thus, not a new subdiscipline of ethics but, compared to the well-known subdisciplines of applied ethics, a categorically different field of ethical deliberation.

## 4.3 Problem-Oriented Ethics of Nanotechnology

However, the results of the previous section do not mean that, nanoethics will be simply dissolved and distributed among the established subdisciplines of ethics. For even if most of the philosophical and ethical questions of nanotechnology are not really new, their co-occurrence in nanotechnology is in many cases new or leads to new constellations of ethical reflection. Analogous to the well-known fact that the nanosciences and nanotechnology are fields in which the traditional borders between physics, chemistry, biology, and the engineering sciences are blurred, the ethical questions surrounding nanotechnology bring together various traditional lines of philosophical and ethical reflection. These include (and this is not an exhaustive list by any means) the philosophy and ethics of engineering, medical ethics, anthropology, bioethics, information ethics, philosophy of mind, theory of justice, and the philosophy of sustainability. This does not imply that there will be new ethical questions (see above) but suggests that questions about overcoming normative uncertainties will be dealt with in a different setting than by applied ethics.

The challenge for nanoethics consists accordingly in permitting the ethical aspects of nanotechnology to be handled across the borders of the classical subtypes of ethics and in a dialogue with natural scientists and engineers. In this way, the expression nanoethics can be salvaged, but not in terms of designating a new subdiscipline of applied ethics. It is hereby understood as a *terminological and conceptual platform* where the individual paths of ethical reflection mentioned above can be brought together. These then reflect — in a problem-oriented and definitely technology-related fashion — the progress made in nanotechnology and products, processes, or systems that are thus made possible or given potential. *The term nanoethics then denotes the interdisciplinary, ethically reflective dialog in which the normative uncertainties associated with nanotechnology and their consequences for society are debated.* The problem-oriented ethics of nanotechnology finds its place in this way within — and not alongside — the established subdisciplines of ethics, complemented by nanoethics as a linking platform.

This proposal for attributing meaning to the notion of nanoethics differs from current usage of the term. The word nanoethics is currently being used in an almost inflationary fashion and given an extremely broad meaning. For instance, on closer inspection the book *Nanoethics* (Allhoff *et al.*, 2007) turns out to be an informative collection of essays on almost all imaginable topics in the context of social aspects of nanotechnology that is worth reading. It covers, for instance, research programs, the public (U.S.) debate on technology revolutions and their predictability, issues of democratic theory, public involvement, educational reform, developing countries, and military developments. The idea of elaborating the ethically interesting aspects of these topics was not considered of interest from the outset, since the term nanoethics was in no way correlated with ethics in a terminological sense. Rather, the term nanoethics serves there as a generic term for diverse social aspects of nanotechnology, including political, social, and methodological ones. This broadening of the term ethics makes the specific solutions contributed by philosophical ethics unrecognizable for specific problem situations and mixes them up with contributions from other areas. Philosophical ethics as a professional handling of normative aspects of science and engineering — particularly in situations involving normative uncertainty — does not appear. "Nanoethics" understood in this way is paradoxical. It is the abrogation of the term "ethics" in and for nanotechnology.

This is aggravated by the fact that in current nanoethics the economic aspects are usually shut out. The frequent use of the acronym ELSI (ethical, legal, and social implications) itself separates out the economic aspects and stands in the way of a comprehensive evaluation. Economic aspects may, however, not be missing from an ethical consideration, at least not in the frequent cases in which the opportunities and risks must be compared and weighed against each other. Nanoethics without consideration of the economic dimensions cannot provide any comprehensive and balanced evaluation of nanotechnological directions or products, but does risk, on the one hand, becoming merely a polemic against certain aspects of nanotechnology and, on the other hand, being caught off balance with the sole aim of gaining approval for nanotechnology.

For these reasons, repeated reference is made in this book to publications within the range of nanoethics, even if they overwork the term ethics, but in so doing in each case specifically ask the question of the relevant ethical challenges in the sense of a problem-oriented ethics (cf. Chapter 3) of nanotechnology. The term nanoethics itself is used in the sense of the platform described above for discussing ethical issues which cross the borders of subdisciplines and are triggered by nanotechnology. These considerations lead to the following key points, premises, and directions of the remarks on the philosophical and ethical aspects of nanotechnology:

- The key point is the search for normative uncertainties (Section 3.1.3), which occur or can occur through developments in nanotechnology. Normative uncertainties create the motivation for ethical reflection, and overcoming them constructively is its main task.
- In order to define existing or possible future normative uncertainties, it is necessary to define the valid "normative framework" for a field of action. Then it can be evaluated whether this is a "standard situation in a moral sense" or not.
- If the answer is yes, it is possible to use the deviations, conflicts, or indifferences to elaborate exactly what constitutes a need for reflection and possibly also action. For this, analyses, defining positions, and if necessary also recommendations can be given.

In this way, a step-by-step systematic analysis is provided of where ethical challenges lie and how they can be overcome, and it is done in a manner that is understandable and amenable to criticism. The ethical aspects of nanotechnology result from challenges that the emerging research and innovation in nanotechnology represent to the established normative frameworks, including the social values they represent. The task of identifying the ethical challenges emerging from nanotechnology is equivalent to the search for the normative frameworks and social values that are affected. No nanoethics discipline of any kind is required for this purpose, but "just" the awareness of the underlying ethical questions of modern science, technology, and engineering (Section 3.2) and the analysis of specific manifestations in nanotechnology.

## Chapter 5

# Ethics of Nano(bio)technology: An Overview

The objective this chapter is to create a bridge between the preparatory and conceptual work done in the preceding chapters and the in-depth case studies in the following chapters. It consists of an application of the understanding of ethics presented in Chapter 3, as set against the background of the concept of an ethics of nanotechnology presented in Chapter 4, to the broad spectrum of innovation and research in nanotechnology in a comprehensive way. In addition to its objective to paint a full picture of the ethical aspects of nanotechnology, it also intends to identify promising and relevant fields for the in-depth analyses to be presented in the remainder of this volume. Drawing a "map" of the "landscape" of ethical questions caused by nanotechnological innovation will answer the questions:

- What are the ethical aspects of nanotechnology and the related innovations in the sense defined above?
- Which of the identified ethical aspects of nanotechnology are *specific* to nanotechnology and *novel* to ethics?
- Where are relationships to recent or ongoing ethical debates in other fields of technology, if any?

The resulting "map" of ethical aspects described in this chapter further develops recent work by the author (Grunwald, 2005, 2008b, 2011a) and also draws upon other explorations of "nanoethics" (e.g., Kushf, 2004a; Allhoff *et al.*, 2007; Jotterand, 2008a; Schummer and Baird, 2006; ten Have, 2007). It is organized with reference to existing ethical debates, for example on privacy, equity, or human nature. This classification has the advantage that it automatically

*Responsible Nanobiotechnology: Philosophy and Ethics*
Armin Grunwald

ISBN 978-981-4316-80-4 (Hardcover), 978-981-4363-33-1 (eBook)
www.panstanford.com

allows me to refer to normative frameworks, making it possible to investigate the following:

(a) Whether the existing frameworks are sufficient to deal with the value problems involved, or
(b) Whether there are new challenges to the frameworks that are specifically caused by nanotechnology.

Following this rationale the chapter is structured as follows. An overview of the relevant literature (Section 5.1) will create the basis for presenting a map of the ethical aspects of nanotechnology. This is followed by consideration of ethical aspects emerging from specific fields of application (Section 5.2) as well as of more cross-cutting ethical issues (Section 5.3). The purpose of the last section is to underpin the selection of the case studies presented in the following chapters.

## 5.1 Literature Overview

Different types of publications on the social and ethical aspects of nanotechnology have been prepared in the last 10 years (ITA, 2010). The first direct work on nanoethics accomplished hardly more than to proclaim a need for ethics in and for nanotechnology (Mnyusiwalla *et al.*, 2003; Khushf, 2004a). Furthermore, in the first years of the past decade the ethical, legal, and social implications (ELSI) of nanotechnology were elaborated by commissions and interdisciplinary study groups (e.g., NanoForum, 2004; Royal Society, 2004; Paschen *et al.*, 2004). These efforts mostly provided checklist-like collections of ethical issues rather than well-structured and systematic approaches. Their main impact was to identify relevant ethical issues and to stimulate further debate. After these first steps, the issues of nanoethics were taken up by conferences and workshops in science and the humanities, which led to the first book publications consisting of collected papers. Today we can see increasingly professionalized publication behavior: nanoethics has entered the sphere of international scientific journals and monographs.

### 5.1.1 Interdisciplinary Expert Studies

The early phase of identifying and classifying the ethical issues of nanotechnology was characterized by interdisciplinary approaches involving either technology assessment (TA; Grunwald, 2009a) or ELSI studies provided by expert groups. Although ethical aspects are implicitly involved, they are mostly not made explicit. In this sense the ethical, legal, and social analysis period of nanoethics might be considered an exploratory stage, which contributed decisively to setting the agenda of the emerging field of nanoethics. In this early work, the nanoethics debate was separated into two branches, which is still valid. On the one hand there are far-ranging, futuristic, and perhaps speculative developments under consideration such as nanorobots and radical human enhancement, while, on the other, more "down to Earth" issues are tackled, such as the possible risks emerging from nanomaterials included in products already available in the marketplace or equity issues. Some examples of technology assessment and studies of the ELSI are briefly described in the following.

The first comprehensive examination of the social effects of nanotechnology was presented by the Office of Technology Assessment at the German Bundestag (TAB) in 2003. As early as 2000, TAB was commissioned by the research committee of the German Bundestag to carry out a technology assessment study on nanotechnology. The results were presented to the committee in 2003 and published (Paschen *et al.*, 2004). In connection with the debate about nanoparticles that was emerging at the time and following the demand for a moratorium from the ETC group (ETC, 2003; cf. Section 2.5 and Chapter 6) this probably includes the first comment on regulating nanotechnology (cf. on these topics, Section 5.3 and Chapter 6).

> Political decisions on the necessity of regulations specific to nanotechnology will have to be made in the near future. A factual basis must be created for such decisions. This includes — in addition to a significantly improved data base on the effects of nanotechnological processes and products on the environment and human health — a systematic and comprehensive analysis of the currently relevant legal framework for applications in nanotechnology. Setting up a monitoring program to support decisions and accompany the further application process of nanotechnology should be considered. (Paschen *et al.*, 2004)

Ethical questions are hidden behind this problem complex but these were not specifically made a topic. The rather futuristic aspects of nanotechnology were however also addressed in the wake of the Bill

Joy debate. There were also comments on converging technologies, which became a topic in later TAB reports (such as brain research, bionics, and enhancement technologies in everyday life).

An intensive debate on nanotechnology issues also took place rather early in the United Kingdom. The widely perceived study by the Royal Society and the Royal Academy of Engineering (2004) resulted in many statements and recommendations in the field of regulation aimed at closing the knowledge gaps and at minimizing the risks possibly resulting from the production and use of nanoparticles before this had happened. In the meantime the UK government responded to these recommendations (UK Gov, 2005) and promised to promote more research on the risks posed by nanotechnology, to conduct reviews of existing regulations with respect to their applicability to nanotechnology, and to insist on public debate.

Expert recommendations on the handling of nanomaterials were issued for the European Commission (Community Health and Consumer Protection; Haum *et al.*, 2004; see also NanoForum, 2004). Special emphasis was given to developing a nomenclature for nanoparticles, to assigning a new Chemical Abstracts Service (CAS) registry number to engineered nanoparticles, and to grouping and classifying nanomaterials with respect to categories of risk, toxicity, and proliferation. Furthermore, many requirements apply to advancing science and developing the instruments for risk assessment methods, screening, and testing. Another important first step was the postulate for establishing institutions to monitor nanotechnologies and to assess continuously our knowledge about them. A dialogue with the public and with industry should be initiated. Freely engineered nanoparticles are, according to the recommendations, to be contained, and the production and unintentional release of waste of nano-sized particles are to be avoided whenever possible and otherwise minimized.

The goal of a study conducted by the European Academy for the Study of Consequences of Scientific Technical Advance was to derive recommendations that took the wide range of societal implications into consideration and that reflected the different views in an integrative manner (Schmid *et al.*, 2006). In accordance with the insight that technology assessment should be carried out in an interdisciplinary manner, the expert group consisted of chemists, physicists, engineers, biologists, toxicologists, philosophers, and economists. While desirable in any scientific endeavor, this mode of operating was *essential* for the assessment of nanotechnology, where the traditional borders between established disciplines are disappearing. The group's starting point was first to ask which properties and functions of materials are changing at the nanoscale in accordance with its own definition (cf. Section 2.3). The focus on potential applications was specifically on emerging technologies in information storage and major biomedical challenges. The final report (Schmid *et al.*, 2006) includes chapters on ethical issues, risk management, and — this might be unique — on aspects of the philosophy of science touched by nanotechnology.

### 5.1.2 Position Papers from Nongovernmental Organizations

Nongovernmental organizations (NGOs) entered the social debate on nanotechnology rather late, with the exception of the ETC group. Possible side effects such as the toxic effects of nanoparticles or the more futuristic expectations and concerns were first expressed by scientists and science writers. NGOs used these early reflections as material for their own approaches and positions regarding nanotechnology, starting with the famous report of the ETC group, "The Big Down" (see below). The focus of most of the NGO positions was threats posed by nanomaterials to the environment and our health, issues of controllability and power, and questions of access and equity. Many of these positions support a strong precautionary principle (Chapter 6) and conclude with postulates for a moratorium concerning the use of nanomaterials in products.

The ETC Group is an international civil society organization based in Canada dedicated to the conservation and sustainable advancement of cultural and ecological diversity and human rights. In its first report on nanotechnology (ETC, 2003), the tools and techniques around nanotechnology are called "atom technologies." They include nanoparticles, nanobiotechnology, nanofabrication, and molecular manufacture, which are regarded part of a process where the human capacity to manipulate matter is extended to the dimension of atoms (the parallel to the title of the National Nanotechnology Initiative "Shaping the World Atom by Atom" is quite obvious), and where "an impressive array of industrial enterprises is targeting a scientific revolution that could modify matter and transform every aspect of work and life." ETC diagnoses that

"Atomtech" (this notion did not find resonance in the nanotechnology debate) poses both tremendous opportunities and horrendous social and environmental risks:

> Atomtech will allow industry to monopolize atomic-level manufacturing platforms that underpin all animate and inanimate matter. The present-day bulk production of materials and new forms of carbon with unknown and untested characteristics is a major concern. In the future, mass production of unique nanomaterials and self-replicating nanomachinery pose incalculable risks. Atomtech could also mean the creation and combination of new elements and the amplification of weapons of mass destruction. [...] Even the production and use of today's nanoscale materials could have breathtaking societal implications and the environmental impacts are unknown due to insufficient data and study. In the future, molecular manufacturing poses enormous environmental and social risks and must not proceed — even in the laboratory — in the absence of broad societal understanding and assessment. (ETC, 2003)

This diagnosis led the ETC group to postulate that products may not be released to the marketplace and to customers until comprehensive knowledge about consequences and risks are available and to call for a moratorium (Section 2.5 and Chapter 6), which has found many successors.

The NGO "Friends of the Earth" published a position paper on the use of nanoparticles in cosmetics (2006). The field of cosmetics is one in which many applications of nanoparticles are expected and some have already been realized, such as the well-known sunscreens including titanium dioxide nanoparticles. As is also the case in other fields of application, there are large gaps in our knowledge of the possible consequences for our health resulting from nanoparticles entering the human body. Friends of the Earth followed the argumentation of the ETC group and proposed a moratorium on the use of nanoparticles in cosmetics until these gaps are overcome.

The ETC group was also active very early in the field of nanobiotechnology and synthetic biology (ETC, 2007; Chapter 7).

> Today, scientists aren't just mapping genomes and manipulating genes, they're building life from scratch — and they're doing it in the absence of societal debate and regulatory oversight. (ETC, 2007)

Starting with this diagnosis the ETC group was concerned about the possible dangers and abuses of biotechnology, building on very optimistic assumptions about scientific progress: "Scientists predict that within 2–5 years it will be possible to synthesize any virus; the first *de novo* bacterium will make its debut in 2007."[1] The group fears that synthetic biology means cheaper and more widely accessible tools for building bioweapons, virulent pathogens, and artificial organisms that could pose grave threats to people and the planet. Another concern is that — despite calls for open source biology — the power to make synthetic life could be concentrated in the hands of major multinational firms. Rejecting proposals for the self-regulation of synthetic biology by a small group of synthetic biologists (Chapter 7), the ETC group demands widespread debate on the social, economic, and ethical implications of synthetic biology, for example, on biosecurity (bioweapons/bioterrorism) and biosafety (worker safety and environment) issues. The report concludes with the proposal for an "immediate ban on environmental release of *de novo* synthetic organisms until wide societal debate and strong governance are in place."

The area of food stuffs is particularly sensitive in the public perception, as has been shown by the reception of genetically modified food in Europe (Meyer *et al.*, 2009). Research concerning the use of nanotechnology in the food sector (nanofood) ranges from new packaging materials to additives for realizing a large range of characteristics for which nanoparticles are to function as the vehicles. In Germany in the past few years, the question has arisen again and again as to whether and which synthetic nanoparticles are already contained in food. According to a study by the Friends of the Earth Germany (Bund für Umwelt und Naturschutz Deutschland, BUND), this is the case. Nanomaterials, they report, are already in various ketchups, vegetable stocks, and powder sugar in order to improve their flowing and pouring qualities. Food storage boxes, refrigerators, and packaging, they say, are increasingly coming onto the market with an antibacterial effect provided by nanosilver ions (BUND, 2008, 2009). The demands of BUND extend above all to transparent information for consumers in the framework of a requirement to label the synthetic nanoparticles contained in them (Chapter 6).

[1] This prognosis did not come to pass. The artificial bacterium produced by Craig Venter in 2010 was not real synthetic life either but only had a synthetically produced genome — which was, however, a large step towards creating artificial life.

### 5.1.3 Selected Edited Books

From 2004 on, several books have been published on the ethical and social aspects of nanotechnology (e.g., Baird *et al.*, 2004; Nordmann *et al.*, 2006; ten Have, 2007; Allhoff *et al.*, 2007). At that stage of the development of nanoethics, such publications were adequate because they allowed for exploring the field from different perspectives without the force of being too systematic or monolithic (as in a monograph). Most of these books put the relations between nanotechnology and society into the foreground, with ethical issues being part of them. In the following the content of some of these books will be described briefly on the basis of a subjective selection. The aim is to collect issues and topics for the classification of ethical aspects of nanotechnology in the remainder of this chapter.

The book *Nanoethics: The Ethical and Social Implications of Nanotechnology* (edited by F. Allhoff, P. Lin, J. Moor, and J. Weckert in 2007) claims to be a comprehensive *anthology*. In fact, even a glance at the book's table of contents shows that the entire field of nanotechnology and society (except the economic dimension) is covered to a large extent. The first part serves as an introduction into the field, starting with an approach to defining nanoscience and nanoethics. The coeditors P. Lin and F. Allhoff use "nanoethics" as an umbrella term for bringing together a lot of different issues in the field of nanotechnology and society. The anthology included the famous paper by Bill Joy, *Why the Future Doesn't Need Us*, as well as an analysis of the policy of the U.S. Congress towards the societal implications of nanotechnology, authored by the futurist writer Ray Kurzweil.

The second part is dedicated to the broader background of the nanotechnology debate, which in the early stage of the debate was characterized by high expectations for nanotechnology to overcome humanity's more pressing challenges. In a more modest version, this conviction is also behind the description of the U.S. National Nanotechnology Initiative (NNI, 1999) in this part. It also led to the necessity of nanotechnology being debated socially. The third part deals with the revolutionary aspects of nanotechnology. It consists of techno-optimistic papers pointing to the benefits of nanotechnology as a revolution, as well as of critical papers considering the applicability of the precautionary principle in coping with the revolutionary situation created by emerging nanotechnologies. The main objective of this part is to present extremely different approaches to and conclusions about nanotechnology as a revolution.

In the fourth part the book is partially more down to Earth, addressing the environmental, health, and safety (EHS) problems possibly posed by nanotechnology. It presents an analysis using patterns from the sociology of uncertainty to identify the nanotechnology issues that could and should be interpreted and managed as risk, primarily with respect to human health and the environment. Challenges to personal choice in emerging nanomedicines are addressed, and the question of whether we are perhaps already "playing God" is raised. Human enhancement is a topic in the context of governance and democracy, which leads to the fifth part dedicated to democracy and policy issues. These papers deal with different forms of participation and deliberation from a stakeholder's perspective.

The *Broader Societal Impact* (part VI) addresses more familiar questions in the field of nanotechnology and society: the privacy issue, building on the case of radio frequency identification (RFID), the challenge of nanotechnology to military applications, educational issues, and the impact of nanotechnologies on developing countries. The final part is, not surprisingly, dedicated to the "distant future" and includes essays on exponential manufacturing, on the use of nanotechnology as set against the background of space travel and settlement, on ethics for artificial intellects, and on life extension. In this way the debate comes back to the high visions often associated with nanotechnology that served as the book's starting point.

The book *Nanotechnologies, Ethics and Politics* (edited by H. ten Have in 2007) is the product of the World Commission on the Ethics of Scientific Knowledge and Technology (COMEST) of UNESCO, which was founded in 1998 to deal with emerging ethical questions at the global level. Its mandate is to be an intellectual forum for the exchange of ideas and experience, to detect early signs of risk situations, to advise decision makers, and to promote dialogue between science, policy makers, and the public. COMEST established an ad hoc group to analyze the ethical aspects of nanotechnology by addressing two questions:

(a) What is the state of the art of nanotechnology and what are the ethical issues related to it?

(b) What opportunities are there for international action regarding these issues?

In response to these questions the first part of the book is dedicated to the state of the art of nanotechnology, giving specific attention to the role that nanotechnology plays for developing countries, including a quick course on the history of nanotechnology. The second part, "Ethical Issues," provides an overview of ethical issues in nanotechnology, building on an analysis of the problem to define nanotechnology and concluding with recommendations for regulation and for further research. In the specific case of nanomedicines, ethical issues are differentiated between short-, medium-, and long-term issues. More generally, the handling of the broad field of nanotechnology and (public) health also goes beyond nanomedicine, including considerations of patient care and clinical innovations and concluding that early stakeholder involvement will be necessary to prevent undesirable developments and to harvest the promised benefits. The third and last part focuses on policy issues in the context of UNESCO as a global organization. The (frequently discussed) impact of nanotechnologies for developing countries is analyzed, and the fields of renewable energy sources, promoting health, reducing hunger, improving water and sanitation are identified as main targets for the use of nanotechnology in developing countries. The consideration of what these countries are doing in nanotechnology is focused on the countries striving to quickly catch up with the industrialized countries (e.g., China and India). The issues of the public's role and education for ethics in nanotechnology are also covered, with a focus on early assessment and policy making (including participatory measures).

The book *Nanotechnology Challenges: Implications for Philosophy, Ethics and Society* (edited by J. Schummer and D. Baird in 2006) unravels the

> philosophical underpinnings of nanotechnology, its metaphysical and epistemological foundations, and its conceptual complexity. It explores the ethical issues of nanotechnology, its impact on human, environmental, and social conditions, and the options for reasonable risk assessment. (Schummer and Baird, 2006, p. 1)

A basic observation is that the international community working on the ethical and social issues of nanotechnology is "growing almost as fast as nanotechnology itself" (Schummer and Baird, 2006, p. 2). The volume is divided in separate parts on the philosophical, ethical, and social issues posed by nanotechnology. The first part includes papers from the perspectives of the philosophy of science, metaphysics, and epistemology, starting with an analysis of the famous Smalley–Drexler debate on the possibility or feasibility of molecular assemblers (Smalley, 2001) and continuing with the relations between nanotechnology and nature, issues of technology determinism, the epistemological status of images created by scanning electron microscopy, and measures of the complexity of nanotechnology. The second part on ethical issues is divided into sections on social and environmental ethics and on risk issues. The challenge of dealing with large uncertainties forms the main and cross-cutting topic in this part. The third part focuses on the public discourse surrounding nanotechnology. It deals with risk perception, communication patterns using nano futures (see also Grunwald, 2007a), and conflicting meanings and interests.

The book *Emerging Conceptual, Ethical and Policy Issues in Bionanotechnology* (edited by F. Jotterand) provides an overview of the nature of nanotechnology, focusing on applications in the biomedical sciences and in humans (bionanotechnology) as well as on the philosophical and ethicolegal issues that are involved. It consists of thirteen articles by scholars from philosophy, anthropology, law, the social sciences, psychology, and the natural sciences. The book is organized in four major parts:

1. Knowledge production in nanotechnoscience
2. Ethics and (bio)nanotechnology
3. Public policy and (bio)nanotechnology
4. Human enhancement and (bio)nanotechnology

In the first part, the nature of nanotechnology as a scientific project is examined, including a critical reflection on its philosophical underpinnings. The second one explicitly addresses the ethics of nanotechnology and bionanotechnology by examining the theoretical framework(s) necessary to sustain ethical reflection at the core of the development of nanotechnology. The third part focuses on legal and public policy issues and how the public perception of nanotechnology could ultimately shape policies and regulations, including regulatory options such as the precautionary principle. The final part focuses on how scientific progress could affect humans, such as through enhancement technologies, and on assessments of whether such progress actually contributes to human welfare and quality of life.

This brief glance at the topics shows that an informal "canon" has evolved in the mapping of ethical questions in nanotechnology (cf. also Kushf, 2004a; Grunwald, 2005a, 2008b; Ach and Siep, 2006). One encounters in the anthologies — and also in essays in journals — the same or similar topics on ethical questions in nanotechnology. This canon has evolved in a somewhat natural way and has acquired dignity and rationality from the fact that it can and could be annexed from several sides: from the developments in the natural sciences, which provide the basic knowledge for technical applications, from the social debates and the concerns communicated there, and also from the discussions conducted in applied ethics. These are presumably also the reasons why this canon has achieved a high degree of stability and — in a structured form — also forms a basis of this book (cf. Sections 5.2 and 5.3).

### 5.1.4 The Journal Nanoethics

The international journal *Nanoethics*: *Ethics for Technologies that Converge at the Nanoscale* was founded in 2007 and has developed into the main organ of the nanoethics community and related fields of research. The self-description (see http://www.springer.com/social+sciences/applied+ethics/journal/11569) follows a "problem-oriented approach" and starts by pointing to social controversies surrounding nanotechnology:

> Nanoscale technologies are surrounded by both hype and fear. Optimists suggest they are desperately needed to solve problems of terrorism, global warming, clean water, land degradation and public health. Pessimists fear the loss of privacy and autonomy, "grey goo" and weapons of mass destruction, and unforeseen environmental and health risks. Concern over fair distribution of the costs and benefits of nanotechnology is also rising.

*Nanoethics* aims at providing a forum for informed discussion of ethical and social concerns related to nanotechnology, and a counterbalance to the fragmented popular discussion. In this way, the journal wants to give voice to reflections on ethical and social concerns rather than to being purely an academic journal. It gives space to a rigorous examination of the ethical and social considerations and policy concerns raised by nanotechnology. While the central focus of the journal is on ethical issues, discussion extends to the physical, biological, and social sciences and to law. "Nanoethics" is, in this sense, more an interdisciplinary journal than an organ of professional philosophical ethics. This judgment is supported by viewing the issues dealt with and the authors submitting papers. A broad interdisciplinary spectrum can be found, with centers in the field of science, technology, and society studies (STS) and in applied ethics. The list of issues covered also demonstrates the broad spectrum. Papers are called for on any area of ethical and social concern to nanotechnology, including:

- Environment and health issues (e.g., nanoparticles)
- Privacy
- Human enhancement, implants, and cyborgs
- Human–machine interface
- Social consequences, for example, redistribution of wealth, un-employment, the nano divide
- The role of intellectual property
- Global social justice
- Risk assessment
- The precautionary principle and cost–benefit analysis
- Nanotechnology in war
- The role of values in design
- Research ethics

This list of issues corresponds very well to the catalogue of ethical issues of nanotechnology that was identified by looking at the edited books on the ethical and philosophical aspects of nanotechnology (see Section 5.1.3).

## 5.2 Ethical Questions Related to Nano(bio)technology Applications

In the debate over ethical issues in nano(bio)technology, it is common to find lists that contain application-related questions (such as the use of nanotechnology in the military) and cross-cutting issues (such as problems of accessibility and justice) next to one another at the same level. This seems unsatisfactory in terms of categories. For this reason, in the following we will first examine fields of application of nanotechnology with relation to ethical aspects.[2] There are direct

[2] The structure of this book, which combines a general overview in this chapter with in-depth studies in the following ones, implies that there are some unavoidable redundancies. I accept this as inevitable in order for both this overview chapter and the in-depth chapters to be independently readable.

connections to applied ethics such as to medical ethics or bioethics so that it seems simple to fit this into the debates underway there and to estimate the degree of novelty caused by nanotechnology. Cross-cutting issues are addressed in Section 5.3.

### 5.2.1 Nanomedicine: Risks and Benefits

Great hopes are placed in the potential of nanotechnology to provide better medical care (Freitas, 1999). Miniaturization is an essential means of progress in many areas of medicine. The use of smaller samples for in vitro analysis allows less invasive and less traumatic methods of extraction. Better interfaces and biocompatible materials create new opportunities for implants and attempts to restore damaged organic facilities. The field of medical applications might develop into the largest field of future nanotechnology applications (Section 2.3). According to an early and widespread definition, which is admittedly unnecessarily stilted, nanomedicine is:

> (1) the comprehensive monitoring, control, construction, repair, defence, and improvement of human biological systems, working from the molecular level, using engineered nanodevices and nanostructures; (2) the science and technology of diagnosing, treating, and preventing disease and traumatic injury, of relieving pain, and of preserving and improving human health, using molecular tools and molecular knowledge of the human body; (3) the employment of molecular machine systems to address medical problems, using molecular knowledge to maintain and improve human health at the molecular scale. (Freitas, 1999, p. 418)

The use of nanotechnology is expected, according to more or less realistic scenarios, to produce improvements in medical diagnostics and therapy. With the aid of diagnostic instruments based on nanotechnology, illnesses or dispositions for illnesses could be recognized earlier than is now possible (Sattler, 2010, part IV). With regard to therapy, treatments might be developed that are dedicated and (more or less) free of side effects. The use of nano-based dosage systems could lead to very precise drug delivery in the human body and thus to substantial advances in treatments (Sattler, 2010, part V). The principle of hyperthermia using magnetic fluids is to inject nanoparticles containing iron into a tumor and to cause them to vibrate by means of alternating magnetic fields, which could heat and kill the tumor tissue without damaging the surrounding tissue (Paschen *et al.*, 2004, p. 225ff.; see also Sattler, 2010, Part VII on nanotechnologies for cancer therapy). Furthermore, the biocompatibility of artificial implants can be improved considerably by means of nanotechnology (Freitas, 2003).

The European Union has set up a technology platform on nanomedicine to address ambitious, responsible research, development, and innovation in the use of nanotechnology for health care. This is intended to strengthen the competitive position of science and industry in Europe in the area of nanomedicine and improve the quality of life and health care of its citizens.

> The ageing population, the high expectations for better quality of life and the changing lifestyle of European society call for improved, more efficient and affordable health care. Our improved understanding of the functioning of the human body at the molecular and nanometre scale as well as our ability to intervene at pre-symptomatic, acute or chronic stages of an illness are of utmost importance to meet these expectations. Diseases like cancer, diabetes, Alzheimer's and Parkinson´s diseases, cardiovascular problems, inflammatory and infectious diseases and depression are serious challenges to be dealt with. Nanotechnology applied to medical problems can offer impressive solutions. Early diagnosis, "smart" treatments and the triggering of self-healing mechanisms are crucial targets on the way to regained health. (ETPN, 2005, p. 1)

A special area that is of interest both practically and ethically is that for making direct connections between technical systems and the human nervous system. Intensive work is being conducted on connecting the world of molecular biology with that of technology. An interesting field of development is nanoelectronic neuroimplants (neurobionics), which compensate for damage to sensory organs or to the nervous system, or increase the performance capacity of these organs and broaden the spectrum of human perception. Microimplants could restore the functions of hearing and eyesight. Simple cochlear or retinal implants, for example, are already possible now. With progress in nanoinformatics, these implants could approach the size and capabilities of natural systems.

The potential of these developments is considerable, even though this is currently only potential and some of the conditions of its realization are uncertain. While the goals are undoubtedly positive, ethical reflection can in this case concentrate above all on the definition and prevention of misuse. Technical access to the nervous system is a particularly sensitive issue because of the possibilities for manipulation and control that it opens. As an enabling technology for many future medical applications, nanotechnology touches on issues such as the sensitivity of genetic information, the gap between diagnosis and therapy, health care resources, and tensions between holistic and functional medicine (Baumgartner, 2006). A comprehensive analysis, however, must also include the possible side effects, especially risks. Risks must also be considered since they might be present in any new treatment:

> There is an urgent need to improve the understanding of toxicological implications of nanomedicines in relation to the specific nanoscale properties currently being studied, in particular in relation to their proposed clinical use by susceptible patients. In addition, due consideration should be given to the potential environmental impact and there should be a safety assessment of all manufacturing processes. Risk-benefit assessment is needed in respect of both acute and chronic effects of nanomedicines in potentially pre-disposed patients — especially in relation to target disease. A shift from risk assessment to pro-active risk management is considered essential at the earliest stage of the discovery, and the development of new nanomedicines. (ESF, 2005, p. 9)

There is probably no field of science in which dealing with risks is so natural and performed so reliably as in medicine and pharmaceutics. In medical research, it is a self-evident fact that new medicinal treatments involve risks and that, for this very reason, animal experiments are often conducted first and then experiments are carried out with human volunteers. For new medicines, there are established and officially supervised approval procedures. Advances in medicine (diagnosis and therapy) are evidently associated with risk, and there are many established mechanisms for dealing with them.

Against this background, it is not probable that direct applications of nanotechnology for medical purposes (e.g., drug delivery) will lead to *completely new* ethical questions. The application of nanotechnology in medicine comprises another step in medical progress but does not pose a new moral challenge in the medical applications that are of interest in this context.

The purpose of the normative idea of healing continues to be to develop approaches to improving healing while reducing the side effects (e.g., drug delivery) and maintaining or further developing established ethical standards for dealing with risks in research, diagnosis, and therapy. Some existing challenges to medical ethics may become more dramatic in nanomedicine, for example the well-known gap between capabilities for diagnosis and those for therapy might be widened in some fields. This would be a gradual shift, demanding ethical attention but not leading to truly new questions in medical ethics. The standards and procedures of medical ethics seem to be sufficient for dealing with new methods of diagnosis and treatment that are based on nanotechnology. This is therefore a standard situation in a moral respect (Chapter 3). This can also be seen from the postulates and the ethical principles derived from them that the European Technology Platform NanoMedicine (2005, p. 34) formulated — not surprisingly — to ensure human dignity:

- *Noninstrumentalization*: The ethical requirement of not using individuals merely as a means but always as an end in their own right.
- *Privacy*: The ethical principle of not invading a person's right to privacy.
- *Nondiscrimination*: People deserve equal treatment unless there are reasons that justify difference in treatment.
- *Informed consent*: The ethical principle that patients are not exposed to treatment or research without their free and informed consent.
- *Equity*: The ethical principle that everybody should have fair access to the benefits under consideration.
- *The precautionary principle*: This principle entails the moral duty of continuous risk assessment with regard to the not fully foreseeable impact of new technologies such as in the case of information and communication technology (ICT) implants in the human body.

The boundaries of such a standard situation would, however, be transgressed in some of the more visionary scenarios such as in the vision of extended longevity, the abolition of aging, or any form of "human enhancement." Nanotechnology could, in connection with biotechnology and, perhaps, neurophysiology, build the technological basis for making such visions become reality. Technical interventions into humans beyond healing would go beyond the classical ethos of healing and curing towards improving the capabilities of healthy persons and would, therefore, lead to normative uncertainties (Chapter 9).

### 5.2.2 Nanoelectronics: Surveillance and Privacy Issues

Information and communication technologies are among the main driving forces in all advanced economies. Miniaturization is a key to further progress, and nanotechnology is expected to provide the basis for the next major step in this direction. Miniaturization is accompanied both by huge promises of technical progress and economical innovation and by concerns about surveillance and privacy issues.

From the very beginning, the threat posed to one's privacy by new surveillance and supervision technology has been viewed as one of the ethical features of nanotechnology that needed to be taken into consideration (e.g., Moor and Weckert, 2004; Nanoforum, 2004). Nanotechnology provides a number of opportunities for personal data to be collected, stored, and distributed on a large scale. It is conceivable that miniaturization might lead to the development of sensor and storage technology that drastically increases the possibility of collecting data from victims without them noticing it (Moor and Weckert, 2004; van den Hoven, 2007). Such miniaturization, in connection with the limited human capacity to see, might make nanotechnology nearly invisible and could contribute to "ubiquitous computing" with the frequently discussed threats to one's privacy (EPTA, 2006). Mehta (2002) considered "nanopanopticism" to be a considerable danger to social trust and the involvement of citizens in public life. Furthermore, the combination of miniaturization and the networking of surveillance equipment could substantially obstruct the present control options and regulations for protecting personal data or even make them completely obsolete. The passive surveillance of humans might be supplemented in the more distant future by active surveillance (that the targets themselves might not notice) such as if it proved possible to construct direct technical access to the nervous system or the brain.

Some of developments in these scenarios are not only considered realistic but even certain: "But what is not speculation is that with the advent of nanotechnology invasions of privacy and unjustified control of others will increase" (Moor and Weckert, 2004, p. 306). Such technology determinism has been proven false by many case studies in research concerning the social construction of technology (SCOT; Bijker *et al.*, 1987). However, in many cases the mere availability of technologies creates some kind of self-dynamic, suggesting that statements similar to the one quoted above — though false — should be taken seriously in order to create awareness.

Health services are particularly sensitive with regard to the private sphere. The development of small units for preparing diagnoses — "lab on a chip" (van Merkerk, 2007) — can make it possible for comprehensive personalized diagnoses and prognoses to be prepared on the basis of a person's health data. The demands placed on data privacy and thus on the protection of one's privacy must therefore be high. The rapid decoding of an individual's genetic disposition can move into the reach of normal hospital activity or of nonclinical service providers. Anyone could then undergo an examination for a genetic disposition to certain illnesses or could be pressed by an employer or insurance company to do so. In this way, individuals could be exposed to social pressure that would limit their capacity to act freely.

All of these questions about privacy, surveillance, and data protection are not specific to nanotechnology. Surveillance technology has attained a substantial state of the art even without nanotechnology, which itself raises questions about the survival of individual privacy (e.g., EPTA, 2006). So-called smart tags are already making use of RFID technology in access control, in ticketing such as in public mass transportation, and in logistics. The size of such tags is currently several tenths of a millimeter in three dimensions, which means that they can hardly be noticed by the human eye. Further miniaturization will permit a further reduction in their size and the addition of further functions. Although this will be possible even without a contribution by nanotechnology, nanotechnology will certainly support and accelerate such developments.

The discussions of ethical issues in information science, the life sciences, and medicine have long included ethically relevant questions such as about a right to knowledge or ignorance, a personal right to certain data, and a right to privacy. They have also encompassed discussions about data protection; for example, a consequence of a drastic extension of genetic and other tests might be undesired social eigendynamics. The developments in nanotechnology can accelerate or facilitate the realization of technical opportunities, and they can extend the technical limitations that currently exist, thus increasing the urgency of the possibly problematic consequences. In this field, however, they do not lead to *new* ethical questions even though the handling of this issue is doubtlessly relevant. Building on recent work in information ethics, privacy and data protection have evolved a standard situation with specific standards for data handling, storage, and evaluation. Nanotechnology does not in principle pose a challenge to these standards. What is indeed challenged are the possibilities for ensuring that these standards will continue to be guaranteed in future developments despite new types of surveillance technologies being developed. These challenges are not challenges to *ethical* reasoning (in the sense discussed in Chapter 3) but challenges to data and privacy protection policies and regulations and to the development of technical facilities for the implementation of what is societally and politically desired in this sensitive field.

### 5.2.3 Using Processes of Life for Technological Purposes

Biotechnology is regarded as one of the most important key technologies to have arisen for decades. Technologies such as genetic modification are used for many purposes and form the basis of many branches of industry. Nanotechnology could, in the form of nanobiotechnology, help expand the technological use of biological processes to a considerable extent. A frequently mentioned field of application could be the development of new forms of microalgae that could be used as a new resource for the energy supply in the field of biomass-based fuels.

Basic life processes take place on a nanoscale because life's essential building blocks (such as proteins) are precisely this size. Nanobiotechnology (VDI, 2002; Sattler, 2010; Part I) makes — or are supposed to make — it possible for nanotechnology to control biological processes. The molecular "factories" (mitochondria) and transport systems that play an essential role in cellular metabolism can be models for controllable bionanomachines. Nanotechnology at this level could permit synthetic biology to engineer cells or even create artificial forms of life (Chapter 7). An intermeshing of natural biological processes with technical processes is already ongoing. The classical barrier between technology and life is increasingly being breached or even crossed.

Biosensors and biomembranes, for instance, could be employed in environmental technology, and the use of photoenergetic processes could provide biological support for photovoltaics. For example, it would evidently be economically and ecologically significant for synthetic biology to be technically able to copy photosynthesis and to exploit it in an industrial context. Scenarios for similarly far-reaching applications can also be found in biomedicine (Section 5.2.1).

New ethical issues are *prima facie* to be expected in this field. Their concrete specification, however, will only be possible when research and development can provide more precise information on the fields of application and the products. The corresponding discussions of risks will be structurally similar to the discussion of genetically modified organisms. It could come to discussions about safety standards for the research concerned, about the risks posed by experiments outside the laboratories, and release problems. The danger of misuse will be made a topic of debate, such as technically modifying viruses in order to produce new biological weapons. A wide range of future ethical discussions is opening up, for which at present there is insufficient practical background to support concrete reflection.

The debate on genetically modified organisms serves as a background (Meyer *et al.*, 2009). Many arguments pro and con have already been discussed there, with the consequence that ethics and society are better prepared for the present and future debates about synthetic biology. In the European Union, the precautionary principle has been installed and serves as a normative framework for challenges of this kind. However, creating life from scratch seems to be much more dramatic than modifying life by means of genetic technology. At the moment, classical patterns of argumentation, such as biosafety and biosecurity, are at the focus of the emerging debate (de Vriend, 2006). Other arguments such as human hubris or the fear of "playing God" are also involved. It is an open question whether the existing normative frameworks will suffice to provide the orientation needed to deal with ethical challenges from nanobiotechnology and synthetic biology. An indicator that this might perhaps not be the case is the severe criticism that has been put forward against the precautionary principle (Dupuy, 2007). Therefore, ethical reflection in the sense of Chapter 3 seems to be urgently needed (see Chapter 7 for an in-depth case study).

### 5.2.4 Human Enhancement

In combination with biotechnology and medicine, nanotechnology offers perspectives that go beyond the traditional medical tasks of healing illnesses, such as "improving" the human body or radically transforming or redesigning it. Although the examples of medical applications of nanotechnology (Section 5.2.1) are still within a, so to speak, traditional framework because the goal (like the classical goal of medicine) is to heal deviations from a healthy state, "enhancement" belongs to a category beyond healing and beyond the traditional ethos of medicine. For example, new or extended sensory functions could be implemented, such as through the extension of the electromagnetic spectrum that an eye is capable of perceiving. It is also possible for new interfaces between man and machine to be created by means of a direct coupling of ICT systems to the human brain ("brain chip"). Another focus of the debate about human enhancements is the lengthening of the human lifetime up to visions of abolishing death.

By extending these lines of development into the futuristic and speculative dimension, it would be possible to study topics such as the increased integration of technology in the human body, the increasing convergence of man and technology, and the creation of cyborgs as technically extended humans or technology that is given human traits. These perspectives raise anthropological questions as to our image of man and to the relationship between man and technology (Jotterand, 2008b) and, at the same time, lead to the ethical question of the degree to which humans *may*, *should*, or *want* to go on transforming the human body. Initially, ethical analysis has to confront the semantic and hermeneutic problems that are tied to the concepts of healing, doping, and improving since the latter are factors that play a determining role regarding each of the relevant normative parameters (Jotterand, 2008b).

The practical relevance of such ethical questions may at first glance appear minimal. Two factors should however be considered. First, the vision of a technical improvement of humans is presented in a rather serious manner, not as science fiction. Research projects are planned in this vein, and milestones to achieving this goal are defined, with nanotechnology playing the role of an enabling technology. The idea of a convergence of nanotechnology, biotechnology, information science, and the cognitive or neurosciences, which is presented at the level of atoms and which is supposed to become possible by means of the technical improvement of humans (Roco and Bainbridge, 2002), has already influenced research funding. Second, technical improvements are by no means completely new. On a rudimentary scale, at least, they are established, as is shown by the examples of cosmetic surgery as a technical improvement of body features that are perceived as being unaesthetic, or of the way in which at least some of the drugs used for improving performance (e.g., Ritalin) are already being employed in the framework of what is called everyday doping (Farah *et al.*, 2004). It is not difficult to assume that the options and practice of technical improvements of humans will increase. A demand for them is probable given our experience with the shortcomings and failings of humans. There can be no doubt that there is a need for ethical reflection on this issue in view of the normative uncertainty and potential for moral conflict that is linked to it.

Ethical reflection thus does not seem to be premature in spite of the speculative nature of the subject. In particular, advances in brain science and developments in converging technologies (Roco and Brainbridge, 2002) lead to this expectation — and would justify ethical reflection "in advance." In the combination of the emergence of enhancement technologies and a social situation with competition strongly influencing every domain of life it seems probable that enhancement technologies will be used if available. The pressure to get advantages over competitors is large, and the doping problem in sports could be a forerunner to other social areas. Ethics is, on the one hand, prepared to deal with such issues, considering for example issues of fairness, equity, and human rights. On the other, however, the magnitude of the challenge and the close relations with societal organization at large, with political philosophy, and with principles of capitalism lead to the conclusion that this is a field where a new type of ethics might be needed — an "ethics of enhancement" in a situation where humans start changing their own nature to an extent unknown and unimaginable in earlier times (Chapter 9). This diagnosis shows that we are far from a standard situation in moral respect in this field.

### 5.2.5 Animal Enhancement

The question of enhancement technologies in relation to humans immediately suggests consideration of whether and how similar technologies can also be used in animals, perhaps even sooner and with less intense or at least different ethical controversies.[3] Scientific and technological convergence processes on the basis of nanotechnology can have various effects in animals. They offer new options for exploiting animals, as for instance the creation of transgenic animals for biomedicine and agriculture.

The difficulties in transferring the enhancement concept from the debate on human enhancement arise from a simple but fundamental fact: the moral and legal status of animals is differently composed and regulated than that of humans. Although according to many laws animals are currently not simply mere objects, humans are permitted to use them in different ways as, for example food, clothes, models, or experimental subjects in research (Ferrari *et al.*, 2010). Due to breeding and husbandry, the very existence of animals depends in many cases strongly on human exploitation. This fundamental difference in moral status and the human–animal relationships that derive from it represent an obstacle to the parallel consideration of animal and human enhancement. Since the value of animals is dependent on human interests, even what is understood as an "enhancement" depends on human interests. In many cases, "animal enhancement" often means improved options for exploiting animals for human purposes, for example in agriculture, in sports, or as experimental animals. Only occasionally is there reference to enhancements "for the animal's sake."

Animal enhancement is not a development for the far-distant future. In many forms it is part of current practice and in many other areas the subject of research (Ferrari *et al.*, 2010). This usually does not trigger dramatically new ethical questions. As a result of the underlying relation to the contexts of human use, the "big" ethical questions are the same as the already familiar ones, such as: For what purposes may animals be instrumentalized? Under what conditions are animal experiments permissible? How should one deal with stress and suffering in the animals? Nanotechnology and converging technologies produce gradual shifts and changes in accent in all of these well-known issues.

However, a speculative but deep-seated ambivalence can be seen in the following example. Animal enhancement is rooted in the

[3] This section owes much to the cooperation with Christopher Coenen, Arianna Ferrari, and Arnold Sauter in the context of a study commissioned by the Federal Ethics Committee for Nonhuman Biotechnology (EKAH; Ferrari *et al.*, 2010).

tradition of genetic interventions in animals. A hypothetical scenario concerns the aim of reducing or completely eliminating an animal's ability to suffer. If animals could no longer experience suffering, they would drop out of the sphere of influence of pathocentric ethics. Animals that do not suffer under human exploitation could accordingly be exploited in any way without any qualms. Cruelty to animals would even be an impossibility because the animals involved could not suffer from it, for example, because their pain perception had been switched off. To call such a change a "technical enhancement" might initially sound cynical. It could, however, indeed be interpreted as an improvement if it meant that human exploitation were made simpler (i.e., no need to consider the animal's well-being) and thus, for instance, more efficient in an economic sense. From one perspective one could claim that if animals are to be employed for human use anyway, then it would be better for humans to use the kind of animals that did not suffer. However, questions do arise regarding the dignity or integrity of the animals, which would be affected by such massive interventions (Ferrari *et al.*, 2010). The option to specifically alter animals by switching off their ability to suffer may be speculative at present. However, it seems advisable to conduct ethical reflection here now (i.e., in advance) in case there is rapid technological progress.

Animal experiments make up a large portion of animal enhancements (Ferrari *et al.*, 2010; Section 2.1), often in preparation for human enhancement. The question whether animal experiments are ethically justified in preparation for a technical enhancement of humans is not readily clear. The very concept of human enhancement is by definition not therapy. If one classifies human enhancements as luxury, animal experiments for this purpose would be just as problematic as animal experiments for producing new cosmetics. In particular if — as in the Swiss regulation of animal experiments — there are legally regulated conditions for the admissibility of animal experiments, there is an urgent need for clarification here.

There is a shift in accent with regard to industrial-scale agriculture. Due to the further increase in economic pressure, there are incentives to further intensify the use of animals. Animal enhancement in this manner could on the one hand lead to increased stress levels for livestock. It could on the other hand open up possibilities for conducting advocatory animal enhancement in the presumed interests of the livestock in order to permit further use for human purposes without creating more stress for the livestock, perhaps even with less.

In conclusion, it should be noted that animal enhancement increases the portion of an animal that is technical and can be manipulated, and reduces its natural portion. It is still an open question whether this will lead to changes in the human-animal relationship (Ferrari *et al.*, 2010; Section 3.4). Careful observation is called for due to the far-reaching cultural significance of the human–animal relationship. Admittedly, this is not really an ethical task but rather one of anthropology, and technical and natural philosophy. Overall it is apparent that nanotechnology — considered in the context of converging technologies — gives us by all means cause for some new or some more strongly accentuated ethical considerations. This is no longer a standard situation in a normative regard due to the far-reaching and in part somewhat speculative developments in animal enhancement. The relationship between humans and animals is the subject of a very fundamental discussion that goes beyond the classical approaches to this issue in animal ethics and animal protection.

### 5.2.6 Military Applications

The military applications of new technology are frequently at the focus of ethical consideration — at least of those expressed in public — in the work of some scientists (e.g., Altmann, 2006) and in the position papers of NGOs. The use of nanotechnology in armaments opens the door to improved weapons, innovative materials, and new areas of application (Lau, 2010). One example is the development of a nanoscale powder for use in propellants and explosives in order to increase the energy efficiency and speed of the explosion. Nanoscale electronic, sensory, and electromagnetic components could improve our capacity to steer and control military vehicles and make them more robust. This could further strengthen the current trend to unmanned and autonomous systems in the air, at sea, and in space. There are numerous potential applications in military intelligence that are based on the use of nanotechnological components for sensors, sensory systems, and sensory networks. Even the field of weapons and munitions will be influenced very immediately by improved sensory capacities and the improved computer power and storage capacity that nanotechnology influences. Developments in nanotechnology will presumably have substantial consequences on military personal, even at the level of personal equipment (soldier as a system). In the forefront is the effort to equip soldiers with additional functionality without substantially increasing the weight of the equipment (Moore, 2007). All of these developments can be viewed as an increase in efficiency compared to previous military technology.

A politically and ethically relevant secondary consequence of these developments could be proliferation, given the assumption that the customary moral standards for the application of technology continue to apply in war. Not all countries observe the standards equally well, and some dictatorships could be tempted to use the new technologies internally (such as for surveillance). These more efficient weapons could even fall into the hands of terrorists. The smallness of nanotechnology is occasionally given as a reason for concern that nanotechnological techniques could lead to the construction of substantially smaller bombs and explosives, which would be considerably easier for terrorists to build and use. Another secondary affect might be a new arms race, which would be driven solely by the speculative concern that potential opponents could arm themselves with nanotechnological weapons to (Altmann and Gubrud, 2002).

In view of these secondary effects, it is not to be expected that dramatically new ethical questions caused by nanotechnological advances in military technology will be raised. The reason is that the types of secondary effects are all known to us from previous developments in military technology. If really new ethical questions are raised in this field, then only in connection with developments in military technology that are rather speculative and visionary. Lines of development that would certainly challenge existing moral orientations would be the creation of entirely new types of weapons. There are no limits on speculation of this nature in connection with visions about nanotechnology. Objects of such speculation include biological weapons manufactured by means of artificial or technically modified viruses and swarms of nanorobots. An emphatic reservation regarding such speculation is appropriate here; it is possible that such ethical reflections might be diagnosed as premature (Keiper, 2007).

The ethical issues involved are in principle not new. Technological progress has at all times also been used for military purposes. To restrict this use to an extent which would be compatible with ethical standards and human rights is a steady challenge and has to do more with political negotiations than with ethical reasoning.

## 5.3 Cross-Cutting Ethical Issues

Some ethical issues in nano(bio)technology are of a fundamental nature and not related to individual fields of application, indeed not even to nanotechnology alone, but arise in every field of technological progress. These are health and environmental effects (Section 5.3.1), questions of fairness of access (Section 5.3.2), and aspects of intergenerational responsibility (Section 5.3.3).

### 5.3.1 EHS: Environment, Health, and Safety

Nanotechnology can be used to create new chemical substances or to study known chemical substances, such as particles whose form and/or size has been changed in order to introduce these substances into the economy (Chapter 2). Generally by either adding nanoparticles or attaching them to key targets, it is possible to improve the properties of the materials or sometimes even to create new ones, such as for treating surfaces, in cosmetics, or in sun creams, but also in the particularly sensitive area of groceries.

The application of nanotechnology in products and systems is expected to produce a significant relaxation of the burden on the environment: a saving of material resources, a reduction in the mass of by-products that are a burden on the environment, improved efficiency in transforming energy, a reduction in energy consumption, and the removal of pollutants from the environment (Fleischer and Grunwald, 2008). Initial studies appear to confirm the tendency of these expectations, yet they also point out the limits (e.g., Fiedeler, 2008). A number of studies on precisely the issue of the sustainability of nanotechnology have been published in the meantime (e.g., JCP, 2008).

The unintended release of nanomaterials into the environment, however, can lead to previously unknown effects. Synthetic nanoparticles are foreign bodies in the biosphere. They can reach the environment by means of emissions during production, during the everyday use of products, or during disposal (Steinfeldt *et al.*

for case studies). It is, for example, possible for the titanium dioxide particles in sun creams to be released during bathing and to reach the sea. The nanoparticles in sprays remain reactive — intentionally — for a long time and can even be transported long distances and diffusely distributed by air, or accumulate at certain locations. Issues such as mobility, responsiveness, persistence, pulmonary penetration, and solubility have to be taken into consideration to evaluate the potential spread of nanoparticles (Colvin, 2003). The ambivalence of technology with respect to sustainable development applies to nanotechnology, too (Fleischer and Grunwald, 2008). The production, use, and disposal of products containing nanomaterials may lead to their appearance in the air, water, soil, or even organisms. Nanoparticles could eventually be transported as aerosols over great distances and be distributed diffusely. Despite many research initiatives throughout the world, very little is known about the potential environmental and health impacts of nanomaterials (see ENRHES, 2010 for a recent review). This applies also and above all to substances which do not occur in the natural environment, such as fullerenes or nanotubes. The challenge to us to act under circumstances of high uncertainty, yet with the nanoproducts already in the market is at the heart of the ethical challenges posed by nanoparticles.

This raises the fundamental question of the acceptability of the possible risks, of the possibly necessary precautions, of a comparison of the expected advantages and the possible risks, and of limits in the form of environmental or health standards. In accordance with the different phases in the life cycle of technology, it is necessary to distinguish the production (e.g., for job security), use (e.g., for consumer protection), and disposal (e.g., the long-term storage of the materials). Despite intensive toxicological study, the state of our knowledge about how nanoparticles spread and about their toxic impact on the environment and health is very incomplete. Because of the specific difficulties in following nanoparticles in the human body and because of their diversity, even the development of appropriate measurement techniques is an extremely complex and time-consuming task. From animal experiments we have discovered that certain nanoparticles can be biologically active. The state of toxicological research does not permit us, however, to characterize the risks of nanomaterials well qualitatively or even to record them quantitatively.

For this reason, classic risk management working with quantitative risks (Section 3.3.1) does not work. "The tools developed in that discipline cannot be used when so little is known about the possible dangers that no meaningful probability assessments are possible" (Hansson, 2006, p. 316). This is an example of an "unclear risk" (Wiedemann and Schütz, 2008), namely, a risk whose magnitude cannot be indicated, for which the nature and extent of the damage are unknown, and where it is even unknown whether a danger exists *at all* (Section 3.3.2):

> Much of the public discussion about nanotechnology concerns possible risks associated with the future development of that technology. It would therefore seem natural to turn to the established discipline for analyzing technological risks, namely risk analysis, for guidance about nanotechnology. It turns out, however, that risk analysis does not have much to contribute here. (Hansson, 2006, p. 315)

Precautionary considerations — such as are foreseen for environmental issues within the context of European framework legislation (von Schomberg, 2005) — must instead be continued until toxicological research makes conventional risk management possible. Ethical reflection is required in several respects: in clarifying and justifying assessment criteria, in assessing the possibility of comparisons to other types of risk, and in assessing the acceptability of risks depending on justified criteria (Section 3.3).

Questions on the toxicity of nanoparticles to the ecology or humans, on nanomaterial flow, on the behavior of nanoparticles in spreading throughout various environmental media, on their rate of degradation or agglomeration, and on their consequences for the various conceivable targets are, however, not ethical questions. Empirical scientific disciplines, such as human toxicology, ecotoxicology, or environmental chemistry, are competent for such questions. They must provide the knowledge to enable us to draw practical consequences for working with nanoparticles and for disseminating products based on them. However, as the debate on the environmental standards of chemicals or radiation has shown, the results of empirical research do not determine how society reacts. Safety and environmental standards — in our case for dealing with nanoparticles — must be based on sound knowledge but cannot logically be derived from that knowledge. In addition, normative standards, for example, concerning the intended level of protection, the level of public risk acceptance, and other societal and value-laden issues, enter the field (Gethmann and Mittelstraß, 1992). Because of this situation it is not surprising that there are frequently conflicts about the *acceptability* of risks. The work of determining the acceptability and the tolerability of the environmental and health risks of nanoparticles is thus an ethically relevant issue (see Section 3.3.2 for general considerations and Chapter 6 for the application to the field of nanoparticles).

### 5.3.2 Distributive Justice: Nanotechnology and Developing Countries

The issue of distributive justice was even mentioned in one of the first publications on the ethical issues of nanotechnology (Mnyusiwalla *et al.*, 2003). In the meantime, it has been discussed in two directions, namely, with regard to developing countries in the sense of intragenerational justice and with respect to our responsibility for future generations in the sense of intergenerational justice (cf. Section 5.3.3). Both branches of the debate have led to the publication of journal articles and book chapters on the social issues of nanotechnology. Distributive justice belongs to the standard repertoire of studies on the ELSI studies in this field. However, its resonance in the public debate is still rather low compared to issues such as the possible risks caused by artificial nanoparticles or the creation of artificial life by synthetic biology (Chapters 6 and 7).

The positive potentials of nanotechnology for developing countries were even emphasized in the first and more or less fantastic writings on nanotechnologies. Drexler (1986) expected the solution of all of the problems of developing countries, such as poverty and hunger. These nanotechnological utopias were confirmed by the developments promised by converging technologies (Roco and Bainbridge, 2002).

Even some ethicists and scientists have expressed high expectations that developing countries would benefit considerably from nanotechnology, such as if nanotechnology would make it possible to pursue a cheap mass production of products needed in these countries (Mnyusiwalla *et al.*, 2003). Especially energy and water technologies were mentioned as carrying high potentials for improving the situation in developing countries.

However, such high expectations were expressed in the early periods of the debate on the social issues of nanotechnology. In the meantime the optimism has disappeared. There is no tendency apparent that nanotechnology will help diminish the technology gap between rich and poor, even if at least a number of developing countries have invested heavily in research on nanotechnology (Schummer, 2007). Optimistic hopes that nanotechnology will provide a rapid solution to the great developmental problems must be relativized: "The selected issues discussed in this chapter allow drawing mostly pessimistic conclusions on the impact of nanotechnology on developing countries"(Schummer, 2007, p. 303; see also Court *et al.*, 2007). Instead, nanotechnology goes for different goals:

> [A]mong the first products with nanotechnological content were: thermal shoes [...], dust and sweat-repelling mattresses, more flexible and resistant golf clubs, personalized cosmetics [...], disinfectants and cleaning products for planes, boats, submarines etc., cream that combats muscular pain, dental adhesives that set the tooth crown better. (Foladori, 2008)

This indicates that demands in the sphere of lifestyle and luxury in rich countries will be fulfilled first, rather than the existential needs of hunger and poverty in developing countries.

An example for similar concerns is nanomedicine (Section 5.2.1). Nanotechnology-based medicine will, in all probability, be rather expensive. Questions of equity and of access to (possible) medical treatments could become urgent in at least two respects.

1. *Within* industrialized societies, existing inequalities in the access to medical care could be exacerbated by a highly technicalized medicine making use of nanotechnology.
2. With regard to *less developed societies*, the particularly dramatic inequalities that already exist between industrialized and developing nations could be further increased (Foladori and Invernizzi, 2007).

The problems posed by an unfairly distributed access to new technology frequently appear particularly clearly in medicine, where inequality often means unfairness:

> The elements outlined above lead us to think that without stringent regulation by the state it is difficult for the important advances in nanomedicine to reach all levels of society and ease the load on the poor, especially in developing countries. (Foladori, 2008, p. 211)

A specific future field of debate with respect to equity is the issue of human enhancement (Siep, 2006). If technologies for improving human performance were available, then the question would arise regarding who will have access to those technologies, especially who will be able to pay for them, and what will happen to persons and groups excluded from the benefits. A separation of the population into the enhanced and the normal could develop where a situation is imaginable where normal is used as a pejorative attribute and coercion towards enhancement might occur (Siep, 2006):

> Merely competing against enhanced co-workers exerts an incentive to use neuro-cognitive enhancement, and it is harder to identify any existing legal framework for protecting people against such incentives to compete. (Farah *et al.*, 2004, p. 423)

Furthermore, we must not only look at the just distribution of opportunities from nanotechnology in order to get the full picture, but also at the possible negative impact directly on developing countries. Nanotechnology might indeed be tied to substantial economic problems in some developing countries. If, for example, cotton were replaced on a large scale by nanofibers or the rubber in automobile tires by nanoparticles, then the economic development of the countries that depend on the export of these natural resources would be endangered.

What becomes clear from such ambivalence is that developing countries will not automatically benefit from nanotechnology (Court *et al.*, 2007). On the contrary, the eigendynamics points in the opposite direction of a widening gap between rich and poor. There is no technology determinism at work in the sense that nanotechnology *as such* could contribute to distributive justice. Instead we are thrown back to the *political dimension* of distributional justice. Although intragenerational justice may profit from technology, it cannot be reached by technology alone. Instead, it requires political measures. One of the early demands in this direction was that developing countries should be involved in the examination of the ethical aspects of nanotechnology (Mnyusiwalla *et al.*, 2003), in order not to exclude large portions of the world's population from the benefits of the expected potentials and, in consequence, to make them victims of discrimination.

At any rate, hopes that nanotechnology could contribute to increasing distributive justice instead of lowering it will not come true as a result of nanotechnology as such, but require political regulative measures that take the forces of civil society into account (Invernizzi, 2008). Technological determinism is sometimes a naïve hope (or concern): while political effort, complex deliberation, and socially aware decision making are complex and require great effort, it would be much easier if technology could bring them about automatically. But this is a simple fallacy and an unjustified expectation. What is needed are political measures lowering the thresholds for developing countries to employ their limited resources and to participate in nanotechnology research and development programs, allowing them to influence the agenda of research with regard to their needs, and promoting the transfer of basic nanotechnology knowledge and skills to them in order to empower developing countries to conduct technology development that takes their own needs into account. Without a doubt, nanotechnology offers great potential to developing countries, but this potential will remain mere potential unless there are political measures that attempt to make this potential a reality.

### 5.3.3 Responsibility for Future Generations

Technology is of major importance for the sustainability of humankind's development. On the one hand, technology determines to a large extent the demand for raw materials and energy, the need for transport and infrastructure, the mass flows of materials, and emissions as well as the amount and composition of waste. Technology is, on the other hand, also a key factor in innovation and influences prosperity, consumption patterns, lifestyles, social relations, and cultural developments. Therefore, the development, production, use, and disposal of technical products and systems have an impact on the ecological, economic, and social dimensions of sustainable development. In most cases, this impact is ambivalent with regard to sustainable development (Fleischer and Grunwald, 2008): there are positive contributions as well as negative consequences. The overall concept of *sustainable development* requires production and consumption patterns to be shaped so that the needs of the current generations can be satisfied in a way that they do not limit or threaten the opportunities of future generations to satisfy their needs. Of special relevance concerning the ecological dimension of intergenerational distributive justice are (following Steinfeldt *et al.*, 2007):

- The *limited availability of many natural resources* such as clean water, fossil fuels, and specific minerals highlights the importance of their efficient use, of recycling, and of replacing nonrenewable resources by renewable ones.
- The *limited carrying capacity of the environment* (atmosphere, groundwater and surface water, oceans and pedosphere, ecosystems) emphasizes the necessity of limiting or reducing emissions and of regenerating damaged environments.

Many scientists and engineers expect nanotechnology to make manufacturing consume less material and energy, and to lead production to generate less waste and pollution. Nanotechnology is also expected to enable new technological approaches that reduce the environmental footprints of existing technologies in industrialized countries. Areas associated with high expectations are energy technologies and water technologies; it is hoped that nanotechnology can provide new materials and (e.g., catalytic) processes, leading to increased productivity and diminished environmental burdens.

As has been described above (Section 5.2.2), these developments might have a price. The consequences of the use and release of nanomaterials into the environment are unknown. Although it is not very probable that synthetic nanoparticles in the environment will have long-term effects because of anticipated agglomeration processes, there is no proof available. We do not know about possible long-term effects comparable to the HCFC problem that created the hole in the ozone layer. This situation of high uncertainty and ignorance places a burden of possible risk on future generations while we are exploiting the benefits of nanotechnology today.

This situation is well known from an ethical point of view. On the one hand, there are positive expectations with regard to sustainable development, which gives legitimacy to a moral position that these potentials be further explored and utilized. But on the other hand, there are risks and uncertainties. This situation is the basic motivation for technology assessment (Grunwald, 2009a). There are two basic challenges in this situation:

1. To *shape* the further development of nanotechnology in the direction of sustainable development
2. To *apply* the precautionary principle in dealing responsibly with the inevitable issues of uncertainty and ignorance.

The questions of the acceptability and comparability of risks, the advisability of weighing up risks against opportunities, and the rationality of action under uncertainty are, without doubt, of great importance in nanotechnology, and they show close ties to the issue of distributive justice.

Shaping nanotechnologies for sustainable development requires anticipatory sustainability assessments in order to permit distinctions to be made between the more and less sustainable technologies (Fleischer and Grunwald, 2008). Anticipatory assessments of nanotechnology have to cover the entire *life cycle* of the respective technological products or systems. They should include a *temporal integration* and *balancing* of all sustainability effects which might occur during the complete life cycle. For such analyses to contribute to shaping nanotechnologies for more sustainability, they must provide reliable *prospective* life cycle information, such as on health and environmental implications, consumption and production patterns, future developments of lifestyles and markets, and the political and economic framework conditions for the later usage of new technologies. These are only some examples of aspects of the future that need to be known *in advance* in order for reliable life cycle analyses to provide sustainability assessments.

A start has already been made toward addressing this challenge for the creation of prospective life cycle assessment. The increasing focus on life cycle assessment as a tool for example in strategy and planning processes, including for long-term issues, and in scenario processes has triggered methodological developments that try combine traditional technology foresight methods with life cycle assessment methods (Schepelmann *et al.*, 2009). Decisive for a comprehensive assessment of nanotechnology or of the corresponding products from a sustainability point of view is that the entire course of the products lifetime is taken into consideration. This extends from the primary storage sites to transportation and the manufacturing processes to the product's use, ending finally with its disposal. In many areas, however, nanotechnology is still in an early phase of development, so that the data about its life cycle that would be needed for life cycle assessment are far from being available. Empirical research on the persistence, long-term behavior, and whereabouts of nanoparticles in the environment as well as on their respective consequences would be necessary to enable us to act responsibly in accordance with ethical criteria of long-term responsibility and with regard to our use of and contact with our natural environment.

One result of these considerations is that it is difficult to operationalize the issue of intergenerational distributive justice because prospective analyses are necessarily involved, with all the well-known problems associated with that type of knowledge and reasoning. Uncertainties cannot be eliminated despite any amount of effort. Therefore, shaping technology by observing the guiding idea of sustainable development and by considering the issue of distributive justice cannot be a simple planning process but must adapt to emerging new knowledge about the respective technologies and the affected areas of society (Grunwald, 2004, 2007b).

This is an ambitious field for shaping technology for the sake of approaching sustainable development. Ethical issues are not directly involved but they form and influence the normative background of shaping activities. This means, for example, that the relevant criteria and procedures must be identified to make the guiding idea of sustainable development work (Grunwald, 2006b). The indicators of what sustainable should mean cannot be derived from a logical deduction. The determination of indicators is itself influenced by values, and any such determination within a given normative framework will not satisfy all of the criteria of a standard situation in a moral respect. On the contrary, controversies will abound, related to conflicting interests on the one hand and to conflicting values on the other. While procedures of negotiation and mediation are needed for the first source of conflicts, ethical reflection is required for the second one. This, however, is generally the case in shaping technology towards sustainable development and is not specific to nanotechnology.

## 5.4 Selection of Issues for In-Depth Studies

The aspects mentioned above have established themselves as an informal "canon" of the ethics debate on nanotechnology. They provide structure to the discussion and are not questioned even if — as we have shown — there is a substantial redistribution of the content of ethical issues in general, of *new* ethical questions, or of issues linked to nanotechnology *in a specific way*. The preceding presentation of ethical issues in nanotechnology does at least achieve something:

- It is orientated towards well-known topics of ethical reflection. Privacy, risk, and justice are classic topics in the ethics of technology. It seems rational to employ these well-known issues and lines of argument to confront the new field of nanotechnology. "Starting from scratch" (Ebbesen *et al.*, 2006) is not necessary, and this is expressed particularly well when links can be shown to existing debates.
- It is thus also amenable to being linked to existing social debates on technology. It is possible to create connections to the main lines of these debates, ranging from nuclear energy, through genetically modified organisms, to chemical regulation and medical technology.
- It both differentiates according to the fields of application in nanotechnology, such as medicine or military technology, and considers cross-cutting issues.
- It allows concretization and "unraveling" (Nordmann, 2007b) of the ethical debate on nanotechnology in terms of the technology and applications and thus makes it possible to take a step away from generalized considerations of nanotechnology as a whole.

The distinction made here in ethical issues between fields of application in nanotechnology and cross-cutting issues is new. It overcomes a weakness in the types of structuring previously suggested in which application issues (for instance, military applications of nanotechnology) and cross-cutting issues (such as distributive justice) are for the most part simply listed additively (also in Grunwald, 2008b).

The assessment of the ethical aspects of nanotechnology using the criteria of urgency and novelty now permit the focus in the further analyses (Chapters 6–9) to be set on the following points:[4]

- *Synthetic nanoparticles* (Chapter 6): This topic is already highly relevant today and is recognized as the field of nanotechnology that has progressed furthest in applications. Here, as explained, there are normative uncertainties in terms of dealing with unclear risks even while the fields of application are simultaneously highly relevant. Ethical reflection can draw

[4] This explains the roadmap for the next four chapters. There is some overlap with the descriptions in Section 5.2 and 5.3, which are unavoidable to keep each of the texts readable on its own.

upon existing considerations of risk ethics (e.g., Shrader-Frechette, 1991; Wiedemann and Schütz, 2008).

- *Synthetic biology* (Chapter 7): Nanotechnology, on the basis of nanobiotechnology, is helping to create a new field of biology. Also being created, although hardly visible yet, is a new cross-cutting topic of ethics at the interface between bioethics and the ethics of technology. There are options for aligning these issues with the ethical debates on genetically modified organisms and with considerations of artificial life in the philosophy of technology.
- *Animal Enhancement* (Chapter 8): The interest of humans in exploiting animals, for example in agriculture, sports, or as laboratory animals in science, leads to these animals being "optimized" for human use. Economic pressure strengthens this trend. Converging technologies on the basis of nanotechnology can, on the one hand, be a means to further intensify the human exploitation of animals. On the other hand, technological intervention "in the animals' interests" is also possible. The relationship between humans and animals will be changed through the further technicalization of animals.
- *Human Enhancement* (Chapter 9): Human enhancement on the basis of converging technologies poses the most far-reaching demands on ethical reflection. Further miniaturization offers, for instance, new chances for implants, particularly in connection with the nervous system and brain (Stieglitz, 2006). In addition, far-reaching (and in part distinctly speculative) visions of future neuroimplants are under discussion. The "technical enhancement" of humans on the basis on nanotechnology is becoming the program for many scientists and raises new issues regarding the future of human nature (Roco and Bainbridge, 2002).

The focus on these points makes it possible to cover the spectrum of nanotechnology applications, from those close to the market to speculative ones. The goal is to view the new opportunities for application made possible by nanotechnology as an "enabling technology" at the same time as considering the normative uncertainties resulting from these opportunities.

# Chapter 6

# Synthetic Nanoparticles

For many years, special attention has been paid to synthetic nanoparticles in the public debate on nanotechnology. A vast potential market is seen for nano-based products. New products based on new properties of nanomaterials can result from either admixtures of nanoparticles or their specific applications, for instance, in a surface treatment, in cosmetics, or in sunscreens. The production, use, and disposal of products containing nanomaterials may lead to their appearance in the air, water, soil, or even organisms. They might, however, produce adverse effects on the environment or on human health (cf. Monteiro-Riviere and Tran, 2007).

In analyzing this situation with respect to conclusions related to responsible risk management strategies, careful normative reflection (including ethical "enlightenment") is required (von Schomberg, 2005). Questions of the acceptability and comparability of risks, the advisability of weighing up risks against opportunities, and the rationality of action under uncertainty are, without doubt, of great importance in the field of nanoparticles (cf. on the more general challenge of dealing with risk and uncertainty (Rescher, 1983; Shrader-Frechette, 1991; Wiedemann and Schütz, 2008). Close cooperation between ethics and toxicology, the social sciences, and jurisprudence is necessary (Renn and Roco, 2006).[1]

[1]This chapter builds on earlier work of the author (Grunwald 2008c, Grunwald 2008d) and develops it further.

*Responsible Nanobiotechnology: Philosophy and Ethics*
Armin Grunwald

ISBN 978-981-4316-80-4 (Hardcover), 978-981-4363-33-1 (eBook)
www.panstanford.com

## 6.1 Synthetic Nanoparticles: Fields of Application and Expectations

Synthetic or man-made nanoparticles are nanoscale particles that are *manufactured precisely* to provide certain functions. Nanoparticles that have been frequently employed are titanium dioxide particles in sunscreen lotions, carbon particles in automobile tires, and silver particles to combat bacteria. Man-made nanoparticles are also employed in sprays,[2] pastes, cosmetics, toner for printers and copiers, paints, varnish, and glues as well as in products for the impregnation of surfaces. Nanoparticles based on carbon, such as fullerenes, lend themselves for use in living systems, such as for nanocapsules. Nanoparticles as such are not consumer products. They are either combined with other substances or applied to a surface in order to improve the properties of products.

Some nanoparticles occur naturally in our environment and originate in different ways. Volcanic eruptions and forest fires enrich the atmosphere with particles of very different sizes, some of which are nanoparticles. The positive health properties of sea air are to a large extent due to nanoscale salt crystals. Organisms and ecosystems have in the course of evolution learned to somehow deal with these different forms of natural nanoparticles, i.e., they have developed strategies for adjusting or for developing defensive mechanisms. The presence of natural nanoparticles can play a role in the assessment of possible newly manifested risks posed by synthetic nanoparticles, analogous to the fact that natural radioactive radiation is taken into consideration in the assessment of the acceptability of man-made radioactivity.

The release of fine and ultrafine particles, extending to the size of nanoparticles, is a side effect of many anthropogenic combustion processes (Oberdörster *et al.*, 2005). This is just as true of the popular pastime of grilling or of an open fireplace at home as it is of the generation of soot in a diesel motor or emissions from industrial combustion processes or incinerators. The pleasure some find in cigarette smoking results to a large measure from the complex micro- and nanoparticles that originate during the slow burning (it

[2]The spray MagicNano reached a degree of familiarity because health problems arose after (accidental) inhalation, forcing the product to be removed from the market in 2006. The use of the prefix "nano" in the name was however for purely advertising purposes. There were no synthetic nanoparticles in it (Grobe, 2007).

is really more of a smoldering). These particles are however also responsible for the toxic and carcinogenic properties of smoking. The issues surrounding fine and ultrafine particles are also the topic of a wide-ranging and intensive debate, which as a consequence of European Union (EU) regulations has even led to measures such as driving restrictions on a sliding scale in cities when the legal limits are exceeded. Epidemiological research is being devoted to the consequences for our health from exposure to aerosols (e.g., Neuberger, 2007 and the references cited there). Regulations published by the EU set thresholds for the maximal concentration of specific classes of dust particles as unintended emissions from industrial processes.

Synthetic nanoparticles, in contrast, are *intentionally fabricated* in order to achieve certain functionalities (Section 2.2.2). They are newly emitted into the environment by man and sometimes inserted into the human body, analogous to synthetic chemicals. During the transition to the nanometer scale, many of the properties of the materials change — although their chemical composition is unchanged — especially depending on particle size, its structure, and its surface properties. With declining size, particles as a rule become more and more reactive since their surface area is relatively larger. This increased reactivity is on the one hand the basis of the desired functionality, but on the other hand also the reason for concern about its toxicological effects since the higher reactivity can lead to an increased responsiveness in the human body or in the environment. This demonstrates the general ambivalence of technology (see Chapter 3): the very same properties that are used for certain purposes can also be responsible for undesirable effects.

### 6.1.1 Surface Treatment

One of the economically significant and earliest applications of nanotechnology consists in the coating of a surface. In our modern industrial society, surfaces play an important role, for example, the surfaces of buildings in the form of walls and windows or the surfaces of motor vehicles. Nanotechnology is supposed to provide the desired surface properties, such as by means of ultrafine coatings. Among these properties are above all the repellant effects against water, oil, and dirt, which make it possible to reduce the ecological and economic costs of cleaning the correspondingly treated surfaces.

Paint for external use on the basis of the lotus effect (Barthlott, 1990), which is given over and over again as an example of nanotechnology, has been available in stores for years. Surfaces are also supposed to be made less sensitive to damage, to smearing with paint by sprayers, and to dampness. Finally, there are also plans to treat surfaces nanotechnologically in order to substantially slow the weathering of concrete. This could also produce substantial ecological and economic savings in the area of infrastructure maintenance (streets, bridges, etc.).

### 6.1.2 Food

The use of nanoparticles in food is an area of application that is frequently discussed and certainly particularly sensitive. The word "nanofood" is already being used for applications in which

> nanotechnology techniques or tools are used during cultivation, production, processing, or packaging of the food. It does not mean atomically modified food or food produced by nanomachines [...]. Instead nanotechnologists are more optimistic about the potential to change the existing system of food processing and to ensure the safety of food products, creating a healthy food culture. They are also hopeful of enhancing the nutritional quality of food through selected additives and improvements to the way the body digests and absorbs food. (Joseph and Morrison, 2006, p. 7)

Research extends on the one hand to new packaging materials and on the other to additives for achieving different types of properties, for which nanoparticles are to serve as the carriers. For example, with regard to packaging, one idea is to equip the packaging material with sensors that detect negative changes and thus provide better protection against signs of deterioration. This could be a sensor in the form of a nanoparticle film that causes the package to discolor if the food degenerates and that works completely independently of any expiration date. This could prevent premature removal of food from shelves, i.e., removal based only on an expiration date and with no regard for the actual condition of the food. A goal of research that is focused both on the better preservation of food and on a simultaneous reduction in preservatives (Bachmann, 2006, p. 39) is "active packaging," which might work, for example, antimicrobially or by absorbing the oxygen that could lead to manifestations of oxidation in food. The point is here to use nanoparticles (e.g., titanium dioxide or silicon oxide) to prevent food from coming into contact with oxygen, carbon monoxide, a liquid, or some other material that would contribute to its degeneration. Furthermore, preservatives could be released upon detection of the first signs of a degenerative process in order to prevent spoilage (release on command, ETC, 2004, p. 42).

While the packaging material of food is something external, nanocapsules could be inserted as containers for food additives that could be used directly to achieve lifestyle or health effects. This opens the door for speculation, for example, that certain capsules could be induced to release flavored substances by being shaken more or less vigorously, so that an individual could determine which taste a drink was supposed to have by shaking it. Microwave technology could also be employed to break open certain nanocapsules to release specific colors or flavors depending on the frequency. This would be development in the direction of individualized foods. The same technology — capsules are always involved that are, as a rule, based on fullerenes — can also be used to send additives to certain locations in the body. The capsules could, for example, be so arranged that they would pass through the mouth and stomach, not being released until they reach the intestines. This can optimize the use of functional food. Less spectacular but according to a BUND study already on the market are additives such as silicon dioxide nanoparticles that influence the flow properties of food such as ketchup:

> Nanomaterials are present, for example, in various ketchups, vegetable broths, and powdered sugar in order to improve their flow and trickle properties. (BUND, 2008)

The food industry disputes this. Strong competition between these companies means, however, that little is known publicly about what items of this kind are already being developed in the laboratories or what is even already being produced. Our knowledge about the nanoparticles that may already be present in food is minimal since there is no requirement for products containing nanoparticles to be registered or labeled. There is not even an accepted method for proving whether they are present. Materials that have so far been added to food as particles larger than the nanoscale do not need re-approval when reduced in size to the nanometer scale since they are chemically identical.

### 6.1.3 Cosmetics

The best-known cosmetics containing nanoparticles are specific sunscreen creams. They contain titanium dioxide nanoparticles that possess excellent properties for absorbing ultraviolet light and are also transparent, meaning that they do not create the aesthetically messy impression of a white smear as do common sun creams. In the process, there is a trade-off between the transparency of the cream and the protective effect against ultraviolet light, which is correlated with the size of the particles in varying ways. Nanoparticles are also used in lipsticks because of certain color effects they produce. Furthermore, nanoparticles of various metal oxides are employed as wrinkle lighteners, as pigments, or for improving the skin sensation (Paschen *et al.*, 2004). This area of application has become a stronger focus of attention since the demand for a moratorium on the use of nanoparticles (Friends of the Earth, 2006).

## 6.2 Possible Risks and Types of Risk

The purposes of most of the applications mentioned above have a positive connotation regardless of the fact that they entail the manufacture of synthetic nanoparticles that are included in the products. The fundamental issue is whether undesired or even catastrophic side effects are possibly linked to these nanoparticles. These synthetic nanoparticles are, after all, foreign bodies in the biosphere, with which man and the environment itself have absolutely no experience.

The unintended release of nanomaterials into the environment can have previously unknown effects. Synthetic nanoparticles can reach the environment by means of emissions during production, during the everyday use of products, or during disposal. It is, for example, possible for titanium dioxide particles in sun creams to be released into sea water during bathing. The nanoparticles in sprays intentionally remain reactive for a long period of time and can even be transported long distances and distributed diffusely by air, or accumulate at certain locations. The nanoparticles used in food packaging could also be released in the trash, and those in food itself might have unintended effects when they enter the human body. Consequently, issues such as mobility, responsiveness, persistence, pulmonary penetration, and solubility have to be taken into consideration when the potential spread of nanoparticles is evaluated (Colvin, 2003; see Steinfeldt *et al.*, 2007 for case studies; ENRHES, 2010 for a recent review).

Concerns of this nature are not entirely plucked out of thin air, as shown by several examples in history (Harremoes *et al.*, 2002). Asbestos fibers are minerals that have only reached the biosphere as a result of human use and that have led to significant health dangers, especially among the people who work with them. There are estimates that in Europe alone there have been some 250,000–400,000 deaths from various types of cancer (mesothelioma and lung cancer) caused by the inhalation of asbestos fibers (Gee and Greenberg, 2002). There is an ongoing debate about possible similarities between asbestos fibers and carbon nanotubes (Pacurari *et al.*, 2010). Although the processing of asbestos was stopped in many countries in the 1980s, a substantial number of deaths are still expected to occur because of the long latency period of these cancer diseases. Hydrochlorofluorocarbons (HCFCs) are another example. The desirable properties of these anthropogenic manufactured chemicals — chemically inert and therefore neither toxic nor inflammable — had led to their use as refrigerants in refrigerators and air conditioners and as a propellant in sprays since the 1930s. For decades now, however, they have accumulated in the troposphere, where in catalytic processes they degrade the ozone layer that is essential for the survival of life on earth. This damage was not anticipated originally, and it may not even have been possible to anticipate it based on the scientific means of the time (Harremoes *et al.*, 2002).[3] Further examples of artificial materials that have led to unanticipated dangers are dioxins from earlier incineration plants and certain wood preservatives, the inhalation of which led to some severe damage to people's health and even to deaths. Although the numbers affected in these cases are small, these examples demonstrate that caution is in general appropriate when new materials are introduced into the biosphere.

[3]This is another good example for the ambivalence of technology. It was precisely the property of chemical inertness that made HCFCs so attractive and that now leads to the fact that the ozone hole will not disappear even decades after the use of these chemicals have been discontinued. HCFCs are not degraded or "used up" by the catalytic reaction.

### 6.2.1 Health Risks

Nanoparticles can enter the human body in very different ways, such as via the lungs after inhalation (Geiser and Kreyling, 2010), via the digestive tract, and through the skin. There is hardly any information available about their effects on public health and the environment, especially about their long-term effects (Schmid *et al.*, 2006). The small size of nanoparticles has made them nearly impossible to track in the human body, or the tracking employs radioactive markers that can change the particles' properties and effects (Marquis *et al.*, 2009). It is known that natural nanoparticles — most of which are soluble in water — embed in liquids upon contact with human tissue, congregate into larger units, and ultimately are excreted via the customary routes. For fine and ultrafine dust from anthropogenic sources (such as diesel soot), toxicological information is available according to which the following mechanisms conceivably produce damage (Moshammer and Wallner, 2007, pp. 172ff. and the references cited there). Some of these mechanisms have already been confirmed by animal experiments with nanoparticles (ENRHES, 2010).

- *Oxidative stress*: Free oxygen radicals can be produced by nanoparticles, whose reactivity can induce inflammatory modifications and thus damage tissue and possibly contribute to arteriosclerosis.
- *Effects on the respiratory tract*: Fibroses and inflammatory reactions depend strongly on the size and shape of the nanoparticle. Animal experiments have shown carbon nanotubes to cause damage to the lungs.
- *Mutagenic changes*: Damage to a cell's nucleus can trigger tumors, particularly in the respiratory tract, but also stomach cancer.

The toxicity of materials can be determined to a large degree in animal experiments. They have demonstrated, for example, that nanoparticles deposited in a lung or the intestines of an animal can reach its bloodstream and thus other organs (such as the brain or liver). A central step toward improving the toxicological test procedure and in particular reducing the number of animal experiments is for in vitro procedures to be validated specifically for nanoparticles. This will enable us to better predict the in vivo effects (Oberdörster, 2010).

Inhalation experiments on rats have become especially well known in which it was empirically demonstrated that carbon nanoparticles can cause considerable lung damage. Their toxic potential increases as particle size diminishes and as the relative surface area of the particle increases. At high concentrations, high mortality is the consequence. The cause however is obstruction of passages in the lungs due to the agglomeration of particles, not the toxicity of the particles themselves. These results do not necessarily indicate any potential consequences for humans since the rats were exposed to unrealistically high concentrations of nanoparticles in the air they breathed, which ultimately led to asphyxiation. The fear has also been frequently expressed that the small size of nanoparticles might enable them to cross the blood–brain barrier. What might ensue is completely unclear. In general, it is important to distinguish carefully between the simple presence of nanoparticles in the human body, their biological endpoints, and their biological action (see Krug and Fleischer, 2007 and the references cited there; cf. Geiser and Kreyling, 2010)).

For numerous reasons, there are still no epidemiological studies that quantify the risks posed by nanoparticle dust. First, there still are hardly any suitable measurement procedures. Second, nanoparticles themselves are substantially more diverse with regard to their chemical composition, size, and structure than, for example, the particles released by road traffic, which substantially impedes empirical studies. Third, synthetic nanoparticles have not been released so far in such large amounts that an epidemiological study would produce reliable results in the one or other direction. Yet the ethical question is precisely whether the attempt should not be made to avoid the situation that an epidemiological study might lead to a positive result. A condition for a positive epidemiological result is, after all, that there has already been damage to the health of a statistically significant number of people or that there has been a sufficient number of premature deaths.

Overall our knowledge today about the means by which nanoparticles are distributed and about their toxic consequences is still very incomplete (Xia *et al.*, 2010). It has been shown, for example, that the titanium dioxide nanoparticles in sun creams presumably do not diffuse through healthy skin into the human body (Choksi *et al.*, 2010). What is still unknown, however, is their behavior on skin that is already damaged (e.g., by sunburns). And even if the nanoparticles could enter the body, it would be unclear whether and what kind of damage they might cause. They would probably rapidly be taken up by other (larger) particles and lose their nanoproperties, analogous to natural nanoparticles. This is still, however, to be demonstrated in individual cases. Intensive toxicological studies are currently being conducted, but this has proved to be a complex and protracted task considering the specific difficulties of tracking nanoparticles in the body as well as their diverse forms (Fadeel and Garcia-Bennett, 2010; Oberdörster, 2010).

### 6.2.2 Environmental Risks

Nanoparticles can be released in the environment during the production, use, or disposal of materials that contain them. Even some of the intended uses of nanoparticles would directly expose the environment to nanoparticles, and precisely to such nanoparticles that are intended to affect biological systems and organisms. Examples are the addition of silver particles to water for fighting legionella (the bacteria that causes legionnaires' disease), the use of fullerenes in the form of soluble crystals with antibacterial properties as a disinfectant, or the use of nanoparticles in agriculture (which are largely still in the planning phase; Bachmann, 2006). The last case raises questions regarding how nanoparticles are distributed, the consequences of their interaction with other elements in the environment, their whereabouts, and their degradability or long-term persistence (Neuberger, 2007). All of these aspects have hardly been studied. Similarly, little is known about the lifespan of nanoparticles in the environment and about their life cycle. Special caution is called for with regard to nanoparticles that have an intended biological effect. Attention must be paid precisely in this context to the ambivalence, as noted above, that an intended function might have unintended consequences.

As a rule, environmental risks are strongly linked with quantitative effects and concentration (see the example of HCFCs). As long as nanoparticles are only emitted into the environment in small or minimal quantities, a first-order approximation would indicate with a high probability that the emission poses no significant danger. And, in analogy to natural nanoparticles, there is the expectation that a large portion of the synthetic nanoparticles in a natural environment would not have a long lifespan. The expectation is that they would be taken up by other particles and lose their specific nanoproperties. Of course, this must be empirically demonstrated. Large-scale interventions in the environment using nanoparticles are problematic as least as long as "sufficient" knowledge of this nature is not available.

### 6.2.3 Nanoparticle Risks as "Unclear Risks"

There is a long tradition of different risk management strategies accompanying the implementation of new technologies and the introduction of new materials. Initially, and sometimes still today, a "wait-and-see approach" is taken, mostly implicitly. New substances have been introduced under the assumption either that problematic developments and impacts would not occur at all or that *ex post facto* repair and compensation strategies would suffice for any adverse effects. The experience with partially dramatic hazards to human health or the environment caused by new materials, by radiation, or by new technologies has, however, led to the establishment of risk regulations in different fields, in order to prevent any further negative impact on health and the environment. Important areas are (following section 5.1 in Schmid *et al.*, 2006):

- Regulations on workplaces with specific risk exposures (e.g., nuclear power plants, chemical industry, and aircraft) to protect staff and personnel
- Procedural and substantive regulations on nutrition and food to protect consumers (e.g., regulations concerning conservation procedures, and the maximum allowed concentrations of undesired chemicals such as hormones)
- Environmental standards in many areas to sustain environmental quality (concerning, e.g., ground water quality and the maximum allowed rate of specific emissions from fabrication plants, power plants, or heating in households)
- Safety standards and liability issues to protect users and consumers (e.g., in the field of automobile transportation and for power plants, engines, and technical products used in households)

There are established procedures for risk analysis, risk assessment, and risk management in many areas of science, medicine, and technology (Renn *et al.*, 2008), for example, in dealing with new chemicals or pharmaceuticals. Laws such as the Toxic Substances Control Act in the United States (Wardak, 2003) constitute the framework for dealing with such situations. Such "classical" risk regulation is adequate if the level of protection is defined and if the risk can be quantified as the result of the probability of the occurrence of the adverse effects multiplied by the assumed extent of possible damage (von Schomberg, 2005). In such situations, thresholds can be set by law, by self-commitments, or by following participatory procedures. In as much as such mechanisms are able to sufficiently cover existing challenges, there is a "standard situation in a moral respect" and the respective normative framework is able to cover the respective situation (Chapter 3).

Classical risk management cannot be applied to synthetic nanoparticles since it is evident that the conditions for its use are not satisfied in this case (Renn and Roco, 2006).

> The tools developed in that discipline cannot be used when so little is known about the possible dangers that no meaningful probability assessments are possible (Hansson, 2006, p. 316).

The lack of empirical knowledge about the cause–effect relationship means that a quantitative determination of the risk is in principle not possible. As a consequence, we have so-called unclear risk, i.e., a risk whose size cannot be given, where the nature and extent of any damage are unknown, and where it is even unknown whether any danger at all is even present (Grunwald, 2008d). For synthetic nanoparticles this means, "It turns out, however, that risk analysis does not have much to contribute here" (Hansson, 2006, p. 315).

This diagnosis continues to be accurate even though a large amount of toxicological research has been conducted on nanoparticles in the last few years, reducing somewhat the gaps in our knowledge (Sattler, 2010, Part II). The question is which concepts for dealing with risk outside classical risk management can provide orientation. Nonetheless, the task of toxicology is to raise the level and reliability of our knowledge as rapidly as possible. Yet the question remains as to which aspects of responsibility have to be observed during the period of time until our knowledge is complete enough for classical risk management to be possible. To answer this, I will first consult the relevant positions of risk ethics and then move on to operative approaches such as the precautionary principle.

## 6.3 Approaches to Dealing with Unclear Risk

One's thoughts go first to philosophical suggestions to risk ethics, which are not linked to a quantitative concept of risk but are more general in nature. However, suggestions that attempt to solve the problem of dealing responsibly with unclear risk in a procedural manner also need to be considered. At the end of this section, I want to make a grounded decision regarding which of these approaches is appropriate for the case of nanoparticles. For this reason, the short presentation of each of these approaches is immediately followed by a critical evaluation of this case.

### 6.3.1 Philosophical Approaches

Although much ethical work is available on risk in general (Keeney, 1984; Shrader-Frechette, 1991; Rescher, 1983; Jonas, 1984), only a few activities have focused specifically on unclear risk. In the following section we will describe and assess the consequentialist approach and its limits, the principle of pragmatic consistency, the ethics of responsibility, a recent approach on "projected time," and deontological advice.

#### 6.3.1.1 *The Consequentialist Approach*

Consequential ethics focuses on the expected consequences and impact of actions and decisions rather than on their intentions. Insofar as unclear risk is a consequence of actions or decisions, a consequentialist approach would be the ideal point of departure in cases requiring ethical reflection. Even in the case of a Kantian approach — which would look primarily at an action's intentions rather than at its expected consequences — the assumed consequences would also have to be taken into account in order to get a comprehensive picture of the situation under consideration. Ethical reflection on what should or not should be done is closely related to consideration of the consequences of actions and decisions, independent of the specific ethical approach. In our case, an action or decision would create a situation involving unclear risk which itself would be a consequence — and possibly a risk in its own right.

However, the notion of "consequentialism" is used most commonly in a more narrow sense (Keeney, 1984), combining an assessment of the consequences of an action with a utilitarian approach concerning a specific type of ethical inquiry and judgment. The aim of the utilitarian consequentialist approach is to investigate all the relevant outcomes and the full impact of a decision in advance, namely, the expected benefits as well as the possible negative effects. This investigation addresses the probability of these outcomes and impacts as well as their extent. Ideally this investigation would lead to specific actions or decisions being related to the expected costs and benefits in monetary terms. Ethical analysis would be transformed into the balancing of risks and benefits using a quantitative calculus (Keeney, 1984). By aggregating these data, the utility of the decision-making options under consideration can be calculated and the option showing the highest expectable utility can be identified, as in a cost–benefit analysis. Often the multi-criteria decision-making approach (MCDA) (Caplan, 2000) is used, which consists of an operationalization of the utilitarian principle of maximizing the utility. Risks are taken into account by denoting a possible damage as negative utility.

In the case of nanoparticles, however, a basic precondition of this approach, the availability of quantitative data on the probability and extent of possible damage, is not fulfilled (see above). The consequentialist approach in the narrow sense with its close relation to an economic assessment is methodologically analogous to the classical risk management approach but does not work in a case of unclear risk. In the absence of accepted quantification procedures, the utilitarian consequentialist approach is not applicable, preventing it from providing any guidance in a case of unclear risk.

#### 6.3.1.2 *The Imperative of Responsibility*

With regard to the fact that quantified risk values must be applied in both of the approaches mentioned above — which by definition are not available in situations of unclear risk — the question is whether ethical approaches are available precisely for this situation. The proposals presented in this and the next section start from the diagnosis that determining the product of the probability of the occurrence of harmful effects and the extent of expected damage might, independent of its methodological problems, be problematic in an ethical regard. The famous work of Hans Jonas (1984) started with the observation that particular technical developments might have an apocalyptic potential that threatens the future existence of mankind. According to his normative presupposition "that mankind should be" and that the existence of mankind may not be endangered, Jonas formulated a new "categorical" imperative: "act in a way that the consequences of your actions are compatible to the permanence of real human life on Earth."

To return to the situation of unclear risk, at first glance it seems that this imperative could provide ethical guidance. Especially in cases of apocalyptic potential (in the sense that the unclear risks include a danger to the further existence of humankind), Jonas' position seems to provide orientation in the direction: stop developing or implementing technology if the result of such situations could be apocalyptic. To make his principle more operable, Jonas postulated a "heuristics of fear" in order to get an impression of possible negative developments and an obligation to use the worst-case scenario as the orientation for action. If the worst-case scenario has apocalyptic potential, then the action should not be taken.

Much criticism has however been leveled against Jonas' approach. Major points have been the naturalistic premises and the supposed teleology of nature used in his derivation of his central imperative as well as the arbitrariness of his conclusions. It might be possible in nearly every case to construct a worst-case scenario showing an apocalyptic dimension. As a consequence, no action would be ethically justified at all, yet doing nothing might also cause situations of unclear risk. An aporetic situation could be the consequence, demonstrating that the "imperative of responsibility" does not provide applicable and operable guidance.

The reason for this diagnosis is that Jonas did not formulate any requirement concerning the evidence or probability of the worst-case scenarios. Such scenarios play a decisive role in his argumentation because they determine whether an action might have apocalyptic potential or not. For Jonas the mere conceivability of the worst case was sufficient for him to use such scenarios in an argumentation for rejecting a whole line of technology. Exactly this is the weak point of his approach because this "low level" opens space for arbitrary speculation about worst cases, and this arbitrariness leads to the aporetic situation outlined above. Thus while Jonas' approach might be very appropriate for raising awareness with regard to situations involving unclear risk, it is completely inadequate as ethical guidance or as orientation in regulatory debates.

#### 6.3.1.3 *The Principle of Pragmatic Consistency*

Ethical guidance in dealing with risk could consist in relating new situations of risk to risk situations that we are familiar with. In this case we could — under specific preconditions — transfer our experience from the established risk strategies to newly emerging situations. A "principle of pragmatic consistency" has been proposed (Gethmann and Mittelstrass, 1992), which takes as its point of departure the observation that the idea of rationally acting persons is related to the request for consistency. We would have difficulty ascribing the attribute "rational" to persons who act in an obviously inconsistent way. The "principle of pragmatic consistency" transfers this observation to the problem of risk, postulating that the de facto behavior of people in existing situations of risk (e.g., in their lifeworlds) could be taken as a measure of what types and extent of risk could rationally be assumed to be acceptable to them. This argument intends to establish a link between empirical risk behavior and normative questions of the acceptability of risks.

This approach faces numerous problems, even in cases of "standard situations" of risk (Grunwald, 2008d). In case of unclear risk, the specific problem arises that the comparison that would have to be made between the existing (and established) risks and the new one would not be possible because of the lack of quantifiable data (see above). Therefore, this approach is not able to provide guidance in cases of unclear risk.

One could try, however, to generalize the basic idea of this principle — that of deriving conclusions regarding the acceptability of risks by combining knowledge about people's empirical behavior and their normative standards — to include cases of unclear risk. It seems promising for us to look at the probably numerous unclear risks that each of us deal with in everyday life, and to ask what this situation would or could imply for newly emerging types of unclear risk with regard to normative standards of rationality. In spite of the foreseeable difficulties in making such a principle work, this would provide an interesting relation between the *is* (empirical behavior with respect to risk acceptance) and the *ought* (normative standards of rational behavior) that should be explored in more detail (see below, Section 6.4).

#### 6.3.1.4 *Deontological Advice*

Deontological ethics does not primarily look at the consequences and impact of actions and decisions but at human rights and obligations on the basis, for example, of a Kantian approach. Concerning risk, Nida-Rümelin (1996b) has proposed an approach which questions and partly rejects the utilitarian-style consequentialist approach. According to this position, there are ethical presuppositions hidden in the decision to follow a consequentialist approach, specifically in the weighing and balancing of risks and benefits. Such presuppositions address the question of whether a balancing approach can be ethically justified at all. The deontological position (Nida-Rümelin, 1996b) emphasizes that there might be strong arguments for or against a particular decision or action that would make a balancing procedure a morally problematic approach or even an immoral one. Take for example a situation in which benefits for a lot of people concerning health, welfare, or even life itself could only be achieved if some people were made victims. The result of a purely balancing approach that applies the principle of maximizing utility might even be that some people would be allowed to die in order to achieve the benefits for the large number of other people (for a famous example, cf. Harris, 1975).

This type of argumentation leads to the conclusion, that there are specific limits to risk management in the consequentialist approach, in particular its utilitarian version even if quantitative data on risk were available (following (Nida-Rümelin, 1996b):

- Human right to life
- Human rights in general (following, for example, the UN declaration of human rights)
- Civil rights (for participation, for example)
- Rights of nonhumans, such as animals (as far as agreed upon due to ethical positions and legal codifications)
- Rights of possession (weaker than the others but established in modern societies)

Such rights lead, following the deontological approach, to restrictions on the balancing and weighing of risks against expected benefits. The deontological position makes the point that there might be ethical problems even in the preliminary phase of a risk management procedure. Instead of the simple use of a balancing approach, the people affected should first assess whether such a procedure would be ethically legitimate.

What the deontological position could contribute to unclear risk is to provide a link to the normative dimension of rights and duties. In the case of unclear risk, there might be situations where a decision creating the unclear risk might even be ethically problematic merely because of the fact that the unclear situation might raise threats to human rights which would not be the case for alternative decisions. In such a situation it could be ethically prudent to follow this argument. Most situations, however, will be more complex. Unclear risks of different types might be the consequence of any of the possible alternatives. In this case the deontological position does not provide orientation because it lacks an operable procedure for comparing different types of unclear risk. In these situations, it might lead to the same problem of aporias that we observed in Jonas' approach (see above).

#### 6.3.1.5 *Projected Time*

The risk debate on nanotechnology has motivated some authors to think even more radically about ethics in the case of unclear risk. Dupuy and Grinbaum (2004) even went beyond Hans Jonas' "heuristics of fear" (see above) and formulated a "duty to expect the catastrophe" in order to prevent a catastrophe. The uncertainty of our knowledge about nanotechnology and its consequences, especially in the face of immense potential for damage, such as from possibly catastrophic effects (e.g., from losing control over self-replicating robots (Joy, 2000)), is here considered a reason to even view the precautionary principle (see below) as being insufficient. The authors put forward the following argument for a "projected time" approach, which makes use of strong assumptions about the future:

- Nobody can know anything about the future of nanotechnology, except that it is the utter catastrophe.
- If everyone could be convinced that nanotechnology is *the* catastrophe, there could be a general renunciation of nanotechnology.
- The catastrophe could consequently still be avoided.

This line of argument is pedagogical: if everyone were to believe in the assertion that nanotechnology is simply the catastrophe, and if they then were to renounce nanotechnology, then the catastrophe would not happen even though it is presently claimed to be inevitable. To put it concisely, the catastrophe will not happen because everybody is convinced that it is certain that it will happen. The assumption of the inevitability of the catastrophe has no validity in the sense of a discourse between opponents and proponents, but only serves the didactic purpose of motivating the ultimate renunciation of, in this case, nanotechnology. The projection of time in this argument employs the well-known mechanisms of self-fulfilling or self-destroying prophecies:

> The predictor, *knowing that his prediction is going to produce causal effects in the world*, must take account of this fact if he wants the future to confirm what he foretold. (Dupuy and Grinbaum, 2004, p. 16)

The assumed projection, however, fails for two reasons. First, it is highly risky to use predictions in order to intentionally influence the further course of developments (here, to motivate renunciation of nanotechnology in order to prevent the catastrophe). Projection, in the sense of determination of the future, runs into the same problems as all other planning approaches. Consequently, there is no guarantee that the catastrophe will be avoided. Secondly, the whole argument relies on the premise that nanotechnology is "the" catastrophe. Though some arguments for this thesis are given (Dupuy and Grinbaum, 2004), it is highly improbable that most people will be convinced, which means that the whole chain of argument fails. For these two reasons, the proposed model of "projected time" will not be able to provide ethical guidance in our case.

## 6.3.2 Operational Approaches

Beyond the philosophical concepts but partly motivated by them, approaches have been developed that attempt to develop procedures for determining socially accepted and legitimized processes for dealing with unclear risk.

### 6.3.2.1 *The Precautionary Principle*

Debates about precautionary regulatory measures have been motivated by the observation that in many cases severe adverse effects had not been detected at an early stage of the introduction of new materials and that these effects had led to immense damage to human health, to the environment, and to the economy (cf. impressive case studies in Harremoes *et al.*, 2002). Wide international agreement on the precautionary principle was reached during the Earth Summit (United Nations Conference on Environment and Development, UNCED) in Rio de Janeiro in 1992 and became part of Agenda 21:

> In order to protect the environment, the precautionary approach should be widely applied by States according to their capabilities. Where there are threats of serious or irreversible damage, lack of full scientific certainty shall not be used as a reason for postponing cost-effective measures to prevent environmental degradation. (principle 15 of the Rio Declaration)

The precautionary principle was incorporated in the Treaty on the EU in 1992. Article 174 postulates:

> Community policy on the environment shall aim at a high level of protection taking into account the diversity of situations in the various regions of the Community. It shall be based on the precautionary principle.

Since the precautionary principle thus establishes a rationale for political action, it substantially lowers the threshold for government action (see von Schomberg, 2005 for the following). It considerably changes the situation compared to the previous context, in which politicians could use (or abuse) a persistent dissent among scientists as a reason (or excuse) simply not to take any action at all. In cases for which the accumulation of relevant scientific evidence can take decades, this would have implied that political action could always be postponed with the argument that scientific knowledge still had to be completed. In this way, political action could simply come much too late.

It is, however, a difficult task to make legitimate decisions about precautionary measures without either running into the possible high risks of a "wait-and-see" strategy (Gannon, 2003) or overstressing the precautionary argument with the consequence of no longer being able to act. The following characterization of the precautionary principle shows — in spite of the fact that it still does not cover all relevant aspects — the principle's complex inherent structure:

> Where, following an assessment of available scientific information, there is reasonable concern for the possibility of adverse effects but scientific uncertainty persists, measures based on the precautionary principle may be adopted, pending further scientific information for a more comprehensive risk assessment, without having to wait until the reality and seriousness of those adverse effects become fully apparent. (modified following von Schomberg, 2005)

Thinking about the application of the precautionary principle therefore generally starts with a *scientific examination*. There is a need to have an assessment of the state of available scientific knowledge and of the types and extent of the uncertainties involved (see Steinfeldt *et al.*, 2007 for case studies). Drawing the line marking the border to the classical practice of risk management or to the situation of purely conjectural risk involves making normative choices that need to be made explicit (Rescher, 1983). In assessing the uncertainties involved, *normative* qualifiers come into play when the precautionary principle is applied (von Schomberg, 2005). It has to be clarified whether there is "reasonable concern" in this situation of uncertainty. The "reasonable concern" qualifier, as employed by the EC guidelines, relates to a judgment on the quality of the available information (von Schomberg, 2005). The assessment of the available knowledge (including its uncertainties) therefore enters the focus of precautionary reflection (for methodical proposals see Pereira *et al.*, 2007). The question of normative guidance is transformed into a procedural challenge of assessing the status of the available knowledge.

A wide debate has underlined many of the shortcomings of the precautionary principle. It has been regarded as ambiguous, as unsuitable for providing orientation for decisions on genuine regulatory options, as capable of increasing people's anxieties, and as too intertwined with a biased social perception of risk (Sunstein, 2005; Marchant *et al.*, 2008; Hull, 2007). There have been fears that it could impose excessive costs on developers and producers, with the result that the precautionary principle might hinder technological innovation and could paralyze development.

These criticisms, however, do not recognize that the difficulties being considered belong to the very nature of any attempt to dealing with unclear risk. It is exactly this situation in which it is by *definition* impossible to rely on scientific evidence. Normative reflection is instead required to an extent that is much larger than in classical risk management, reflecting a "persistent under-emphasis of the ways in which risk assessment inevitably rests on normative commitments," which take the form of

> subjective judgements, influential social values, contestable assumptions and administrative procedures that are open to contingent framings and the tacit or deliberate exercise of power. (Wynne *et al.*, 2007, p. 33)

The precautionary principle was conceived as an approach to contexts that cannot be properly faced by means of the model of acceptable risk, cost–benefit analysis, or the reference to the best available technology (Marchant *et al.,* 2008). Even the fact that it does not in itself contain the criteria and tools to automatically fulfill this task belongs to the situation of unclear risk that the principle has to be designed for. In this sense, the principle is only formally a legal principle; in reality is it much more a political principle for whose application political debate and reasoning are necessary.

A closer look at the precautionary principle in this context (Synthetics, 2010) shows that:

(a) The appeal to the precautionary principle does not necessarily involve the introduction of legally binding measures.
(b) The normative decisions stemming from the application of the principle should meet the requirements of proportionality, be nondiscriminatory, and be based on a cost–benefit analysis that also takes opportunity costs into consideration, not just economic ones.
(c) Such decisions should be open to revision and should ensure a wider spread of the responsibilities for the production of the scientific evidence needed for wider risk assessment. The application of the precautionary principle is part of the risk management process when scientific uncertainty does not allow a full understanding of the risk itself.

#### 6.3.2.2 *The Prudent Avoidance Approach*

The prudent avoidance principle states that reasonable efforts to minimize unclear risks should be taken when the actual magnitude of the risks is unknown. The principle was proposed in the context of electromagnetic radiation safety (in particular, fields produced by power lines). A report for the Office of Technology Assessment of the U.S. Congress (Nair *et al.*, 1989) described the prudent avoidance of power line fields as

> looking systematically for strategies which can keep people out of 60 Hz fields arising from all sources but only adopt those which look to be "prudent" investments given their cost and our current level of scientific understanding about possible risks.

The principle has been adopted in a number of countries, for example, in Sweden, Denmark, Norway, Australia, and New Zealand. It has also been adopted in some form by a number of local regulatory bodies in the United States, for example, by the public utility commissions in Colorado:

> The utility shall include the concept of prudent avoidance with respect to planning, sitting, construction, and operation of transmission facilities. Prudent avoidance shall mean the striking of a reasonable balance between the potential health effects of exposure to magnetic fields and the cost of impacts of mitigation of such exposure, by taking steps to reduce the exposure at reasonable or modest cost. Such steps might include, but are not limited to (1) design alternatives considering the spatial arrangement of phasing of conductors; (2) routing lines to limit exposures to areas of concentrated population and group facilities such as schools and hospitals; (3) installing higher structures; (4) widening right of way corridors; and (5) burial of lines. (Colorado Public Commission, 1992)

Some health departments have also adopted policies or published informational literature that recommends prudent avoidance as a policy tool. They assessed the scientific literature on electromagnetic fields (EMFs) and concluded that adverse health effects from exposure to EMFs have not been established. Most of them agreed that there is some evidence that exposure to EMFs may pose a risk to health and suggested a cautious approach when building new electrical facilities, homes, and schools (including kindergartens and child-care structures) near existing electrical facilities such as power lines and substations. However, all of them have rejected the imposition of arbitrary low numeric exposure levels for EMFs since these are not supported by the scientific literature.

Of great interest in this chapter is what "prudence" means in this approach. Coming from Aristotelian ethics, prudence denotes a pragmatic approach to what should be done. In this case, prudence consists first in balancing two frequently diverging ideas: the idea of precaution and the idea of making use of a technical advance even in the case of unclear risk. Secondly, prudence is asked for in the concrete design of the technical facilities: alternatives are sought which might be similarly useful but would create lower radiation loads for people.

### 6.3.3 Interim Conclusions

The ethical approaches presented above do not provide clear guidance in situations involving unclear risk. Some of them (the utilitarian approach and the principle of pragmatic consistency) need quantified risk values, so that they by definition are applicable only in standard risk situations. Others will lead to aporetic situations (Jonas, 1984; Dupuy and Grinbaum, 2004) because they do not offer an operable strategy of distinguishing between unclear risk and mere speculation. The deontological position provides some hints as to what has to be taken into account but misses the specific challenge of unclear risk.

This conclusion might sound disappointing but it is not. It would be an exaggerated desire to expect ethics to be able to solve the problem of decision-making involving unclear risk solely by its own means. Instead, in the following I will look for ethical support to identifying problem-solving strategies but without expecting the solution to come from ethical guidance alone. In this way, ethical guidance contributes to a more comprehensive picture, allows for structuring the normative issues involved, and offers analytical work concerning the structure of the respective problem in a normative regard. Problem-solving itself cannot, however, be expected from these concepts, but must be the outcome of socially accepted and democratically legitimized processes. This is the main reason in the following to continue working with the precautionary principle in the version accepted by the EU (von Schomberg, 2005) and to include the implications of the ethics of risk in the implementation as far as possible.

Both the precautionary principle and the prudent avoidance principle do not have an explicit ethical foundation. While the precautionary principle might be understood as an attempt to make Jonas' work (which stands in the Kantian tradition) more operable by involving procedures for assessing the available knowledge about the unclear risk under consideration, the prudent avoidance principle is obviously motivated by the Aristotelian approach. With respect to outcome and consequences, both approaches converge strongly. They lower the expectations with respect to ethical guidance and transform them into questions of "prudent" knowledge assessment, which is to serve to guide the search for adequate actions and decisions. They consequently aim to solve the paradoxical challenge by enabling assessments where assessments do not seem to be possible (cf. Colorado Public Commission, 1992). This challenge, however, can only be met by a procedural approach where the involvement of participatory processes seems plausible (Pereira *et al.*, 2007) but which also should be ethically informed. Following this assessment, I will use the precautionary principle to clarify the risk situation of nanoparticles and to derive some orientation for further operationalization.

## 6.4 Dealing Responsibly with Nanomaterials

Since the precautionary principle as such cannot be readily applied, a normative analysis has to introduce distinctions on the basis of which different cases can be handled separately.

### 6.4.1 Conditionally Normative Reflection

There is a remarkable consensus — not seriously contested by anyone — about the fact that we have hardly any knowledge about the possible health and environmental impact of synthetic nanoparticles (e.g., Schmid *et al.*, 2006). This evaluation is even shared by nanotechnologists and the companies in the relevant industries. This consensus even extends to the fact that toxicological research is in particular needed to close these gaps in our knowledge. The consensus does not, however, cover the question of what should be done in the period of time until these gaps are closed:

- Representatives of a strong precautionary principle (see Rippe, 2002; Bachman, 2006, pp. 55ff.) demand a moratorium on the commercial use of nanoparticles until there is clarity about their possible consequences. In a normative sense, this means that a higher priority is assigned to the protection of those possibly affected than to the use of the potential expected of nanoparticles. Deontological arguments, in particular, are useful in grounding this position since the risks might violate the rights of individuals, such as those to physical integrity and life. The burden of proof is placed on the side of those promoting the introduction of a new technology. They have to ensure that the introduction of the new technology will not violate anyone's rights.

- Representatives of a weak precautionary principle (Bachmann, 2006, pp. 55ff.) argue that while it is not appropriate to introduce an immediate moratorium and while there must be more research, special caution must be exerted until synthetic nanoparticles have been proven to be harmless. As in the liberal tradition customary, the burden of proof here dictates that regulatory intervention by the government is not permitted until there is ample evidence of a danger or risk. Caution is nonetheless appropriate even though evidence of a danger or risk is not yet available. Thus "the proactive introduction of protective measures [is demanded] in view of the possible risks, which science today — because of a lack of knowledge — can neither confirm nor disprove" (Swiss Re, 2004, p. 47).
- The representatives of a wait-and-see strategy[4] refer to the anticipated utility, to the fact that all progress is fundamentally fraught with risk, and to the fact that we simply do not know whether a decision not to use synthetic nanoparticles ultimately poses more risks or whether the unused opportunities also violate people's rights. The burden of proof here is seen to be on the advocates of precaution. In line with a libertarian attitude that only something can be forbidden or regulated that has been proven to be injurious, new technology must be presumed to be harmless until there is proof to the opposite.

Behind these different positions are ultimately assumptions about societal risk disposition and about the freedom to weigh alternatives, i.e., about de facto moral sensitivities. If these were shared assumptions in the sense of a predeliberative agreement (Section 3.1), or if it could be assumed that they were shared, then

[4]There is no theoretical background for this position. It rather belongs to the attitudes toward the lifeworld that can be heard at coffee breaks at conferences, not in the presentations themselves.

it would be simple to draw a conclusion about measures in the one or the other direction. The consequence in the case of presumed harmlessness with regard to unclear risk — in addition to the need for research, which would be expected for each of the positions — would be to make rapid technical use of synthetic nanoparticles, at least until clear damages might possibly be empirically detected. The immediate consequence in the case of the strong precautionary principle would, in contrast, be a moratorium since otherwise the rights of those possibly affected would be violated.[5]

It is evident that there cannot be any talk of a predeliberative agreement shared by all the members of society in either the one or other direction. There is thus a conflict that gives us reason to diagnose a nonstandard situation in a moral respect (Section 3.1). The analytical questions then to be posed as part of the further analysis and to prepare alternative possible solutions aim to determine *who* would either benefit or possibly be exposed to risk by synthetic nanoparticles. Since ethics is related to social contexts and cannot be blind to who is an actor and who is affected, who is an agent and who a victim, who benefits and who suffers, an ethical analysis may not refer to chances and risks in a purely *abstract* fashion. Just the opposite, it has to ask who the involved persons and groups are and what possible influence they have on the exposure to synthetic nanoparticles. Further questions are how those possibly affected by the risks posed by nanoparticles can behave toward the risks. At this point, I propose distinguishing the following levels of response:

1. *Impositions that can be controlled individually*, such as riding a motor cycle, doing risky types of sports, or in the future possibly taking a vacation in space. In these cases, the individual affected determines him- or herself which risks he or she is willing to accept. Among these *impositions with simple alternatives* are the possible health dangers posed by certain foods or cosmetics. A labeling requirement gives a customer the opportunity to avoid these dangers in as much as there are sufficient alternatives.
2. *Impositions where the alternatives are onerous*: While it is possible to avoid some impositions in principle, it is often

[5]Bachmann (2006, pp. 55ff) argues accurately that — despite the fundamental differences between the strong and weak versions of the precautionary principle — there is not much difference normatively between the corresponding positions of the ETC Group (strong) and the Royal Society (weak).

impossible in practice. Those affected are exposed to real (or unclear) risk. Classical examples are the problems of determining the sites of incineration plants, permanent repositories for radioactive waste, or chemical factories. Evasive action is possible in principle — one could move away — but as a rule this would be fraught with substantial social and economic burdens.

3. *Impositions with no alternatives*: Some of the consequences of technology are diffusely distributed: the hole in the ozone layer, global warming, creeping groundwater pollution, soil degradation, the accumulation of pollutants in the food chain, noise, and ultrafine dust. Nobody can avoid them, but the amount of exposure varies strongly, such as by residential area.

Ethical analysis leads to different results depending on which of these categories is applicable. While the issue for the first item is to provide the necessary information to those affected (Section 6.4.2), for the third it is have regulations that are binding and democratically endorsed (Section 6.4.3). The second item is frequently the object of controversy regarding whether the specific case is closer to the first or the third item.

## 6.4.2 Informed Consent and Consumer Freedom

First I will consider the case that nanoparticles might pose unknown risks to the current *users or consumers* of such products. The example of food is very illustrative (nanofood, Section 6.1). The purpose of the demand for a labeling requirement for food containing nanoparticles is to put consumers in a position to themselves determine if they are willing to take a possible risk in order, for example, to enjoy certain advantages or lifestyle features. Formally, this is sufficient to satisfy the demand for informed consent:

> Yet because there is no labeling requirement, you do not know which products contain nanomaterials. Even the responsible authorities do not have sufficient information. And consumers have hardly a chance to avoid consuming nanoparticles included in food. (BUND, 2008)

For two reasons, however, the labeling model is problematic or does not confront part of the ethical problems:

1. Labeling would only create or improve the choices of *current* users. The spread of nanoparticles could however also affect *future* persons and limit their rights. It is impossible, however, to ask those possibly affected in the future for their approval. What is involved here is responsibility for the future, together with the relevant normative issues and problems (e.g., Birnbacher, 1988).
2. Synthetic nanoparticles contained in food could also be distributed in the environment diffusely, affecting numerous people in addition to the consumers immediately affected. This case would not be covered by a labeling requirement. Nanoparticles free in the environment — such as from automobile exhausts or residues from incineration — would in principle affect everyone, independent of their approval following any introduction of labeling. At this level, such problems cannot in general be solved according to the model of an individual's informed consent. At this level the issue is the collective acceptability of risks, which has to be legitimized democratically.

A labeling requirement would thus only address some of the normative issues that are involved. This does not change anything about the fact, however, that labeling represents a significant element in helping those who are possibly affected to participate in determining whether they are affected. Without labeling, they are exposed without knowing it and without their approval. Moral autonomy in this context consists in enabling those who are affected to give their approval — or not — to being a participant in a live experiment involving nanoparticles.

Furthermore, when impositions cannot be easily evaded, it is time for government risk policy to become active. This can take the form of operationalizing the precautionary principle, to which ethical reflection should prepare the groundwork with regard to the acceptability of risk (see above). At issue here are generally binding regulations (see below), not the possibility of private reactions.

### 6.4.3 Regulation, Code of Conduct, and the Common Good

Beyond individual freedom there is the level of binding regulation and its relation to manifestations of the "common good." The debate on regulatory issues related to nanotechnology started in the United States and was in response to reports of possible risks. Currently an intensive debate is taking place between representatives of nanotechnology businesses and the relevant regulatory bodies, such as the U.S. Food and Drug Administration. The debates take a product-based approach and, because of the great importance of product liability in the United States, lead to a pragmatic decision in each case. The aim of the Toxic Substances Control Act (TSCA), the law regulating chemical substances in the United States (enacted in 1976), is to regulate the production and use of chemicals which pose a risk or a potential risk to the environment (Wardak, 2003). The main element is the TSCA Chemical Substance Inventory, which is a database of about 80,000 chemical substances, of which about 50,000 have been reviewed and about 5000 have been subject to rigorous testing (Wardak, 2003). Ideas reflecting the precautionary principle are not incorporated in the TSCA. Regulatory action can only be undertaken if any scientific uncertainty about possible hazards has been resolved. The burden of proof requires that there is clear scientific evidence for the existence of adverse effects. With regard to nanotechnology, there are three types of possible exemptions: the low volume exemption, the low release and exposure exemption, and the test marketing exemption. The low volume exemption frees a manufacturer from full reporting in accordance with the TSCA regulations if less than 10 tons of a particular substance are produced per year (in the envisaged but not yet adopted new European system for regulating chemicals, REACH, this threshold will be lowered to 1 ton/year). The production of many nanomaterials will probably fall under this category of exemption. The test marketing exemption might also apply to many nano developments because in many cases small amounts will be produced in order to exploit its market potential (Wardak, 2003).

The European REACH system is based on a different perspective, one that takes a stronger precautionary approach. Its goal is to test all known chemical substances and include the data in a central database while the review of new chemicals will be optional (Haum *et al.*, 2004, p. 36). Although this approach is *prima facie* strange, it is motivated by the diagnosis that the number of new substances is very small compared to that of the complete set of existing chemicals — whose features, side effects, and risks are often only partially known. In the case of nanoparticles, this situation is ambivalent: on the one hand, it provides the chance to carefully check the impact of nanoparticles based on well-known chemicals (differing only with respect to particle size and/or morphology), but on the other hand, there might be a gap if such nanoparticles were declared as new chemicals and thus only reviewed optionally within REACH.

The existing regulatory systems for chemicals have been optimized for large volumes, and they do not take the specific attributes of nanoparticles into consideration (it is possible that this point will be changed in REACH in the course of recent initiatives in the European Parliament, but this is still an open process under political pressure). It is well known that the toxicity of fine and ultrafine particles depends strongly on the size and shape of the particles. The regulation of chemicals, however, does not take these attributes into account (since they are indeed irrelevant for usual chemicals).

An intensive debate on nanotechnology issues also took place rather early in the United Kingdom. The widely read study of the Royal Society and the Royal Academy of Engineering (2004) led to many statements and recommendations concerning regulations designed to close the knowledge gaps and to minimize the risks that might result from the production and use of nanoparticles before this knowledge becomes reality.

Experts have released their first recommendations for handling nanomaterials for the European Commission (Community Health and Consumer Protection; Haum *et al.*, 2004). The establishment of institutions to monitor both the nanotechnologies and our knowledge about them is also seen as an important step. A dialogue with the public and with industry should be initiated. Freely engineered nanoparticles should be contained, and the production and unintentional release of waste containing nanosized particles should be avoided whenever possible and otherwise minimized.

The debate in Germany began on the basis of a study on nanotechnology requested by the Bundestag, the German parliament (Paschen, 2004). It contains what may be the first reference to regulation of nanotechnology, which must be seen in connection with the incipient debate on nanoparticles of the time and with the demand of the ETC Group (2003) for a moratorium (see Section 5.1.1).

The German government (Bundestag, 2007) released an official report on the status of its plans to regulate nanotechnology in 2007. According to this, the government at that time did not see any fundamental need for changes to the existing laws and regulations resulting from the development of nanotechnology. In its view, the existing system of laws and regulations at the national and EU levels offered flexible instruments for recognizing and, if appropriate, reacting to the possible risks posed by the developments of nanotechnology. At the time, the government said it could not determine whether any additional regulations specific to nanotechnology were necessary in order to meet its legal obligations to ward off dangers and take precautions.

According to this report, future regulatory challenges will consist in conceiving regulations that accompany and support the innovative process and that keep pace with the rapid development and distribution of technology. Strategies to minimize the nanospecific risks are supposed to be developed without imposing unnecessary limitations on the technology's chances to grow. The government said it had introduced comprehensive measures to extend the level of knowledge regarding the potential dangers posed by nanomaterials, to initiate an intensive dialogue with the representatives of the social interests that are involved, and to support and stimulate actors to act responsibly. The need for regulation was thus said to already be covered by the laws on the respective fields, for example, chemicals, worker protection, pharmaceuticals, and food.

The European Commission has adopted a code of conduct for responsible research in the nanosciences and nanotechnologies. The Commission has drawn up a voluntary code that covers seven general principles, including sustainability, precaution, inclusiveness, and accountability. With the publication of its "Code of Conduct for Responsible Nanosciences and Nanotechnologies Research," the Commission hopes that universities, research institutes, and companies in the European Union will sign up and ensure the safe development and use of nanotechnologies. It refers explicitly to the precautionary principle:

> "Given the deficit of knowledge of the environmental and health impacts of nano-objects, Member States should apply the precautionary principle in order to protect not only researchers who will be the first to be in contact with nano-objects but also professionals, consumers, citizens and the environment." (EC, 2008)

Particular attention is paid to the food sector: "As long as risk assessment studies on long term safety are not available research involving deliberate intrusion of nano-objects into the human body, their inclusion in food [...], feed, toys, cosmetics and other products that may lead to exposure to humans and the environment should be avoided."

All of these activities build on the precautionary principle, some of them explicitly, others implicitly. Following the introduction of the precautionary principle (see above), the central questions as to whether the precautionary principle could or should be applied to the production and use of nanoparticles given our knowledge about the possible consequences are:

1. Is there a precautionary situation characterized by *epistemic uncertainty* or *unquantifiable risk*?
2. Is there "*reasonable concern* for the *possibility* of adverse effects" in the field of synthetic nanoparticles that would provide legitimacy to the application of the precautionary principle?

These questions concern the diagnosis of the current state of affairs, especially with respect to an assessment of the current state of knowledge. The first question has to be answered *positively* (e.g., Schmid *et al.*, 2006; Haum *et al.*, 2004). There are unknown or uncertain cause–effect relationships and still unknown probabilities of risks resulting from the production, use, and proliferation of nanoparticles. The scope of possible effects, their degrees, and the nature of their "seriousness" (in relation to the chosen level of protection) can currently, even in the best cases, only be estimated in qualitative terms. Therefore, we are witnessing a typical situation of uncertainty where established risk management strategies cannot be employed (in the same sense, see also the Royal Society, 2004, summary, p. 4; Haum *et al.*, 2004).

The answer to the second question, namely, whether there is "*reasonable concern* for the *possibility* of adverse effects" in the field of synthetic nanoparticles, is also to be answered positively. The initial toxicological results from the exposure of rats to high concentrations of specific nanoparticles show severe and lethal consequences. Because the exposure concentrations in these tests are extremely high and it is difficult to transfer any knowledge gained from the exposure of rats to the human situation, the results do not provide evidence of harm. They do however let us conclude that there is "reasonable concern for the *possibility* of adverse effects" caused by synthetic nanoparticles. Please note that evidence of the *possibility* of adverse effects is sufficient in order to legitimate the application of the precautionary principle. This is, on the one hand, an important difference to the imperative of responsibility (Jonas, 1984) because in the Jonas' approach no *evidence* of the possibility of adverse effects is required to reject a certain technology, just the hypothetical conceivability of adverse effects. On the other, there is a strict difference to classical risk management situations where *evidence of adverse effects* is required, not evidence of their *possibility*.

While it is thus simple to answer the question of the applicability of the precautionary principle, it is much more difficult to answer the question as to what would follow in terms of measures from its application. Two levels must be observed (Haum *et al.*, 2004, p. 26): the level of improving the knowledge base in order to enable classical risk management strategies, as well as the level of elaborating regulatory mechanisms to be better able to deal with the uncertainties involved as long as the first level has not been reached. While a consensus is simpler to achieve at the first level — despite the problem of scarce resources to do all the research necessary — the main problems concerning "weak" and "strong" precautions (see above) arise at the second level.

Recalling that the precautionary principle requires a scientific assessment of the state of the art, and that the "quality of the information available" (von Schomberg, 2005) serves as a qualifier in assessing whether concern is reasonable or not, it becomes clear that the assessment of the state of knowledge plays a key function in deciding among the two alternative options. This scientific assessment of the knowledge available is provided in a comprehensive manner in Schmid *et al.* (2006), reaching the frontier of present research.[6] The result is that in some cases there are indications of risks from nanoparticles to health and the environment. Based on the state of the art, there is *no reason for serious concern* about adverse effects, but there is for serious concern about their *possibility*. This result is supported by other studies, such as that by Haum *et al.*:

> Taking into account our present-day knowledge, there is, with regard to nano-specific effects (excluding self-organization effects

[6]Such a comprehensive analysis is usually not included in ELSA studies on nanotechnology (cf. Haum *et al.*, 2004; Royal Society, 2004; Nanoforum, 2004), which endangers the strength and evidence of the recommendations given.

> and cumulative effects of mass production), no reason for particularly great concern about global and irreversible effects of the specific technology "per se," with it being on a par with the justifiable apprehension concerning nuclear technology and genetic engineering (Haum *et al.*, 2004, p. 16).

The mere possibility of serious harm implied by a wider use of nanoparticles, however, does not justify the use of the precautionary principle to argue for a moratorium or other prohibitive measures. While it will always be possible to create speculative scenarios with highly adverse effects, to follow these speculations would not reach the "rationality" of the precautionary principle. The present state of our knowledge therefore provides no reason for a moratorium or similar strict measures. Because the state of knowledge changes permanently, it is urgently necessary for the production of knowledge about the impact of nanoparticles on human health and the environment to be continuously monitored and assessed.[7]

### 6.4.4 Operative Approach: Remarks on the Next Steps

In order to draw conclusions regarding the further nature of our dealing with nanoparticles from the results of ethical reflection in the sense of conditionally normative consultation, it appears to be sensible to first ask whether synthetic nanoparticles could endanger "the entirety" in the sense proposed by Jonas (1984). According to a wide consensus, this is not the case, yet even this conclusion cannot be drawn without scientific and technical assessment of the possible consequences (e.g., Schmid *et al.*, 2006). The most important reason is that these particles do not have the potential for causing apocalyptic damage, even if some of them could be toxic for humans or ecosystems.

It must be noted, of course, that a sentence such as that by Haum *et al.* (2004) that is quoted above can only be expressed given conditions that themselves could be doubted. Decisive is their reference to "present-day knowledge." A similar sentence could have *correctly* been said about HCFCs for several decades until it finally turned out that the consequences of their massive emission into the atmosphere did amount to a global threat after all (Harremoes

---

[7]One of the most dramatic lessons to be learned from the asbestos story is exactly the lesson about the crucial necessity of such a systematic assessment (Gee and Greenberg, 2002).

*et al.*, 2002). This example does not question the validity of Haum *et al.*'s (2004) sentence since the sentence is true in relation to the "present-day knowledge" of the time. Yet present-day knowledge can age or become obsolete. The example thus focuses our attention on the fact that the knowledge that is the basis of the sentence's validity does not include future knowledge, and that any conclusions may have to be changed if new information comes to light. This leads to the demand for a permanent review of the knowledge needed as a foundation for conclusions. The phrase "until further notice" can continue to be used for orientation as long as this constant review does not report any "reasonable concern" or an argument for the plausibility of scenarios resulting in possible global damage. This means that certain patterns of ethical argumentation, such as that from *The Imperative of Responsibility* (Jonas, 1984), do not apply here.[8]

This keeps us in a sphere that has a great deal of experience in dealing with new materials or chemicals. The question is now how we can act responsibly toward nanoparticles in a precautionary situation when the available knowledge is extremely uncertain, the normative bases of a decision are controversial, but the consequences are far reaching (a typical situation in "post-normal science"; Funtowicz and Ravetz, 1993).

Finding a functioning system of analogous social action — i.e., action situated according to explicable criteria in a normative sense that is analogous to the challenge posed by synthetic nanoparticles — would provide a viable standard of comparison. This would offer the opportunity to "normalize" the new situation by providing orientation according to a tried schema of action. First, this would maintain the primacy of the lifeworld (Chapter 3), and second the lifeworld would serve as a role model by providing a functioning system of social action for dealing with unclear risk. The normative portion of this mechanism of a functioning system of social action would provide a framework within which the new situation (in this context, dealing with synthetic nanoparticles) could be considered as a *standard situation in a moral sense* (Section 3.1) even though it is not one. The acceptance of the analogy with a functioning system of social action would transform the issue of dealing with nanoparticles into a moral standard situation, with the consequence

[8]This must be completely re-examined as soon as self-organizing or even living nanoparticles are being considered (on this see Chapter 7).

that the respective rules and procedures could be implemented. In this way, the new quality would be normalized. Since the role model character of the functioning system of social action would first have to be accepted, there could of course be controversies about the moral permissiveness, about the limits, or about the implications of transference from a functioning system of social action to a new situation.

These considerations about normalization remind one of the "principle of pragmatic consistency" (see above), which employs arguments of rationality and consistency to aim precisely at such normalizations. As described above, this principle, however, cannot be applied directly to the problem posed by synthetic nanoparticles since no measurements and criteria are available for comparing the unclear risks posed by synthetic nanoparticles with established unclear risks. In the following, I shall nonetheless attempt to use this principle in another manner that makes the analogy appear plausible. The issue then is not to compare unclear risks with other unclear risks, but to select several different conceptions for dealing with unclear risk and to identify one of them as being responsible for dealing with nanoparticles. In doing so, the enormity of the risks that may be involved will play a significant role. This enormity does not put "the entirety" at risk (see above), yet such statements on risk can only be made on the basis of present-day knowledge.

With regard to the goal of a "normalization of the abnormal," we might think of two role models: on the one hand, the handling of synthetic chemicals, and on the other, that of genetically modified organisms (GMO). The regulation of chemicals has a long tradition of dealing with new materials that appear in the biosphere. This has resulted in numerous suggestions that the regulation of chemicals should be considered as a role model for the handling of nanoparticles (e.g., Schmid *et al.*, 2006), while however taking note of the particularities of nanoparticles (Section 2.2). On the other hand, the precautionary principle is intimately linked with the field of genetically modified organisms (von Schomberg, 2005).

The extent to which genetically modified organisms are damaging our health or the environment has not been demonstrated experimentally yet. Ecologists thought and still think it plausible that the release of individual genetically modified organisms could, for example, have irreversible effects on natural vegetation, for example, with regard to a decline in biological diversity. It is scientifically uncertain whether — compared to traditional forms of agriculture — genetically modified organisms pose an additional risk even though no discernable risk factors have so far been identified. This situation of a lack of knowledge about the possible consequences of genetically modified organisms has been a substantial motive for the introduction of the precautionary principle, which happened to be its first use case. For this purpose, the precautionary principle was transferred into a system of legislation that is based on a *case-by-case* and *step-by-step* concept (von Schomberg, 2005).

This procedural implementation of the precautionary principle requires an ongoing scientific assessment and determination of possible risks. This framework makes it possible to circumvent the inability of science to decide whether genetically modified organisms can be considered fundamentally safe or not. It shifts the discussion to the microlevel of scientific information and to whether a *specific* genetically modified organism — following successful authorization — can be considered safe for a *specific* application, for a *specific* and limited period of time within *specific* requirements on monitoring and management. This implies regulatory requirements regarding monitoring, labeling, and traceability. Temporary approvals are measures that are in accord with and support the precautionary principle. They also promote learning activity that puts us in a position to collect information at the appropriate time and to assess the innovative aspects of genetically modified organisms, specifically their potential utility.

The implementation of the precautionary principle in the European Union can in this sense be considered a functioning system of social action.[9] On the basis of an assessment that corresponds to our current level of knowledge, namely that synthetic (abiotic) nanoparticles pose an unclear risk of presumably less consequence than genetically modified organisms since they are not alive and do not propagate,[10] the application of the precautionary principle for nanoparticles would put us in a responsible framework of action from a normative perspective under the condition that the

[9]This does not mean that everything functions smoothly and noiselessly. The example of the new system of regulating chemicals in the European Union, called REACH, in which the precautionary idea is more strongly anchored, demonstrates how difficult the implementation frequently is in view of massive interests.

[10]The challenges for synthetic biology, in which living nanoparticles may once be produced, look very different (Chapter 7).

above assessment does not change. This would, in accordance with the variations of a weak precautionary principle named above, mean "that there is a need for a cautious step-by-step diffusion of risk-related activities or technologies until more knowledge and experience is accumulated" (Haum *et al.*, 2004, p. 6).

This results in a proceduralization that takes the form of small steps toward extending our knowledge of the possible health or environmental consequences as rapidly as possible and in an ongoing evaluation and assessment of the knowledge situation. The resulting procedure would be an *incrementalism* that is reflective and obligated to constantly question itself. In order to operationalize this incrementalism (even the concept of which by the way collides with the frequent pathos of the revolutionary quality of nanotechnology), there are a number of conceivable measures for proceeding responsibly in the current situation. Among these measures are (extended after Schmid *et al.*, 2006; see also Maynard *et al.*, 2006):

- The development of a nomenclature of nanoparticles in order to be able to classify them according to toxic and other criteria as a means for escaping the pointlessness of risk assessment of individual types of particles.
- The recommendation that nanoparticles be handled like new chemical substances even though their chemical composition is known, with the respective consequences for requiring new approval procedures, for example in the area of foods.
- The development of verification and test methods for gathering necessary data in a rapid and comparable manner. This is lacking in the spheres of human medicine and ecology, which is a substantial limitation on the study of an impact.
- Research activities, in particular in environmental chemistry and toxicology, in order to overcome the deficits in our knowledge regarding the means by which nanoparticles are distributed, their reactions with the respective environmental media, the locations they end up in the medium and long term, and their consequences for the environment and our health.
- The avoidance of economic dependence on specific nanoparticles, which is important should unacceptable toxic properties of certain nanoparticles be proven. In this case, opportunities to replace them rapidly should be available without the replacement itself creating economic difficulties (this is an imperative of economic wisdom).

- The development of good practices and corresponding guidelines for dealing with nanoparticles, for example, at the workplace, such as by minimizing or avoiding the release of nanoparticles.
- The permanent observation and evaluation of the knowledge available worldwide about the effects that nanoparticles have on our health and the environment, in order not to repeat such failures known from the history of asbestos (Gee and Greenberg, 2002).
- An open and transparent discourse with the public.

Such an incremental approach that is accompanied by constant reflection on the level that has been achieved would be a very cautious integration of nanoparticles into the economy and into future material flows in the exchange between the circular economy and the natural environment. Yet this route, despite its cautiousness, is not without risk. The asbestos story teaches us that "'no evidence of harm' must not be re-interpreted in the way of 'evidence of no harm'" (Gee and Greenberg, 2002). Most important in this situation is taking maximum advantage of our learning opportunities. This means that the introduction of synthetic nanoparticles is understood as an open process in which the experience gained while carrying it out will be one factor that determines its future course. This will ensure that the relevant effects — should they occur — will be recognized rapidly, enabling consequences to be drawn, which would be a huge difference to the manner in which this problem was handled in the asbestos industry (Gee and Greenberg, 2002).

Risk ethics in this incrementalist process points to the fact that we must take into account who accepts which risks voluntarily, in which conditions this voluntariness exists, which risks we impose on people, how we decide about impositions and their criteria, how risks and chances are distributed, and what the opportunities are for weighing the chances versus the risks. Also ethically relevant is the question of whether one person's chances are not another person's risks. The example of green genetic technology, in which the chances are predominantly seen to be on the side of the food and agricultural industries while those of consumers are not perceived — yet precisely this group is thought to be burdened with possible risk — points furthermore to the everyday political significance of this dimension. In a normative sense, this significance is given strength by the deontological side (Nida-Rümelin, 1996b). Even so, several suggestions are derived for dealing with synthetic nanoparticles that are concerned with what it means to act responsibly given the great uncertainty in our knowledge of the consequences.

In addition to risk management, risk ethics, and regulation, public risk communication has to be observed carefully because irritation caused by this communication could have a dramatic impact on public acceptance and political judgment. The public perception of risk has to be taken seriously even in the case that it seems to be inadequate compared to the results of scientific risk assessment. Taking it seriously, however, does not imply simply accepting it, but starting an open dialogue about risk perception, patterns and underlying reasons of perception, acceptability, acceptance conditions, trust-building measures, etc. Denying the existence of risks from nanotechnology often causes mistrust and suspicion instead of creating optimism. A frank discussion of possible risks increases the chances for establishing a trusting relationship between science and society: for a public debate to build trust, it must be open about the chances and risks, and often there must be schemes available for comparing different types and amounts of risk (Renn and Roco, 2006). Transparency about the premises of different risk assessment exercises is urgently required. Knowledge about risks includes knowledge about the validity and the limits of that knowledge. The communicative and participative instruments of technology assessment (Decker and Ladikas, 2004) could help to improve mutual understanding and public risk assessment.

### 6.4.5 Epilogue and Reflection: Risk Ethics and Nanoparticles

Whether risk ethics is even possible is a matter of controversy. Some reject the possibility of attributing responsibility to a subject, arguing that personal responsibility is obsolete in view of the division of labor associated with the modern organization of technology, and of the functional differentiation of society into social systems (Bechmann, 1993). According to this view, moral communication does not have a place in the functional systems of society since they are oriented according to their own code (e.g., true/false in systems of science) and because they cannot absorb communication that does not employ this code. And, furthermore, since no moral system of its own has developed as a functioning societal system, moral communication does not have a place in the functional systems (such as politics, economics, law, or science) and thus leads at best to white noise (Luhmann, 1989) or to edifying Sunday sermons of no consequence (Beck, 1988). Finally, according to this view, the object of risk ethics, namely, the uncertain risks, is so uncertain that the object of responsibility disappears together with the uncertainty regarding the consequences (Bechmann, 1993; Chapter 10).

The preceding analysis has indeed demonstrated that ethical and thus conditionally normative reflection cannot solve the problems of dealing with nanoparticles and unclear risks. Yet it would be demanding too much to expect ethical reflection to provide the *solution* to societal and political problems. Valuable advice can at best offer a solution to a problem, but as a rule it may not implement it because of legitimation problems. It can also consist in pointing to ways to reach a solution to a problem and to related issues that in this way should be taken into consideration. And a few pointers to this end have in fact been produced in this conception that is modest yet commensurate to the reflective nature of ethical advice being conditionally normative (Section 3.1).

The initial point is to recognize that ethical reflection draws attention to the shortsightedness of some positions. First, a wait-and-see strategy cannot be carried out without injuring the potential rights of persons who do not have a chance to defend themselves, not even by casting their vote in a democratic process. Second, an argument would not be sufficient that operates with the de facto approval of the use of nanoparticles by today's society. For even if, for example, it would be correct that everyone in today's society would concur with an argument of something's harmlessness (until the opposite is proven true), from an ethical point of view it would be important also to reflect on the fact that this approval is merely de facto and that those affected by these risks in the future might not concur. If *future* actors and people affected by a weighing of alternatives are not per se to be worse off than those alive today, then limits must be set to the potency of arguments built on the facticity of the existing. Third, this implies lastly that an argument about a labeling requirement for products containing nanoparticles cannot suffice since the users who on this basis would approve of their use within the framework of informed consent might not be the ones affected, at least they might not be the only ones affected. This results in at least the following pointers from considerations about risk ethics:

- Frequently the issue is to compare newly developing situations — such as dealing with nanoparticles — with familiar ones in order to learn from the familiar for the new. Since no quantitative comparisons between the various risks from nanoparticles or between the risks and chances can be undertaken, the question turns to qualitative comparisons — i.e., those that are led by analogy — in order to normalize the situation.
- The different concepts in risk ethics point to different aspects of the problem, all of which have to be taken into consideration. Jonas (1984), for example, calls for taking possible apocalyptic scenarios into consideration, Gethmann and Mittelstraß (1992) the de facto approval relationships to risks, and Nida-Rümelin (1996b) the possible violations of rights. The survey of ethical concepts contributes in this way to a comprehensive understanding of the ethically relevant aspects of the situation of interest to us.
- This uncovers analytically the normative factors underlying the decision-making situation with reference to the aspects that at the current focus of reflection. The task of risk ethics is to uncover the implicit ethical and methodological assumptions contained in risk assessments and to subject them to an explicit analysis (Shrader-Frechette, 1991). It should in particular be noted that as a rule normative premises are contained even in the methodological aspects of registering risk (so for example in the assumptions of quantifiability or the corresponding rules).
- It becomes clear that in the case of nanoparticles not only the rights of the currently living could be violated (e.g., of consumers whose autonomy with regard to the risks to be taken could be protected by a labeling requirement), but also those of future generations. Responsibility for the future must be taken into account.

- An essential point proves to be judgment about how obvious and sufficient the available knowledge is with regard to the possible consequences of the release of nanoparticles. The ethical questions themselves often retreat behind the challenges of assessing the extent and the gaps in our knowledge. For this reason, it is to a certain extent justified for people to refer to the significance of evaluating the level of our knowledge regarding this type of problem.[11] At the least, special attention must be given to this point in conditionally normative reflection.

Even if the possibilities of risk ethics are limited and we must warn against exaggerated expectations from providing conditionally normative advice, the results of the studies named above have contributed significantly to a better understanding of the normative reasons for decisions and of the implications of these decisions for the further handling of nanoparticles. This helps satisfy the conditions for being able to conduct an enlightened democratic debate. This is after all where the formation of opinion on such issues — a political problem — ultimately belongs. An essential function of providing conditionally normative advice is in the *hermeneutic* dimension: at issue is the *enlightenment* of the debate on nanoparticles with reference to the normative questions that are involved — and viewing this as a contribution to deliberative democracy (Grunwald, 2008a).

[11] One such demand is: *From the Ethics of Technology to Knowledge Assessment* (von Schomberg, 2007). For an example of such assessment procedures, see Fleischer *et al.*, 2010; see also in general Chapter 10.

# Chapter 7

# Toward Creating Artificial Life

The goal of nanobiotechnology — using nanotechnology to access biological systems — is to employ technology to influence and shape living systems to a greater degree. It even offers the perspective of artificial life being created by means of synthetic biology. The question whether such developments are responsible has been raised in intensified form in view of the ever increasing extent of man's interventions in nature (Jömann and Ach, 2006). This chapter will first describe nanobiotechnology and synthetic biology (Section 7.1) and name the chances and risks that they pose as far as something can already be said about them (Section 7.2). It will then analyze the starting points of ethical reflection, focusing on the normative uncertainty that may be developing (Section 7.3). This is followed by hermeneutic questions as to the changing relationship between technology, nature, and life (Section 7.4) and by conclusions for dealing responsibly with the current situation of research (Section 7.5).

## 7.1 Nanobiotechnology and Synthetic Biology

The concept of "nanobiotechnology" (Section 7.1.1) — the word "bionanotechnology" is also used occasionally (Jotterand, 2008a; Goodsell, 2004) — was coined in the United States in the context of the National Nanotechnology Initiative (NNI, 1999). Synthetic biology (Section 7.1.2) employs the knowledge of nanobiotechnology with the goal of creating new forms of life. In 2010, Craig Venter,

*Responsible Nanobiotechnology: Philosophy and Ethics*
Armin Grunwald

ISBN 978-981-4316-80-4 (Hardcover), 978-981-4363-33-1 (eBook)
www.panstanford.com

one of the pioneers of synthetic biology, announced that he had successfully implanted artificial DNA into a bacterium. Synthetic biology has rapidly become known to the public (Synth-Etics, 2010), and it has already led to a series of activities on its ethical, legal, and social implications (ELSI).

### 7.1.1 Nanobiotechnology

In the last few years, nanotechnology has become the focus of lively interest in science and research policy, and, increasingly, of media and public interest. Fascinating prospects are the possibility of designing materials at the molecular and atomic levels and making completely new product features available, including some at the macroscopic level, and also the further miniaturization of components, products, and methods, extending even to the construction of nanomachines. They open far-reaching possibilities for applications. Nanobiotechnology is characterized by the fact that — in addition to the usual elements of a definition of nanotechnology (see Section 2.3) — biological components are elements of the application (according to the Association of German Engineers VDI, 2002). A distinction is made between "Nano2Bio," in which use is made of nanotechnology for the analysis and production of biological nanosystems, and "Bio2Nano," which refers to the use of biotic materials and structures for the production of technical nanosystems.

The starting point is that basic life processes take place on a nanoscale because this is precisely the size of life's essential building-blocks (e.g., proteins). Nanobiotechnology (VDI, 2002; Roco and Bainbridge, 2002) will make it possible to control biological processes by means of nanotechnology. Molecular "factories" (mitochondria) and "transport systems" can — precisely because they play an essential role in cellular metabolism — be models for controllable bionanomachines. At this level, nanotechnology could make it possible to engineer cells. Networking natural biological processes with technical ones seems conceivable. There are visions of nanomachines in cellular and subcellular processes, which might take the form of mechanisms for producing energy, molecular factories and transport systems, or high-capacity data storage and data reader systems. Functional biomolecules would then act as components of units for gathering and transforming light, or as signal converters, catalysts, pumps, or motors. In nanobiotechnology, the language of engineering is extended to the nano level, just as in biotechnology it was applied to biological systems: "The fact that biological processes are in a way dependent on molecular machines and clearly defined structures shows that building new nanomachines is physically possible" (Kralj and Pavelic, 2003, p. 1011).

A characteristic example is the attempt to create a technical replica of photosynthesis (Cheng *et al.*, 2010). Plants and some forms of bacteria assure their energy supply by means of photosynthesis. Sunlight is used to synthesize complex carbohydrates from carbon dioxide and water, which serve both for energy storage and as energy supply. In contrast to current photovoltaic cell technology, this principle even functions in diffuse or very weak light. The idea of using the principle of photosynthesis, as it has developed in the course of evolution, to technically ensure man's energy supply is exceptionally appealing. Energy supplied on the basis of this principle would be $CO_2$ neutral, would be easily storable, could be produced in a decentralized fashion, would be practically inexhaustible, and would not produce any problematic waste. Nanobiotechnology provides the techniques needed to understand the natural processes at the molecular level and possibly to be able to replicate them. It concentrates on replicating the simpler manner of functioning in bacteriochlorophylls, relying on the principle of *self-organization* for the formation of the corresponding nanoscale structures (nanotubes or tube-like structures; Balaban and Buth, 2005).

There is hope that such research can contribute to the development of artificial antennas that can even function in weak and diffuse light. They could then be useful for the design of hybrid solar cells based on economical polymer technologies (Balaban and Buth, 2005, p. 207). Such research is however still entirely in the sphere of basic research. The point is to understand essential processes in the context of their technical functioning. Designations such as "light harvesting complex" or "proton pump" demonstrate the technical view of photosynthesis (Cheng *et al.*, 2010).

### 7.1.2 Synthetic Biology: Engineering Life

Synthetic biology has recently turned into a vibrant field of scientific inquiry. The combination of engineering with biology promises to make it possible to fulfill many of the goals expected of nanotechnology in an even easier fashion: while nanotechnology involves the development of materials and machines at the nanoscale, synthetic biology builds on the insight that nature already employs components and methods for constructing machines and materials at very small scales. Synthetic biologists hope, both by employing off-the-shelf parts and methods already used in biology and by developing new tools and methods, to develop a set of tools to hasten the advent of the various promises of nanotechnology (Synthethics, 2010).

The knowledge gathered by nanobiotechnology can be used *to produce new* functions in living systems by modifying biomolecules or the design of cells, or designing artificial cells. The traditional self-understanding of biology, which is molded by the natural sciences and which aims to *understand* vital processes, is reinterpreted by synthetic biology (Ball, 2005; Woese, 2004) as a *new invention* of nature and as the creation of artificial life on the basis of our knowledge about "natural" life. This transforms biology into a technical science (de Vriend, 2006) that embodies the dual strands of cognition and design and that is subordinate to the primacy of design goals, just as in the classical technical sciences. In this new form of biology, "the pre-existing nanoscale devices and structures of the cell can be adapted to suit technological goals" (Ball, 2005, p. R1).

Synthetic biology differentiates between an approach that uses *artificial* molecules in order to reproduce biotic systems and one that uses the elements of classical biology and combines them to form new systems that function in a "non-natural" manner (Benner and Sismour, 2005). The force behind this is the thought of creating artificial life or a technically modified form of life that is partially equipped with new functions: "how far can it [life] be reshaped to accommodate unfamiliar materials, circumstances, and tasks?" (Ball, 2005, p. R3). Examples of these efforts range from the design of artificial proteins to the creation of virus imitations or the reprogramming of viruses, and even extend to attempts to program cells to perform desired functions (Ball, 2005; Benner and Sismur, 2005, pp. 534–540).

Various suggestions have been made for defining synthetic biology, all of which point in the same direction despite some differences in nuances (on definitions in general, see de Vriend, 2006, pp. 13ff.). Accordingly, synthetic biology is:

- The design and construction of biological parts, devices, and systems, and the redesign of existing, natural biological systems for useful purposes (LBNL, 2006).
- The design and synthesis of artificial genes and complete biological systems, and the modification of existing organisms, aimed at acquiring useful functions (COGEM, 2006).
- The engineering of biological components and systems that do not exist in nature and the re-engineering of existing biological elements; it is determined by the intentional design of artificial biological systems, rather than by the understanding of natural biology (Synbiology, 2005).

A characteristic feature of each of these definitions is the turn to artificial forms of life — whether they are newly constructed or produced via the redesign of existing life — each of which is associated with an expectation of a specific utility. Since the diameter of DNA, a typical object of technical operations, is approximately 2 nm and since such technical operations are thus part of nanobiotechnology, synthetic biology can be considered a part of nanobiotechnology: "synthetic biology could be considered a specific discipline of nanobiotechnology" (de Vriend, 2006, p. 23). In turn, synthetic biology can be viewed as the continuation of molecular biology with the means of nanotechnology.

The starting point of synthetic biology is for it to model biotic units as complex technical relationships and to break them down into simpler technical ones (deconstructing life according to de Vriend, 2006). While this is still, so to speak, a form of *analytic* biology, it becomes a *synthetic* one when the knowledge about individual processes of life that has been obtained from technical modeling and the corresponding experiments are combined and utilized in a manner so that certain "useful functions" can be achieved as a dedicated result (de Vriend, 2006). "Seen from the perspective of synthetic biology, nature is a blank space to be filled with whatever we wish" (Boldt and Müller, 2008, p. 388). Cells are interpreted as being machines that consist of components. The language and approach of mechanical engineering are transferred to living systems. Following the tradition of the idea of technical standardization, the

Registry of Standard Biological Parts was founded at MIT. Gene sequences are stored there as drafts for different cellular machine components. Using the dedicated design of artificial cells based on such components, micromachines are supposed to be produced that can, for example, process information, manufacture nanomaterials, or make medical diagnoses. Entirely in the tradition of mechanical engineering and electrotechnology, such machines are supposed to be built part by part according to a design drafted top-down. The result is supposed to be a functioning entity:

> Engineers believe it will be possible to design biological components and complex biological systems in a similar fashion to the design of chips, transistors, and electronic circuits. (de Vriend, 2006, p. 18)

There are also approaches for utilizing the principles of evolution to achieve certain new effects. For example, cells could be subject to the pressure of artificial evolution by turning off the genetic sequences responsible for the building of certain amino acids. By adding chemical substances that are chemically sufficiently similar to the missing amino acids, the cell can be brought to use the substitutes in place of the amino acids. The result of this is a cell with modified properties. Here there is a tight interface to systems biology (Bruggeman and Westerhoff, 2006), in which the complex interaction of the many individual processes is to be understood as a complex entity:

> Systems biology is crucial to synthetic biology. It includes knowledge about the natural basic biological functions of RNA and DNA sequences in information storage, energy supply, membrane functions, cell structure, cell-to-cell signalling, gene regulation (gene expression), and metabolic functions in natural systems. (de Vriend, 2006, p. 23)

Decisive for scientific and technological progress is the *combination* of knowledge about molecular biology and genetic techniques with the new opportunities offered by nanotechnology. The prerequisite for a precise design of artificial cells would be a sufficiently thorough understanding of all the necessary subcellular processes and interactions. The current state of knowledge is still far from this. The purpose of much of the research and development in nanobiotechnology is to provide such knowledge (Schmid *et al.*, 2006).

## 7.2 Chances and Risks

It is still too early to provide a comprehensive and concrete description of the chances and risks posed by developments in nanobiotechnology and synthetic biology. Both of these possibilities probably exist most clearly in the areas of environmental science and health: "It [synthetic biology] has potential benefits, such as the development of low-cost drugs or the production of chemicals and energy by engineered bacteria. But it also carries risks: manufactured bioweapons and dangerous organisms could be created, deliberatively or by accident" (Church, 2005, p. 423). Although synthetic biology did not become known outside a circle of experts until 2006, a public discussion developed rapidly (Synth-Etics, 2010). The first projects and activities were started on its social and ethical issues.

### 7.2.1 Chances

The aim of synthetic biology is to design biological systems for numerous useful purposes, such as for the production of alternative fuels or new materials (e.g., biodegradable plastics). "Artificial organisms" might also be used in medicine, for example, as biosensors, or they can help to degrade environmental pollutants. The almost complete absence of restrictions limiting possible applications means there is immense room for thinking about applications. Synthetic biology could thus provide a considerable contribution to solving many urgent problems — with all the problems involved however in statements of the type "it could be" (Chapter 10).

The applications of nanobiotechnology are in medicine, the life sciences, and biotechnology. In medicine, items that deserve to be emphasized (see Paschen *et al.*, 2004 for details) are new diagnostic means (such as biosensors and the possibility of permanently monitoring the status of someone's health), developments in drug delivery (e.g., the procedure for precisely transporting active substances and targeting their deposition), and the use of new biocompatible materials and surfaces (e.g., Freitas, 2003). The following areas of application can be distinguished in synthetic biology:

- Nanomanufacturing and nanostructuring using bio-based methods. The goal is to use the principle of the

self-organization of molecular units to create more complex structures in order to achieve certain functions (biomineralization).
- Production of the desired chemicals by means of bacteria that are manufactured or modified in a dedicated manner (plan drug factories; de Vriend, 2006, p. 30).
- The technical utilization of functional biomolecules and of hybrid systems (e.g., of biomolecular motors and actuators) in technical systems or in combination with nonbiological components.
- The realization of functions at the interface between biological and technical materials and systems (e.g., for neuroimplants or prostheses).

These areas of application and the corresponding functional combinations make it possible for biosensors and biomembranes, for instance, to be employed in environmental technology, and for photoenergetic processes to provide biological support for photovoltaics. For example, it would obviously be economically and ecologically significant if synthetic biology were technically able to copy photosynthesis (see above). Scenarios for similarly far-reaching applications can also be found in biomedicine. Furthermore, the techniques developed by synthetic biology could be used in the production of biofuels.[1] Solving the energy problem in a sustainable manner is one of the great challenges facing today's and tomorrow's energy policy and technology development. The creation of synthetic organisms that could help produce biofuels in an efficient way and that would not lead to conflicts with the food supply is an attractive idea to many people.

The far-reaching hopes placed in the technical use of processes of self-organization, independent of the concrete fields of application, constitute a point that deserves a great deal of attention. The utilization of phenomena of self-organization, including the possibility of replicating "living things" that have been created or modified by technology, is a central aspect of synthetic

[1]The problem of equity (Section 5.3) will also apply to synthetic biology. Issues of just distribution will arise with, for example, the development of novel ways to produce bioethanol. An issue that is not referred to very often is: "In what way are issues of land use and of economic equality or inequality enhanced or decreased by the introduction of synthetic biology in third world countries" (Synthethics, 2010).

biology and leads to self-organization becoming increasingly significant: "The paradigm of *complex, self-organizing systems* envisioned by von Neumann is stepping ahead at an accelerated pace, both in science and in technology [...]. We are taking more and more control of living materials and their capacity for self-organization and we use them to perform mechanical functions" (Dupuy and Grinbaum, 2004, p. 292; this leads the authors to express serious concerns about the risks; see Section 7.2.2).

There are also economic reasons for the fascination exerted by self-organization. If we could succeed in making materials organize themselves in such a manner that technically desired functions or properties would result, this might be substantially less expensive than if we had to specifically create them, such as by means of a nanotechnological top-down procedure in which atoms are given a specific arrangement. The goal would thus be to substitute processes that run by themselves for construction processes that are designed laboriously by manual action. Human intervention would thus be moved to a different level. The builder would become the controller of a nature that carries out the construction:

> The self-assembling properties of biological systems, such as DNA molecules, can be used to control the organisation of species such as carbon Nanotubes, which may ultimately lead to the ability to "grow" parts of an integrated circuit rather than having to rely upon expensive top-down techniques. (de Vriend, 2006, p. 19)

### 7.2.2 Risks

The primary fears of risks connected with nanobiotechnology and synthetic biology concern possible, unintended, negative consequences for our health and the environment, but also the intended utilization of this technological potential for novel biological weapons. "In the case of synthetic biology, specific risks in need of close scrutiny and monitoring are uncontrolled self-replication and spreading of organisms outside the lab, and deliberate misuse by terrorist groups or individuals or by 'biodesigner-hackers'" (Boldt and Müller, 2008, p. 387). These threats stand for two general areas, namely, safety and security. In our context, the former refers to harms that occur naturally while the latter concerns harms that occur intentionally. The World

Health Organization (WHO) defines biosafety as "the containment principles, technologies and practices that are implemented to prevent the unintentional exposure to pathogens and toxins, or their accidental release" (WHO, 2006). Biosecurity in a laboratory context is regarded as "control and accountability for valuable biological materials [...] within laboratories, in order to prevent their unauthorized access, loss, theft, misuse, diversion or intentional release" (WHO, 2006).

Let us first look at biosafety. A well-known scenario from the debate about organisms modified by genetic technology concerns a product of synthetic biology (e.g., an artificial or modified virus) that might escape from a laboratory and cause considerable risks to our health or the environment without there being a way of recapturing it. These are risks that involve biological safety (de Vriend, 2006). Beyond posing an immediate danger to living species and individuals, the genetic pool of certain species might be contaminated as a result of genetic transfer, which would thus lead to ongoing and irreversible modifications.

A major cause for such expectations is that synthetic biology could create living things that are alien to the natural biosphere and that we do not possess any evolutionary experience for dealing with them in ecosystems and organisms (according to de Vriend, 2006). The consequences of the release of such partially or completely invented living things might thus be impossible to anticipate and to calculate. One — accurate — response to this is that the probability is very small that such artificial living things could survive and cause damage in the natural world because they are not adapted to natural processes. But even a small probability is a finite one, which is why problems of this nature do require careful observation. A "green goo" was described as the scenario of such a catastrophe, in analogy to the well-known "gray goo" thought by some to be a possible consequence of our losing control of self-replicating nanorobots (Joy, 2000): "The green goo scenario, somewhat more likely, if only because it is more easily in reach, suggests that a DNA-based artificial organism might escape from the lab and cause enormous environmental damage" (Edwards, 2006, p. 202f).

This scenario shows the ambivalence of referring to the potential of self-organization. Although self-organization has been used above all to refer to capacities that are considered positive and whose technical use promises many advantages, especially economic ones, the capacities of living systems for self-organization and for self-replication also pose a special type of potential threat. As the debate about the gray goo scenario in nanotechnology showed, positive visions can quickly turn into negative ones (Schmid *et al.*, 2006). The technical utilization and organization of living processes lead us to a fundamental and inherent risk, namely that we can lose control of living systems in a much more dramatic manner than we can of technical systems because they pass on their properties and can multiply.

Following the debate about the risks of genetically modified organisms, the precautionary principle was developed and implemented in European law for dealing with challenges posed to biosafety (see on this von Schomberg, 2005 and Synthethics, 2010). Within the framework of a gradual step-by-step procedure and with a commitment to carefully examine at every step the consequences of that step, a layered containment strategy was established. Accordingly, the research initially takes place under high-security conditions, subsequently in "normal" laboratories, then in controlled open experiments, and finally, currently on application, in controlled plantings that keep a minimum distance from fields used for nongenetically modified agriculture. This has led to a successive reduction in our initially complete ignorance about the consequences of genetically modified organisms in our natural environment.

The essence of the biosafety argument is that synthetic biology, by intentionally creating new or technically modified living objects, intervenes in the course of natural evolution to a considerable degree. While natural evolution only takes place in small steps, and large modifications only take place over extremely long periods of time, man is now endeavoring to take control of evolution in just a few years or decades by creating artificial life: "Ponder for a moment the incredible hubris of the entire endeavor of bionanotechnology. The natural environment has taken billions of years to perfect the machinery running our bodies and the bodies of all other living things. And in a single generation we usurp this knowledge and press it to our own use" (Goodsell, 2004, p. 309). In this sense, the massive *acceleration* of natural development by means of nanobiotechnological procedures is a special challenge that even within the field of synthetic biology is viewed as a risk factor. Artificial cells — even if they are based on knowledge gained from natural cells — will possibly only have a few years of experiments in a laboratory behind them, not millions of years of evolution. The new construction of cells or the reprogramming of viruses must be placed under special observation.

Finally, the complexity of the processes in molecular biology per se poses a gateway for potential risks, as has already been shown in the initial experiments with forms of gene therapy:

> A more concrete rationalization of this distrust is the feeling that unpredictable consequences can follow from rearranging a complex system that is not fully understood. The recent history of gene therapy offers us a cautionary tale. To introduce genes into patients, it has been necessary to use what is essentially a piece of nanotechnology — an adapted virus is used to introduce the new genetic material into the cell. But fatalities have occurred due to unexpected interactions of the viral vector with the cell. (Jones, 2004, pp. 214ff.)

The biosecurity argument focuses on the possibility of the *intentional* construction of novel biological weapons on the basis of newly constructed or modified cells (de Vriend, 2006, p. 54). The products or techniques of synthetic biology, or perhaps only the knowledge produced by it, might be misused for military purposes in government weapons programs or by terrorists. Our fantasies have more or less free rein with regard to the concrete possibilities, and in view of the fact that hardly anything is known about military programs of this nature, there is a danger of chasing after conspiracy theories. One consequence of this state of ignorance is that no purely imagined details should be made known, such as those regarding intentionally reprogrammed viruses. On the contrary, the issue should be the ethical aspects of the *possibility* of such developments. In this respect, it should be taken into consideration that high-technology techniques must be applied in order to create synthetic biological weapons: "Contrary to popular belief, however, a biological weapon is not merely an infectious agent but a complex system consisting of (1) a supply of pathogen [...]; (2) a complex formulation of chemical additives that is mixed to stabilize it and preserve its infectivity and virulence during storage; (3) a container to store and transport the formulated agent and (4) an efficient dispersal mechanism" (Tucker and Zilinskas, 2006, p. 39).

This means that it is by no means an easy matter to produce and use such biological weapons. Since the scientific and logistic effort required would be considerable, such a development is rather improbable among terrorists. Nonetheless scenarios have been created even for this risk in which synthetic biological weapons could be created by, for example, a fanatic lone operator, an expert in synthetic biology who could misuse his specialist knowledge to develop such living objects out of personal hate and to use them against the object of his hate, or by a biohacker who, analogous to computer hackers, could construct a damaging virus just to demonstrate that it was technically possible or to rouse public awareness (de Vriend, 2006). And there is also the military, which in some countries has almost limitless resources at its disposal and hardly any logistic problems. In this sense and despite all the uncertainty and improbability involved, the conscious creation of harmful synthetic or technically modified living objects must indeed be considered a potential risk of synthetic biology.

The situation is made more difficult by the dual use dilemma. It "arises in the context of research in the biological and other sciences as a consequence of the fact that one and the same piece of scientific research sometimes has the potential to be used for harm as well as for good" (Miller and Selgelid, 2006). Of special relevance to synthetic biology are the dual use concerns raised by the *publication* of research relating to biological agents with a large capacity for harm. Key examples given in the literature are the publication of the genomes of engineered virulent mousepox virus, polio virus (Selgelid, 2007), and the 1918 influenza virus (Sharp, 2005). The dual use dilemma means that "publishing studies like these [mousepox and polio] both alerts potential bioterrorists to possible biological weapons and actually gives them explicit instructions to produce them [while] the mousepox and polio studies may advance medical science and biodefense preparation" (Selgelid, 2007).

All these arguments show that the risks associated with synthetic biology are clearly "unclear" risks in the sense of Section 3.3: neither quantitative data nor clear knowledge about cause–effect relationships is available.

## 7.3 Ethical Issues

In view of our experience with genetically modified organisms and their regulation, and of the often speculative nature of reflections on the consequences of synthetic biology, it is not immediately clear what the *specific* challenges are that synthetic biology poses to ethical reflection. In other words, what are the situations of normative uncertainty? A general thesis asserting that the products of synthetic biology are value neutral, such as the following assertion

> The technology of Synthetic Biology provides a new set of tools. Any ethical challenges come from the way we use the tools and not from the tools themselves. (Chopra and Kamma, 2006)

is however inacceptable under all circumstances, especially since such assertions of value neutrality are not even accepted in the traditional sphere of engineering (e.g., Jömann and Ach, 2006; van Gorp, 2005). The ethical issues posed by synthetic biology can be classified according to the different normative frameworks that are affected: the question regarding how to deal with risks comes after precautionary considerations (Section 7.3.1), that regarding the moral status of artificial living things follows bioethical considerations (Section 7.3.2), and that regarding the human hubris — or "playing god" — requires an analysis of its own (Section 7.3.3).[2]

### 7.3.1 Dealing with Risks Responsibly

The uncertainties about the applications of synthetic biology mean that it is senseless to go into *concrete* risks at this point. What however is possible and has already provided cause for expressions of concern is to reflect on types of risk that can be linked specifically to synthetic biology. The first characteristic that can be named in this regard is that synthetic biology concerns interventions in or constructions of *living* systems. These systems are, in contrast to abiotic synthetic nanoparticles (Chapter 6), characterized by self-organization, they can (as a rule) reproduce and multiply, they can modify their properties (within limits) as a consequence of evolutionary pressure, and they can exchange substances as well as gene sequences with their environment. Subsequently, such a risk scenario takes on a completely different character than, for example, those about synthetic chemicals or nanoparticles.

The idea of self-organization has already served as an inspiration in the visions of a molecular assembler (Drexler, 1986), namely at

[2]The definitions of synthetic biology (cf. Section 7.1.2) sketch the field as something radically new. Does this also mean that we need a new ethics of such a discipline? Arguments apply here that are very analogous to those in the debate as to whether there has to be a new field of nanoethics (Chapter 4). They show that the fundamental ethical issues are in no way radically new, but rather that they demonstrate differences of degree and urgency.

the beginning of the history of nanotechnology (Section 2.1). This assembler was supposed to be self-replicating and produce as part of its propagation further machines, which in turn could build further machines themselves. This idea, which Drexler considered to be the solution to practically all of mankind's problems, was reinterpreted by Bill Joy as a horror scenario to humankind in which the future no longer needs man because development continues even without man (Joy, 2000). In contrast to the case of synthetic nanoparticles (Chapter 6), it is at least conceivable that the creation of synthetic living things,[3] as described in such scenarios, can pose a danger to "the entirety" in the sense described by Hans Jonas (1984) on the basis of the heuristics of fear. The issue below initially concerns our grasp of how strongly the situation of synthetic biology — as described here — deviates from the challenge posed by dealing with synthetic nanoparticles (cf. Chapter 6). Is it nonetheless possible to infer anything from that analysis about the role that the precautionary principle can play for nanobiotechnology and where its possible limits lie?

Dupuy and Grinbaum (Dupuy and Grinbaum, 2004; Dupuy, 2007) have undertaken a very trenchant interpretation without however referring concretely to synthetic biology. In the following, I attempt first to take this interpretation seriously and then to deconstruct it pursuing a critical goal. This effort follows from my evaluation that the problems of nanotechnology that the authors diagnose and that they react to could most probably arise in nanobiotechnology, and in particular in synthetic biology.

According to Dupuy and Grinbaum, the growing significance of self-organization in nanotechnology means ultimately that it inevitably has to lead to a catastrophe. They maintain that its complexity means that this development can be neither predicted nor controlled. Their argument consists in drawing attention to the necessity of the *deed* before it is possible for us to think about controlling its consequences. "The unpredictable behaviour of nanoscale objects means that engineers will not know how to make nanomachines until they actually start building them" (*The Economist*, cited after Dupuy and Grinbaum, 2004, p. 288). Regarding

[3]Lost in the process is the difference between synthetically manufactured nanorobots as a manifestation of technology and synthetically manufactured living things, such as viruses. The fears presented in the following refer to the shared property of self-organization and are independent of whether we catalog the objects of fear as technology, biofacts, or living things.

options for control, the authors then continue: "The engineers of the future will be the ones who know they are successful when they are surprised by their own creations" (p. 289), and in consequence "it is neither by error nor by terror that Man will be dispossessed of his own creations but by design" (p. 291).

They conclude from this diagnosis that the classic techniques for dealing with uncertainty and unclear risks (Section 6.2) are insufficient: "We believe that none of these tools [utility analysis, scenario technique, precautionary principle, A.G.] is appropriate for tackling the situation that we are facing now" (*The Economist*, cited after Dupuy and Grinbaum, 2004, p. 293). More research will not necessarily increase certainty from our knowledge, but rather create more uncertainty:

> In cases where uncertainty is such that it entails that uncertainty itself is uncertain, it is impossible to know whether or not the conditions for application of the precautionary principle have been met. If we apply the principle to itself it will invalidate itself before our eyes. (Dupuy and Grinbaum, 2004, p. 294)

The issue for the authors is instead to replace the open nature of the future, represented as tree diagrams on a linear time scale, by a projected time in which the connection between today's talk of the future and the bringing forth of a future present is established. The means of a self-fulfilling prophecy is to be used in this manner to "set" the future. "We will call *prophecy* the determination of the future in projected time, by reference to the logic of self-fulfilling prophecy" (Dupuy and Grinbaum, 2004, p. 303). Even exceeding the gloom of Hans Jonas' "heuristics of fear" (1984), Dupuy and Grinbaum assert a duty to anticipate the catastrophe in order to prevent its occurrence:

> A catastrophe must necessarily be inscribed in the future with some vanishing, but non-zero weight, this being the condition for this catastrophe *not* to occur. The future, on this part, is held as real. This means that a human agent is told to live with an inscribed catastrophe. Only so will he avoid the occurrence of this catastrophe. (Dupuy and Grinbaum, 2004, p. 310)

The interpretation presented by Dupuy and Grinbaum fails however with respect to its negative diagnosis on the precautionary principle. The sentence that this diagnosis is built on is not tenable:

> In truth, one observes that applications of the "precautionary principle" generally boil down to little more than a glorified version of "cost–benefit" analysis. (Dupuy and Grinbaum, 2004, p. 293)

The authors fail to recognize that the precautionary principle is a *political* principle, neither a specification for a calculation following a preceding quantification, nor a legal specification that is easily applied (see the detailed presentation in Schomberg, 2005 and in Section 6.4). The argument cited above, namely, that application of the precautionary principle to itself makes it obsolete, is also wrong because the precautionary principle is constructed precisely for this case.[4] Beyond the precautionary principle there is no room for a rational dispute about possibly problematic futures. The limits to the rationally possible and negotiable are already sounded out in the precautionary principle itself. This means that the precautionary principle is the appropriate framework for addressing risks even in the case of synthetic biology.

We can directly build on the experience gained in working with genetically modified organisms (Synthethics, 2010). In view of the facts that compared to traditional gene technology, synthetic biology leads to a further increase in the depth of man's interventions in living systems and that the pace of innovation continues to increase, the precautionary measures will tend to become even stronger, in as much as we operate in the same normative framework. The responsible action of scientists will form a major issue in the run-up to genuine regulation (Section 7.5).

### 7.3.2 The Moral Status of Created Organisms

The production of new living things or technically strongly modified ones by synthetic biology will raise the question of their moral status. Insofar as with respect to its moral status — and various bioethical positions differ on this considerably — a difference in principle is made between the living and nonliving objects of ethical reflection, the question will be whether synthetically produced living things are also accorded this moral status. Would synthetically manufactured living things also be accorded reverence for life in Albert Schweitzer's sense? This implies first the question of whether

[4]This would undoubtedly cause specific problems in operationalization, which would increase the difficulties of management. Yet this is no argument against the precautionary principle.

a living thing in itself (i.e., for no other reason than that it is alive) is accorded a special moral status, such as dignity or a value of its own, as opposed to a value relative to purposes set by man. According to the biocentristic position of bioethics, even the justification of technical manipulation of "lower" forms of life could be considered problematic, or it could be asked if synthetic living things are supposed to have the same moral status as natural ones.

> Given that creations of synthetic biology can qualify as living organisms designed to be useful, can they be regarded solely as having instrumental value and being devoid of any intrinsic value? (Bhutkar, 2005)

Aside from several specific issues, the claim that life has an intrinsic value simply because it is natural faces a simple but severe problem: if life has a value of its own, why should it be relevant whether it is natural or artificial? Simply to insist that, for example, nature earned this value or dignity through the process of evolution is no argument at all (Link, 2010). If life had an intrinsic value solely because it is natural, why should it be questionable to create a living organism lacking such an intrinsic value? One worry could be that we might lose respect for life in general by getting used to considering organisms as living machines (cf. Boldt *et al.*, 2009). But that would be an indirect slippery slope argument based on mere speculation.

This skips the question of whether life itself has an intrinsic value and if and how synthetic biology could come into conflict with it. For many people, the conviction that life has an intrinsic value is deeply rooted in their intuitions. This would request to protect life for its own sake. But from which harm could or should one protect a cell or a bacterium? Provided that a living being lacks any ability for subjective experience, it is impossible to inflict pain on it. In other words, living organisms, as for instance, bacteria, simply fail to have a relevant "basis" for being protected against harm (Link, 2010).

Bacteria and viruses — which themselves or their components are the objects of work in synthetic biology (see numerous examples in ETC, 2007) — are not the proprietors of rights in the customary ethical categories that could be harmed. In this sense, technical manipulations of these living things would not be ethically relevant as manipulations *of them* but only with regard to the possible consequences of these manipulations *for humans* (e.g., in the form of risks; see Section 7.2.2). The same would apply, *a fortiori*, also to synthetic or technically modified living things if they could be produced despite the objections mentioned above. In as much as synthetic biology is concerned with living things that are not accorded any rights, ethical issues therefore are raised at the most with respect to the consequences for humans and the environment, not however with respect to the technical manipulations of such life itself.

It is another question whether synthetic biology might come into conflict with an assumed intrinsic value of life. The goal of synthetic biology is to *create* life, not to destroy it. It is therefore hardly imaginable that synthetic biology could come in conflict with the intrinsic value of life. Maybe the mere act of creating life is morally questionable, but that is different question and does not hinge on the intrinsic value of life (Link, 2010).

More far-reaching ethical questions are posed by the technical modification or creation of higher organisms. Regarding the issue, in particular, of the capacity of animals to experience pain, the focus of ethical consideration is not just on the consequences of technical interventions for humans but also on the affected organisms themselves (pathocentric position). The question of what the consequences of such technical manipulations for individuals are — such as the administration of pharmaceuticals that promote muscle growth or genetic interventions — is a topic that has already been frequently raised. Even if one does not take a strong position on animal ethics issues, this is evidently a field in which normative uncertainties can manifest themselves with regard to how higher organisms are handled as a consequence of measures of synthetic biology (or at least their possibility). The ground has been well prepared, however, with regard to how these uncertainties will be handled in the framework of ethics as a result of the animal ethics debates over how animals are treated in the agricultural industry and of the ethical problems related to animal experiments (Chapter 8).

### 7.3.3 Quasi-ethical Concerns: Humans "Playing God"?

In synthetic biology, man moves from being a modifier of what is present to a creator of something new, at least according to the visions of some biologists: "In fact, if synthetic biology as an activity of creation differs from genetic engineering as a manipulative approach, the Baconian *homo faber* will turn into a creator" (Boldt and Müller,

2008, p. 387). Several voices call it likely that work to create new life forms will give rise to fears, especially that of synthetic biologists "playing God." The accusation that man is playing God is presumably as old as technology itself and is a popular rhetorical maneuver of technophobes against new technologies. Things do, nevertheless, come to a head in the case of synthetic biology. Primarily motivated by mistrust against science, often fed by religious or spiritual beliefs or by quasi-religious ecological convictions, a substantial portion of the public still exhibits some reluctance to passively allow scientists to meddle with life, either by incubating it in a laboratory or by developing it from scratch, as some synthetic biologists promise to do. The question should be scrutinized seriously, especially since playing God is one of the favorite buzzwords in media coverage of synthetic biology.[5]

For the sake of illustration, let us consider a report by the influential German news magazine *Der Spiegel* (following Synthethics, 2010). The title of the article is "Konkurrenz für Gott" (Competing with God), which is a reference to a statement by the ETC Group ("For the first time, God has competition"). The introduction states that the aims of a group of biologists are to reinvent life, thereby raising fears concerning human hubris (Grolle, 2010). The goal of understanding and fundamentally recreating life would, according to the article, provoke fears of mankind taking over God's role and that a being such as Frankenstein's monster could be created in the lab.

Some ethicists and theologians have objected generally to any technology that attempts to create life on the ground that we may not play God. Yet it is unclear what this concern means argumentatively: "There is very little agreement about the precise nature or legitimacy of these concerns, let alone what, if anything, might be done to address them" (Parens *et al.*, 2009). Most of the papers "that mention this class of issues usually do not elaborate them. Most papers are restricted to raising questions in a rather unsystematic sense" (de Vriend, 2006). These issues are often briefly mentioned in reports or articles containing an overview of ethical aspects of synthetic biology, but are rarely discussed in any depth. An important task in achieving a fruitful debate would therefore be to clarify the ethical question: what does "playing God" mean?

[5]The thoughts presented in this section have been elaborated in the Synth-Ethics project of the European Union (Synth-Ethics, 2010). In particular, I will refer to analyses made by Link (2010).

Even the plausible position that religious people might fear "that when the creation of life at human hands happens this will knock down a key theological tenet that only God can create life from non-life" (Caplan, 2009) is controversial. Even if we accept a religious view, why should the capability of creating life threaten to undermine God's dignity or authority, at least as the "prime mover?" "It is hard to credit the view that God would give us the abilities to make new life forms and then argue that to do so crosses a line that God does not want crossed" (Caplan, 2009). It is quite interesting that this view is supported by many theologians.

An analysis of the literature shows that there are two important things to be done. First, it would be ill-advised to simply shove aside the worries and fears associated with phrases like "playing God." This is all true if the aim is to ignite a public debate and to ensure there is social agreement on synthetic biology as a field of research. Therefore, we might first consciously examine what rational arguments can be made out of these fears. What rationale may lay behind them, even if at first glimpse it seems to be just an irrational uneasiness? Second, an examination of political philosophy may be needed to determine how to deal with these fears in democratic societies if it turns out that many of these concerns are not rational. This is because it is advisable to inform the public and to stimulate a discourse about these issues (cf. de Vriend, 2006; van Est *et al.*, 2007). But what if public uneasiness nonetheless increases or at least persists? Clearly, we cannot just disregard these worries. But, on the other hand, a supposedly irrational uneasiness does not seem to justify restricting or overregulating scientific freedom. An approach must be developed that takes such concerns into account, methodically dispels those that are irrational, and seriously addresses those that are serious both philosophically and publicly.

An obvious interpretation is that the creation of life is God's prerogative and that infringing upon His privilege is to be rejected. However, in the view of many theologians, to think that synthetic biologists overstep a crucial boundary rests on a gross misunderstanding or modern transfiguration of the term "creation." Creation, it is argued, may not be considered merely as causal origination. It is not about the first limb of a causal chain but about the continual reason of the whole chain. Creation is, in this view, a hermeneutic category for interpreting the existing world and revealing a specific view of it; causal categories are out of place (cf. Kessler, 2008). We humans may thus participate in God's creation, but merely as *cooperatores Dei*. We simply do not have the potential for a *creatio a novo*, and therefore cannot break into this sacrosanct, divine domain at all: "Man can principally not act like God" (Dabrock, 2009). In conclusion, religion seems to provide no reason to accuse synthetic biologists of playing God since they cannot play God. The only possible sin would lie in the arrogance of presuming that they were able to play God.

A different and important aspect of "playing God" is addressed by referring to human hubris. Because synthetic biologists have a powerful instrument at their disposal, the worry is that they might "play" with nature in a carefree way. Moreover, since the possible impact of new technologies emerging from synthetic biology will increase further dramatically, the risks might seem so high and incalculable that pretending to have them under control nearly amounts to playing God. While this concern is certainly warranted, both the possibility that the techniques for creating life could fall into the wrong hands as well as the harm that might result from an accidental release of engineered microbes are already covered by the debate about biosafety and biosecurity (see above). In this respect the charge of playing God does not add anything new (Link, 2010).

We can conclude that the "playing God" argument is an expression of concern and fear but that it does not touch on any ethical issues themselves in the sphere of synthetic biology. Ultimately there is no accepted normative framework enabling us to distinguish were playing God begins and where it stops. We similarly do not have the means for distinguishing hubris from nonhubris. In the context of problem-oriented ethics, there is no means for operating with the hubris or playing God positions that claim to offer orientation for action.

In summary, we can concede that it is quite understandable to feel unease with regard to human's creating life. And the objections or demurs put forward can be fairly interpreted as attempts to conceptualize this unease. But all things considered, relying on the distinction between the natural and the artificial or claiming that life has an intrinsic value are not arguments apt to put this discomfort on a firm footing. This, of course, does not imply that these qualms are nonrational or irrational. Even if some of our intuitions rest on a "secularized religiousness," we are well advised to explore what the rationale behind them might be (Link, 2010) for two reasons:

(a) Because the speakers that advance these positions must be taken seriously as moral actors.
(b) Because previously unknown arguments might be hidden behind this uneasiness. The task of enlightenment that results is presumably less one of ethics than one of hermeneutic diagnosis.

## 7.4 Hermeneutic Dimensions

The search I undertake in this section has a *hermeneutic* character. I interpret nanobiotechnological developments, especially those in synthetic biology, in relation to possible changes in the relationship between technology and nature (Section 7.4.1) and to changes in our understanding of "life" (Section 7.4.2).

### 7.4.1 Technicalization of the Natural or a More Natural Technology

One characteristic of nanobiotechnology is the extension of classical machine language to the sphere of the living. The living is increasingly being described in technomorphic terms, tied to a change in the self-understanding of biology.

> Although it can be argued that synthetic biology is nothing more than a logical extension of the reductionist approach that dominated biology during the second half of the twentieth century, the use of engineering language, and the practical approach of creating standardized cells and components like in an electrical circuitry suggests a paradigm shift. Biology is no longer considered "nature at work," but becomes an engineering discipline. (de Vriend, 2006, p. 26)

Examples of such uses of language are referring to hemoglobin as a vehicle, to adenosine triphosphate synthase as a generator, to nucleosomes as digital data storage units, to polymerase as a copier, and to membranes as electrical fences (see section V.1.2.4 in Grunwald and Oertel, 2006). *Prior to* considering synthetic biology from the perspective of technology ethics, it appears sensible to ask if and how such changes in the use of language and such reinterpretations modify the relationship between technology and life or modify man's view of this relationship and thus possibly also the way he acts toward it.

Although *prima facie* the transfer of machine language to living systems in this manner undoubtedly appears to be a step toward the technicalization of life, just the opposite is sometimes expected. Nanobiotechnology is sometimes classified as a special form of bionics and referred to as *nanobionics* (see below; cf. Grunwald and Oertel, 2006). *Bionics* refers to a field of research that pursues a *technical* cognitive interest, such as in searching for solutions to problems, discoveries, and innovations, and to this end employs knowledge derived from the observation and analysis of biotic systems (Grunwald and Oertel, 2006; von Gleich *et al.*, 2007). Bionics attempts, as is frequently expressed metaphorically, to employ scientific means to learn from nature in order to solve technical problems (Nachtigall, 2002). The promise of bionics is that the bionic approach will make it possible to achieve a technology that is more natural, more nearly natural, or better adapted than is possible with traditional technology. Examples of desired properties that could be achieved include adaptation into natural cycles, low levels of risk, fault tolerance, and environmental compatibility.

In grounding such expectations, advocates refer to the problem-solving properties of natural living systems, such as optimization according to multiple criteria under variable boundary conditions in the course of evolution, and the use of available or closed materials cycles (von Gleich *et al.*, 2007, pp. 30ff.). According to these expectations, the targeted exploitation of physical principles, of the possibilities for chemical synthesis, and of the functional properties of biological nanostructures is supposed to enable nanobiotechnology to achieve new technical features in hitherto unachieved complexity, with nature ultimately serving as the model. The concept of *nanobionics* was created against this backdrop. The emergence of nanobionics was motivated by the possibility of obtaining increasingly detailed insights into the molecular functioning of biological macromolecules and of utilizing them technically. The principles of bionics find application at the nanoscale (Hampp and Noll, 2003).

This is the classic bionic step from the structural to the functional analysis of living systems or their components in order to employ this knowledge to design solutions to technical problems:

> Nature has made highly precise and functional nanostructures for billions of years: DNA, proteins, membranes, filaments and cellular components. These biological nanostructures typically consist of simple molecular building blocks of limited chemical diversity arranged into a vast number of complex three-dimensional architectures and dynamic interaction patterns. Nature has evolved the ultimate design principles for nanoscale assembly by supplying and transforming building blocks such as atoms and molecules into functional nanostructures and utilizing templating and self-assembly principles, thereby providing systems that can self-replicate, self-repair, self-generate and self-destroy. (Wagner, 2005, p. 39)

If the above-mentioned promise of bionics is transferred to nanobionics — as is suggested by the references also cited above — a completely different impression of a more natural technology might be created to take the place of the initially *technomorphic* view of nature that is clearly visible in machine language. This perspective stirred hopes for a more nearly natural technology on the basis of nanobiotechnology.

A closer look at the research process in nanobionics suffices, however, to prevent one from having hasty expectations and hopes. The cognitive process in nanobiotics attempts to gather knowledge about the structures and functions of natural systems, but from *technical intervention*, not from contemplation or via distanced observation of nature. Living systems are not of interest *as such*, for example, in their respective ecological context, but are analyzed in the *relationship of their technical functioning*. Living systems are thus interpreted in nanobiotics — just as in bionics generally — as *technical systems*. This is probably a significant precondition for it to be at all possible to succeed in transferring knowledge gained from living systems to technical ones. Even in traditional macroscopic bionics, living beings were not considered as living systems, but as a supplier of ideas for technical solutions. A technical cognitive interest is already at work in the recognition of the functional and structural principles of living nature.

From this perspective, nanobionics is linked epistemologically to a technical view of the world and to technical intervention. It carries these technical ideas into the natural world, modulates nature in a technomorphic manner, and gains specific knowledge from this perspective that can then be transferred back to the technical sphere and used there for solving problems. "Bionics chooses a technically oriented access to nature, in order to move from technically understood life to life-optimized technology" (Schmidt, 2002, p. 141). The expression, "Biology is the nanotechnology that works" (Brown, 2004) hits the needle on the head: Nature is seen as technology, both in its individual components and also as a whole.

> This is where a natural scientific reductionistic view of the world is linked to a mechanistic technical one, according to which nature is consequently also just an engineer [...]. Since we can allegedly make its construction principles into our own, we can only see machines wherever we look — in human cells just as in the products of nanotechnology. (Nordmann, 2007b, p. 221)

Bionics casts a specific epistemological glance at nature by considering the latter to be an ensemble of solutions to technical problems reached under evolutionary pressure. This means that the nature that is taken as a role model is itself technicalized by nanobiology. Even Drexler (1981) made this change in perspective by comparing the functions of familiar techniques and examples of cell or organisms, thus ultimately interpreting the biotic world as a machine. In this sense, even the controversially discussed idea of nanorobots is ultimately to be classified as a bionic idea. It is namely the creation of artificial living things on the basis of knowledge gained from studying natural living beings and their components (e.g., viruses).

Instead of eliciting a more natural technology per se as promised by bionics, nanobionics in research must *necessarily* rely on a technicalization of the living, i.e., on its technomorphic modeling. The result of nanobionic research signifies a far-reaching technicalization of what is natural. This observation attributes neither a higher nor lower value to nanobionics and is ethically indifferent. It only points out that we should be very reserved in dealing with normative expectations and promises that are widely tied to the concept of bionics.

This makes it possible to provide a clear answer to the hermeneutic question discussed in this section. In nanobiotechnology there is a technicalization of the living, not a technology that is *per se* closer to nature, when we are supposed to learn from nature. Learning from nature for technical problem solving must of necessity already take a technical view of nature. This is in the nature of the research process. Whether and what this means in an ethical sense cannot be clarified for nanobionics generally. Any clarification requires an analysis of individual cases, specifically with regard to their effects on the environment and their risks (in this sense, also see von Gleich *et al.*, 2007).

## 7.4.2 The Relationship Between Technology and Life

Concepts such as that of life belong to the core concepts of anthropology, biology, and philosophy and, like as a rule most core concepts, are semantically controversial. Life, in the form in which living beings exist and develop in nature, which is only partially accessible to human influence (e.g., via breeding), is open to numerous interpretations. One that gets special attention is the relationship between technology and life. If synthetic biologists were able to create something that can be called "artificial life," how should we consider these entities? This is not only an interesting metaphysical question about the status of artificial living beings — sometimes called "biofacts" (Karafyllis, 2006) — there is also the ethical question of how we are to treat them: Should we treat them as machines or as living beings?

Precisely the increasing degree of opportunities for human intervention in living beings provides the motive for conceptual debates, such as recently in the context of biofacts (see Karafyllis, 2006 and the subsequent discussion section). Karafyllis saw an inherently necessary link between "life" and "growth" resulting from an autonomous drive and concluded that there cannot be artificial life: "We can reject the thesis that man can produce 'life' as an artifact by referring to the fact that aggregation is not growth and gestalt is not habitus" (i.e., a settled tendency of behavior; Karafyllis, 2006, p. 555). If this dictum were taken seriously, it would be easy to clear the table of ethical challenges regarding how to deal with artificial life because the latter would be an impossibility.

In several respects, however, critical questions must be raised as to whether the issue of the possibility or impossibility of artificial life can be determined in a *purely conceptual* manner. The first question is — in view of the challenges posed by synthetic biology — whether the rejection of the possibility of artificial life and artificial living beings is really as clear as the quotation suggests. The dictum that artificial life is in principle impossible must be doubted at least with regard to man's capacity to technically construct artificial *living beings*. Traits similar to those of living beings are already known from some products of software technology, such as computer viruses. The assertion that something is generally impossible appears to be very bold. Even if "a process of life cannot be reduced to chemical processes that have been linked together" (Psarros, 2006, p. 594), we cannot for ever exclude the chance that such linking of individual functions by synthetic biology might result in an overall situation that appears to be alive or very similar to it.

Hidden behind this question is the even more far-reaching one as to whether it is possible to resolve in a purely *conceptual* manner the dispute about whether life or living beings can be constructed technologically, i.e., without reference to the *de facto* level of knowledge and skill in biology. Even very substantial advances in synthetic biology could not change anything on the dictum of the impossibility of artificial life as it is constructed. This type of argument — that asserts the general impossibility of something or makes claims of universal applicability — is in a logical respect prone to being refuted and historically has often proved to have been premature.

Another argument that challenges Karafyllis' conclusions that the technical production of life is impossible consists in doubting that a clear distinction can be drawn between being able to influence living systems by means of technology and to newly invent and construct them by means of technology. More likely is rather a continuum of technical interventions in living systems ranging from minimal interventions to constructions that are either largely or entirely new. Disregarding the fact that the distinction between the natural and the artificial is in itself problematic; it is by no means clear that the (possible) products of synthetic biology are to be considered as artificial life in the sense that synthetic biologists would "create" life (Brenner, 2007). Since there can be no doubt about technology's ability to influence living systems, Psarros and Karafyllis would have to indicate at which point in this continuum it would be appropriate to speak of a new construction of life by means of technology, which their positions claim is not possible. It is difficult to imagine that this can succeed.

Furthermore, assuming finally hypothetically that the thesis of impossibility were correct and that we could in fact decide conceptually about the possibility of artificial life, then this thesis would be *irrelevant* in a practical sense. It would be clear then that the term "life" could not be attributed to artificially created "beings." Yet if advances in synthetic biology would lead to the creation of such beings, we would face the question of how to deal with them independently of whether they are classified as living beings or not. A definition of life that is not sensitive to empirical changes in the state of knowledge and skill in biology would also be of little assistance in the hermeneutic analysis of these changes or in respect to ethical evaluations.

What however is evident from these rather episodic considerations is that synthetic biology provokes precisely this type of question. It does not provoke them simply as academic questions, but rather as questions whose perspective at least has a real background.

> Additionally, synthetic biology forces us to redefine "life." Is life in fact a cascade of biochemical events, regulated by the heritable code that is in (and around) the DNA and enabled by a biological machinery? Is the cell a bag of biological components that can be redesigned in a rational sense? Or is life a holistic entity that has metaphysical dimensions, rendering it more than a piece of rational machinery? (de Vriend, 2006, p. 11)

While many biologists incline toward a reductionist–materialistic view (obvious for example in the context of converging technologies; Roco and Bainbridge, 2002), some take the holistic position, as shown in statements such as, "While machinery is a mere collection of parts, some sort of 'sense of the whole' inheres in the organism" (Woese, 2004). The debate about the relationship between technology and life thus represents another stage in the debate between reductionism and holism.

## 7.5 Responsible Governance of Synthetic Biology

The regulation and governance of synthetic biology are still in their beginnings, and no specific legal provisions have yet been enacted. It is only just possible to foresee most applications, and risk management relates mainly to research itself rather than to products and other innovations (IRCG, 2009, p. 7). This diagnosis focuses the need for governance of synthetic biology on the necessity to debate the responsibility of scientists as individual professionals and of science as a system. The main field of applied ethics that is relevant to this situation is research ethics, which involves more "soft" than "hard" regulation (Synth-Ethics, 2010).

"Soft law" means both "rules of conduct which, in principle, have no legally binding force but which nevertheless may have practical effects" (Mörth, 2004, p. 5) and guidelines and recommendations by national, supranational, or international organizations or by institutional or consultative bodies. The absence of formal binding force in this kind of norm depends on the absence of formal sanctioning mechanisms (Synth-Ethics, 2010), which also applies to moral rules. Soft law in particular tends to be a tool in which legal and ethical principles converge. The normative framework relevant to this field is science ethics, which is a part of professional ethics. It considers issues such as the dissemination and publication of possibly problematic information (e.g., how to build a virus).

Due to the ability of living systems to reproduce, modify, and organize themselves, the risk situation in nanobiotechnology and in synthetic biology is clearly different from that of abiotic synthetic nanoparticles. The very early stage of development in nanobiotechnology and synthetic biology and the distance from the market, however, prevent any discussion in concrete terms of the risks or even of unclear risks resulting from products or processes. There are no specific synthetic living beings or their parts under risk observation in a manner similar to titanium dioxide or silver oxide nanoparticles in marketed products. The situation is just the opposite: the subject of the ethical debate is the *process of research itself*. Thus the focus here is on questions of the responsibility of the scientists and the disciplines involved, of the accountability of certain areas of research, and of the relationship between the self-regulation and self-obligation of science and state regulation.

It is thus not surprising that the well-known conference of Asilomar (1975) is repeatedly cited as a model for future steps in the field of nanobiotechnology (e.g., Boldt and Müller, 2008). That conference took place in a situation in which a global spirit of optimism regarding genetic engineering could be observed while at the same time the first signs of public criticism and demands for state regulation could be heard. The outcome of the conference was that genetic engineers committed themselves to taking responsibility and exercising caution. Interpretations of the conference are controversial. On the one hand, it is praised as a positive example of science proactively assuming responsibility; on the other hand, it mainly served the purpose of pre-empting state regulation so that genetic engineers could carry on conducting their research with as little interference as possible (de Vriend, 2006).

At the second global conference on synthetic biology (2006), there was an attempt to follow up the actions at Asilomar and a corresponding declaration was passed (Maurer *et al.*, 2006). This, however, only refers to the possible *military* use of synthetic biology and puts up a set of self-obligations for possibilities of this kind (http://syntheticbiology.org/SB2Declaration.html):

> First, we support the organization of an open working group that will undertake the coordinated development of improved software tools that can be used to check DNA synthesis orders for DNA sequences encoding hazardous biological systems; we expect that such software tools will be made freely available.
>
> Second, we support the adoption of best-practice sequence checking technology, including customer and order validation, by all commercial DNA synthesis companies; we encourage individuals and organizations to avoid patronizing companies that do not systematically check their DNA synthesis orders.
>
> Third, we support ongoing and future discussions within international science and engineering research communities for the purpose of developing creative solutions and frameworks that directly address challenges arising from the ongoing advances in biological technology, in particular, challenges to biological security and biological justice.
>
> Fourth, we support ongoing and future discussions with all stakeholders for the purpose of developing and analyzing inclusive governance options, including self-governance, that can be considered by policymakers and others such that the development and application of biological technology remains overwhelmingly constructive.

Self-obligations in science have, however, also come in for criticism. On the occasion of the same conference on synthetic biology, 35 nongovernmental organizations (including the ETC Group, Greenpeace, and the Third World Network) wrote a joint letter on the subject, from which a long passage can be cited here in which they refer specifically to this restriction to a possible military aspect of synthetic biology (ETC, 2006; also mentioned in de Vriend, 2006, p. 49; Wolbring, 2006):

> Moreover, the social, economic, ethical, environmental, and human rights concerns that arise from the field of synthetic biology go far beyond deterring bioterrorists and "evildoers." Issues of ownership (including intellectual property), direction and control of the science, technology, processes and products must also be thoroughly considered. Society, especially social movements and marginalized peoples, must be fully engaged in designing and directing dialogue on the governance of synthetic biology. Because of the potential power and scope of this field, discussions and decisions concerning these technologies must take place in an accessible way (including physically accessible) at local, national, and global levels. In the absence of effective regulation, it is understandable that scientists are seeking to establish best practices but the real solution is for them to join with society to demand broad public oversight and governmental action to ensure social well-being.

This letter is characterized overall by:

- Mistrust of scientific self-regulation and self-obligations. Such self-governance is undemocratic as scientists should not be allowed to decide such far-reaching questions affecting their own activities.
- Demands for a broad investigation of the social consequences of synthetic biology instead of restricting any investigation to abuse scenarios, for example, by terrorists.
- Emphasis on the necessity for including social groups in dialogs about the agenda of research and the handling of possible social consequences.

The normative uncertainty and conflict that are symbolized by this letter and the activities at the second conference on synthetic biology concern the *distribution of responsibilities* for the further research process. What influence do scientists, the public, the state, or other social actors or areas have on the further course of events in nanobiotechnology and in synthetic biology in particular? Should there be a "policy of knowledge" (Stehr, 2004) that decides where the acquisition of knowledge is desirable and where it should be prevented? What role might this policy play? And, how should responsibility and accountability consequently be distributed?

The question is *which* responsibility *should* be attributed to scientists in the fields of nanobiotechnology and synthetic biology. The answers frequently demand that scientists are supposed to reflect on the consequences of their actions in a manner that constitutes a complete assessment of the technology. This is often done with the implicit hope that scientists — if they assessed the results of their own actions comprehensively — would make judgments in a responsible manner and act accordingly, and that negative and unintended consequences could be avoided to a large extent or even completely (cf. in this direction also Presidential Commission, 2010, p. 13). These expectations are however doomed to failure because technology assessment in general and in nanobiotechnology and in synthetic biology in particular is characterized by the following:

- The need for *social* consultation (on state sponsorship of research, on government policy toward science and technology, and on regulating the context of technical development by means of legislation, judicial decisions, or economic measures)
- The need for a systemic perspective on the extent of the consequences
- The problem of anticipating and *evaluating an uncertain future* (systemic feedback, long-distance effects in both spatial and temporal terms, the lack of social causal relationships)
- The *evaluation* problem (how to evaluate the consequences of technology with regard to their desirability or tolerability in view of the diversity of interests and opinions in society)

Both *inter- and transdisciplinary efforts* are needed to be able to at least approximate a solution. Neither individual scientists nor disciplines such as synthetic biology or even philosophy can address these questions alone with any prospect of success. Scientists in nanobiotechnology and synthetic biology are experts in their fields, but not in the possible social consequences of their actions and not for the question of the acceptability of uncertain risks and dealing with them. I must thus warn against unreflected optimism with regard to devolving responsibility, regardless of whether biologists commit themselves to these expectations or society tries to commit them to it.

When it comes to attributing responsibility, a broader approach is thus necessary, one that does justice to the realities of an extensive division of labor, citizens' claims for democratic participation, and the specific circumstances in the sciences. The public sphere in democracies should be considered first and foremost. One of the demands for a transparent relationship between science, politics, and the public is for there to be deliberation about the agenda of science, in this case of synthetic biology that is conducted in a democratic manner (Habermas, 1970).

To take demands seriously for participation by a democratic public as well as for decision-making processes that are politically legitimized, however, does not lead to nanobiotechnology and synthetic biology being freed of all responsibility. These fields are justifiably expected to provide transparent information to the public. This is particularly true for potentially worrying developments. Faced by such developments, society might initiate ethical reflection or technology assessment in order to analyze and evaluate the problem systematically. Yet even independent of such developments, society is faced by the question of shaping the scientific agenda. The specific responsibility of scientists to provide information at an early stage lies in the fact that they possess particular cognitive competence in their own area and are the first to have certain information. This responsibility also extends to participation in interdisciplinary and social dialogues and in political counseling. Science, including synthetic biology, is part of society, not something external to it. The expectation toward science is that it reflects on its role in society and actively accepts this role in its many aspects.

A professional code of ethics or declarations of commitment can certainly be useful and valuable in many situations: "A code of ethics and standards should emerge for biological engineering as it has done for other engineering disciplines" (Church, 2005, p. 423). Yet neither of them can replace ethical reflection or the power of democratic opinion. This is particularly true in view of the new options for technically modified or newly created life forms and the associated higher costs of preventive care. The task in the future will be to operationalize the principle of preventive care. This can be achieved through appropriate containment strategies along the lines followed in genetic engineering, where the knowledge base is broadened at each step by systematic investigations of developments and their consequences. Such strategies transform the initially "unclear" risk into risk in the sense of classical risk management (cf. Grunwald, 2008d). The success of such strategies depends on having responsible scientists, an interested democratic public that expects clarification and demands its own participation, and measures at the governmental level. These measures do not immediately have to take the form of regulations but can be expressed, for example, as systematic observation of further developments in nanobiotechnology and in synthetic biology within the framework of technology assessment. Finally, such observations should be accompanied by philosophical interpretation and ethical reflection of the risks and opportunities posed by each case and with reference to the relevant normative frameworks.

Another level of complexity consists in the international dimension. Although the following demands are justified, they are probably (still?) rather unrealistic:

> For synthetic biology a strong case can be made for international dialogue on the appropriate role of regulatory oversight. The difficulties which arise from piecemeal and divergent national approaches to the regulation of innovative technology in life sciences were very apparent in the case of GM crops, and this experience offers lessons for synthetic biology. (IRGC, 2009, p. 17)

The synthetic biology community, which is internationally organized, clearly has the advantage over a nonexistent public at the global level. The lack of an international level at which democratic deliberations could take place must act as an incentive for us to develop at least some forms of such interaction and to utilize the first elements of global governance. In 1998, for example, the UNESCO commission "World Commission on the Ethics of Scientific Knowledge and Technology" (COMEST) was established. Its mandate includes both policy consultation and promoting public consideration (see ten Have, 2007 regarding nanotechnology). Much still remains to be done.

# Chapter 8

# Animal Enhancement

The use of animals for human purposes, such as for nutrition, as working animals, or for sports, has a long tradition in human civilization. In order to help man utilize animals as effectively as possible for these purposes, techniques — such as for breeding and keeping animals — were developed early in human history. Although genetic engineering and molecular biology have already enabled profound steps to be taken toward even further reaching interventions in the nature of farm animals, it appears that nanotechnology will enable man to take additional steps to increase his influence over animals.[1] The term "animal enhancement" is increasingly used in this field. However, its meaning needs to be clarified.

This chapter first describes some of the directions of research that are currently in progress in this field (Section 8.1). It then inquires about the semantics of animal enhancement (Section 8.2), in part in preparation for considerations of "human enhancement" (Chapter 9). Against the backdrop of the existing normative framework, the ethical challenges are identified (Section 8.3) and lead beyond the level of applied ethics to issues from hermeneutics and the theory of science that are related to the relationship between man, animal, and technology (Section 8.4). In conclusion, I present some thoughts about issues of responsibility (Section 8.5).

[1]This chapter is based on a study prepared for the Swiss Federal Ethics Committee on the Non-Human Field (Eidgenössische Ethikkommission für den Außerhumanen Bereich; EKAH) (Ferrari *et al.*, 2010). I would like to thank Arianna Ferrari, Christopher Coenen, and Arnold Sauter for permitting me to present some of the joint results here.

*Responsible Nanobiotechnology: Philosophy and Ethics*
Armin Grunwald

ISBN 978-981-4316-80-4 (Hardcover), 978-981-4363-33-1 (eBook)
www.panstanford.com

## 8.1 (Nano)Technology for Intervening in Animals

In the scope of the "converging technologies" (Roco and Bainbridge, 2002), nanotechnology, biological and genetic technology, information and communications technology, and cognitive science and brain research are said to be converging. This convergence is supposed to open up radical new opportunities, the focus of which in the report named above is on "improving human performance" (Chapter 9). As the science of dealing technologically with matter at the level of the smallest particles, including atoms and molecules, nanotechnology provides key competence required for this convergence. Convergence tendencies that will also be employed in the animal realm can be recognized, such as the enrichment of research and technology in molecular biology from the utilization of methods from neurophysiology and the linking of information and communications technology with genetic procedures. Existing "enhancement" techniques are being further developed with the assistance of procedures and know-how from nanotechnology. Animal enhancement builds on known technologies in order to (even) further improve the usefulness of animals for human purposes.

The utilization of animals by humans takes different forms, such as experimental animals in science, working and farm animals in agriculture, in the military, as pets, in sports, and in zoos and for entertainment purposes. In the following I will name — as examples — several fields of research in which nanotechnology is beginning to play a role in man's interventions in animals in order to better enable him to pursue his own interests.

Agricultural research focuses on improvements in animals — in the sense of maximizing their performance as working animals — primarily for economic reasons. The production and utilization of transgenic farm animals is based on the desire for a — with regard to human interests — perfect design of animals. Animal utility dominates the perception of what constitutes the enhancement[2] of a farm animal. In this area, animal enhancement dovetails with the perspective of the classical forms of "enhancing" farm animals.

[2]The word "enhancement" will as a matter of principle be placed in quotation marks in the following. This is to indicate that it is important to note the substantial conceptual complexity of the word in contrast to its simplistic positive connotation (Section 8.2).

The specific purpose of cloning farm animals, for example, is to reproduce animals that have already been optimized by breeding, in order to further improve their utility. This combination of classical approaches with new (nano)technological means also includes visions regarding nanotechnology, such as the use of intelligent biosensors in farm animals (Roco and Bainbridge, 2002, pp. 5f.) that can check the state of an animal's health. The principal goal is the integration of different technological systems for diagnosis and drug delivery. In this connection, according to the idea of converging technologies, nanoelectronics joins nanobiotechnology and bioinformatics in playing a central role since extremely small instruments have to be produced that can activate and regulate themselves (Scott, 2005).

Increasing animals' competitiveness is the goal of the use of doping medication and of cloning sport animals for breeding. Cloned horses are used as suppliers of valuable reproductive material, such as sperm. The direct doping of animals, in contrast, is forbidden by the regulations of sporting associations, but has been reported, especially with regard to racehorses. Conditions such as strong competition, insufficient regulation, and economic interests can lead to illegal enhancement by doping, just as in doping in human sports. However, it is not well known how widespread such practices are.

New technology also makes enhancements possible in animals kept as pets in response to new desires and needs of humans. Cosmetic surgery is used to modify animals to meet the respective aesthetic desires of humans. The goal might be to make the animals more successful in animal beauty pageants — which would be something analogous to doping — or to make them simply more attractive to their owners. Even though it might be advantageous for a dog to be more attractive if it were therefore treated more endearingly, this does change anything about the fact that the operation takes place according to human criteria. Such interventions have been criticized as "unnecessary" medical actions on the basis of a veterinarian's ethos (Neumann, 2008).

Fish provide examples of the dedicated enhancement of pets. The fluorescent zebra fish Night Pearl from the Taiwanese company Taikong has been on the market in Asia since mid 2003. This is a genetically modified fish, in whose genome a gene for a fluorescent protein has been inserted. The production and sale of these fish met with objections in Europe (Whitehouse, 2003). The trade with genetically modified fish has been forbidden in Europe, Canada, and Australia. Further fluorescent fish products have in the meantime reached the market (Robischon, 2007).

The most frequent technological applications in the area of animal enhancement are animal models for dedicated experiments.[3] In biomedicine, animal models for human diseases or their symptoms are frequently produced in order to test therapies or medications prior to clinical trials. Enhancement of the experimental animals serves to improve the conclusiveness and applicability of the subsequent results for humans. Examples are xenotransplantation, in which animals serve as the source of organs, tissues, and cells for humans, and the use of mouse models for investigating cause–effect chains and possible therapies of diseases such as Alzheimer's. To facilitate this, animals are "enhanced" in different ways. For example, pigs are modified so that they elicit a milder rejection response in primates, or primates are treated pharmacologically so that they accept the porcine organs or tissues more readily. Animals are also used in research in human enhancement (Chapter 9) for testing enhancement medication or implants.

A well-known example, which in a certain sense has led to an increase in animals' abilities, is the mouse model known as *fearless mouse*. In this mouse model, the gene encoding stathmin was knocked out (Shumyatsky *et al.*, 2005). Stathmin inhibits the formation of microtubules, which are responsible for the delivery of information about learned and innate fear to the amygdala, a central area of the brain important for memory. Such mice have a reduced memory of fearful experiences. They cannot recognize dangerous situations because they lack the congenital mechanisms. Transgenic mice have also been created whose olfactory system has been modified so that they possess none of the congenital mechanisms for recognizing bad or dangerous scents. A video is available in the Internet that shows such a mouse approaching a cat and even snuggling with it (Kobayakawa *et al.*, 2007). It is obvious that knocking out fear does not at all constitute a form of positive improvement from the imagined perspective of animals since under normal conditions the mechanisms of fear are decisive for survival.

[3]The term "animal models" does not refer to models of animals, but to real animals that are considered to be models for certain functional relationships and that therefore can serve as objects for dedicated experiments.

Researchers expect these models to produce new knowledge for treating human mental disorders such as panic attacks or the post-traumatic stress syndrome as well as new insights into the mechanisms of fear.

Among other enhancements are cases in which new abilities have been implanted in animals, such as a resistance to illness that is not present naturally. In such cases, it must be determined to which extent there may be advantages for the animals and, possibly, whether additional animal experiments appear to be justified. This also points to a general issue. An ethical judgment must, entirely analogous to the life cycle approach to determining the sustainability of technology (Grunwald, 2006b), take the entire process into consideration, including the research and the animal experiments that must take place on route to creating an enhanced animal. It is not sufficient to observe the properties of an enhanced animal — its well-being, integrity, and similar features — and the related ethical aspects. The route to the goal must also be taken into account (Section 8.4).

Animal experiments are also taking place in the research and development for novel brain–machine interfaces. For rodents and apes it has already been demonstrated numerous times that implanted electrodes make new means possible for the animals to control machines. Following appropriate training, apes with implanted electrodes are able to control robot arms sufficiently that they can grasp objects and feed themselves (Velliste, 2008). Apes whose arms had been temporarily paralyzed have been successfully trained to move their own temporarily paralyzed arm in a targeted manner, using a brain–computer interface (Moritz *et al.*, 2008). Technology has also been successfully developed to remotely control animals. Brains of rats, for example, have been stimulated in such a manner that it was possible to precisely control their movements (Talwar *et al.*, 2002). The remote control of external objects by means of brain activity is thus reality in animal models (Hatsopoulos and Donoghue, 2009). Nonetheless, it is only possible to speak of animal enhancements to a very limited degree. In some cases, animals were previously paralyzed or injured in order to simulate the special situation of human patients, or they were even created as disease models. Examples include rat models of traumatic injuries to the spinal cord (Truin *et al.*, 2009) and of paralysis following stroke

(Pedrono *et al.*, 2010), Parkinson's disease (Fuentes *et al.*, 2009), and Huntington's disease (Kraft *et al.*, 2009). In a certain sense, the primary objective is improved communication between mammals and computer technology.

Animal experiments are also taking place in work on extending the human life span, which is a significant topic in human enhancement (Section 9.1). Of particular interest are mice that have been the object of various types of interventions. While it is true that animal life spans have been extended, it is very difficult to interpret the results or to transfer them to humans (Ferrari *et al.*, 2010) since it is unclear what is being compared.

## 8.2 The Semantics of Animal Enhancement

The meaning of animal enhancement needs to be clarified since it has not yet become a commonly used phrase. The starting point is the fact that the term enhancement is used very generally as a synonym for improvement, while it on the other hand also serves as a symbol for a debate about the goals and limits of modern medicine, specifically with regard to possible "technical enhancements" of humans (see, e.g., Schöne-Seifert *et al.*, 2009). The new, emerging use of the phrase animal enhancement cannot ignore this ongoing debate, but must clarify whether and which elements of the previous use of the term can be transferred to the enhancement of animals.

### 8.2.1 The Semantics of Enhancement

The term "enhance" has fundamentally positive connotations in everyday language as the opposite of "worsen." The word enhancement and its use are, however, much richer and more complex than often assumed and must therefore be examined carefully. Enhancement is not a one-place predicate, but can only be determined relative to certain criteria. Caution must be used to avoid running into the rhetorical trap suggested by everyday language in regarding an enhancement to be intrinsically positive. One must, rather, basically enquire as to the criteria according to which a simple change is evaluated as an *enhancement.*

Enhancement represents an activity or action by which an object is changed in a particular direction: there are *actors* (the subjects of enhancement), who enhance *something* (the object of enhancement) according to *criteria*. In accordance with this, enhancement necessarily includes three semantic dimensions:

1. A *starting point* for enhancement. An enhancement is only plausible as an *enhancement* if the starting point of the change is given.
2. A *criterion* of enhancement. A normative criterion, relative to which the enhancement takes place, must be given. A criterion consists of the declaration of a *parameter* (quantitative or qualitative) and the *direction* in which the parameter will be altered to constitute an enhancement. The direction of change in which something is seen as enhancement depends on the target of the enhancing.
3. A *measure* of enhancement. Measuring the *size* of an enhancement is primarily significant in weighting processes, such as if the enhancement in one place is offset by deterioration in another, and balancing is necessary.

Many ethical conflicts about possible enhancements, regardless of whether they concern humans or animals, flare up due to the very criteria for enhancement. This occurs in at least three directions.

*First*, even something that is desired or accepted by certain actors or groups as enhancement could be considered ethically problematic according to the criteria for the enhancement. Since the criteria for designating a change as enhancement are normative and attribute a value, they can be an object of controversy, and this is frequently the case. The statement that it is sensible to enhance automobiles by constantly improving performance and acceleration may not meet with universal approval.

*Second*, in connection with the criteria employed, there is a *social* dimension to enhancements since the question can be raised as to *for whom* an enhancement is truly an enhancement according to given criteria. An example is the debate about genetically modified plants in the production of food. Enhancements such as resistance to herbicides — and the resulting increase in production efficiency — would primarily benefit agricultural producers and the food industry, but not the consumers, on whom however any undesired effects on health would land.

*Third*, a change that is an enhancement relative to one criterion can be a deterioration relative to another. As is well known from technical development, the enhancement of an automobile from the point of view of sportiness can be achieved at the expense of the environment, and an enhancement in terms of decreased costs can endanger safety requirements. Conflicting goals are everyday occurrences in technical enhancements just as they are in political decision making. Enhancements are frequently linked to deteriorations, making complex assessments of relative value necessary before a decision can be made.[4]

*Enhancing* describes a change compared to a starting point in an intended direction. It is open with regard to the extent of change, and it has no defined end point. *Optimizing* and *perfecting* are different. They are oriented on an envisaged *final* or *target status*. An optimization is at an end when the optimum is reached.

The object of the analysis given in this chapter is not the optimizing or perfecting of animals. The analysis is terminologically about enhancing that includes the direction and criteria of change but does not aim to achieve an "optimal" final condition. Being open with regard to a final condition implies that there is no inherent limit to the enhancement: the result of each enhancement can serve as the starting point for new enhancements (possibly based on different criteria). The process of enhancement is not completed and cannot be completed. Any limits to enhancement that may be mentioned have to be external and can, for example, consist in ethical arguments or in arguments about efficiency or acceptance.

Furthermore, the difference in action theory between the goals of an enhancement measure as envisaged *ex ante* and the actual *ex post* consequences of the measure must be noted if the enhancement is understood to constitute a course of action. A measure intended to be an enhancement can turn out *ex post* to be a worsening, or it can have unintended consequences. The following must therefore be taken into account when enhancements are being evaluated: the *goals* of the enhancement, the *criteria* oriented on these goals, the *deteriorations* that may be linked to a specific enhancement from a different perspective, the criteria and magnitude of the *assessment* of such conflicting goals, and finally the possible manifestation of *unintended consequences* of the enhancement measures.

### 8.2.2 Animal Enhancement

Semantically, the application of the enhancement concept to animals initially appears to be almost trivial. On closer examination, however,

[4]This is also familiar from breeding. For example, the highest production cows are as a rule more susceptible to disease and less robust against environmental factors.

it does contain several noteworthy problems which can be addressed by adopting the semantic introduction of enhancement given above: "someone enhances something according to a criterion," and by using the references to the starting situation, the criterion, and the magnitude of the enhancement. The actors currently involved in enhancement are mainly researchers, who react in part to a demand from science and in part to a nonscientific demand for enhanced animals. In a more general sense, those responsible for this demand must also be included among the actors involved in animal enhancement since they determine to a large degree the criteria for enhancement. The *objects* of enhancement are animals or, more specifically, certain features of animals' performance. Both of these considerations are not difficult. More interesting, however, is the question as to the criterion or the multiple criteria for enhancement and the motivations and goals related to them.

The fact that the actors involved in animal enhancement are always humans makes it immediately clear that we must look to human interests regarding animals to find answers to the question as to the origin of the goals of these enhancements. With animals it is impossible to obtain an informed consent, which could legitimize the desire for an enhancement or at least ease the legitimization. Animal enhancement is thus the product of human decisions. These can however be looked for in two directions that are normatively very different:

- *Utilitarian perspective*: From this perspective, the current or anticipated applications of animal enhancement can consist in the fact that animals are supposed to be utilized or better utilized for humans. It follows from this that animal enhancement must be seen in relation to the human interests in utilizing the animals.
- *Advocatory perspective*: It is possible that humans might argue from the "imagined perspective" of animals. Animal enhancements could be undertaken whose goals — as determined by humans — consist in realizing an animal interest — as imagined by humans — such as reducing pain.

Animal enhancement can thus be understood, on the one hand, as a direct enhancement of the utility of animals to humans, but on the other as enhancement in the animal's interest within the practices of human usage. Both goals are set by humans but against completely different normative backdrops. This duality permeates the ethical issues of animal enhancement.

This can even be seen in the role of veterinary medicine. Frequently (not always, of course), those animals receive medical treatment that present health problems *resulting from human use* or that are in danger of no longer sufficing human interests. Here, in contrast to human medicine, there is no ethos of healing for its own sake but a "should" that is influenced by human interests in animals. Because of the different moral and legal statuses of humans and animals, human and veterinary medicine are subject to different underlying normative conditions. It is true that veterinary medicine includes aspects of "should," such as avoiding suffering. The basal normative structure is, however, different from that in humans and can differ according to the respective national legislation. While for example the concept of the "dignity of the creature" is anchored in the constitution in Switzerland and that of "intrinsic value" occurs in animal protection legislation in The Netherlands, there are no equivalent concepts in German law. In France, animals are even considered to be things. Analytically, the weaker the rights of animals are anchored, the more the character of veterinary measures is tied to the "desires" and interests of animal owners.

Overall we take animal enhancement to refer to the following:

- Measures that are supposed to enhance the performance of animals compared to what is "customary," where "performance" can be recognized by the human purposes, i.e., user interests in the animals.
- Measures that permit or facilitate the human use of animals in which animals are exposed to fewer or no heavy burdens than would be the case without the enhancement measures.
- Measures that include the creation of *new* properties in animals such as resistance to diseases that in a strict sense go beyond improving already existing properties or performances and thus constitute a change (on human enhancement, see analogously Jotterand, 2008b).

To enable us to speak of enhancement, there has to be a reference value, a criterion, and a goal (see above). For animals, the goals of enhancement — and thus also the reference parameters on which an enhancement is measured — are dominated by utility, as shown by the examples. The ambivalence of the concept of enhancement quickly becomes apparent, however, since interventions can cause health and other problems. According to a utilitarian perspective, one cannot assert that enhancement measures can be equated automatically, as it were, with enhancements in an imagined interest of the animal. Ethical evaluations must at any rate always precisely enquire in which sense and for whom a change constitutes an enhancement.

Interventions to improve disease resistance or a genetically engineered change to optimize certain cognitive abilities of animals could appear to be an exception or a different case. Although the creation of disease resistance should, on the face of it, be in the imagined interest of animals, the situation is more complex. As a rule, the motivation to create disease resistance is not to save the animals from illnesses but to improve the opportunity for human use. Here is an example from agriculture. Among dairy cows, mastitis (inflammation of the udder) is, after fertility disturbances, the most frequent reason for premature slaughtering. That a typical Holstein Friesian cow, the most widespread race, today lives less than 5 years is an unnaturally short live span. This poses a problem for their owners since these cows do not reach the age of maximum performance (Reuter, 2007). Taking advantage of performance potentials has in this case increased susceptibility to illness. If the attempt were made to counteract this by means of a technical enhancement, i.e., to find a resistance to mastitis, this would — if successful — spare animals illness from the disease. Yet the motivation would continue to be the interest in using farm animals. Animal enhancement would ultimately be employed to compensate for negative developments (increasing mastitis) that themselves are caused by the (intense) use of animals by humans in the first place. The talk about the enhancement of animal abilities therefore appears to be ambivalent in a subtle manner that is not the case for humans.

## 8.3 Relevant Ethical Challenges and Normative Frameworks

Animal enhancement is not some future dream but both a part of current practice and an object of research. Handling animals is a classic topic of ethics and is subject to underlying normative conditions and moral standards and their concrete manifestations in animal welfare legislation. In the past decades a movement has become apparent that contains two different and contradictory features. On the one hand, animal welfare is often attributed a high level of social significance, as expressed in tougher legislation. On the other, however, the pressure for using animals has risen generally, above all in the field of agriculture and for experimental animals. Developments in the life sciences, especially in genetic engineering, have led to a greater depth of human intervention in animals. Nanotechnology and the converging technologies that build on it do not add anything radically new to this development, but they increase the possibilities for human intervention and thus lead to more critical questions as to the limits of such intervention. Because of the fundamental relationship to the context of usage by humans, the "grand" ethical issues are similar to the known ones, such as for which purposes may animals be instrumentalized? Under what conditions are animal experiments permitted? How should we deal with animal stress and suffering? In the following sections I examine the newer issues in the areas of animal experiments (Section 8.3.1), the extensive change in regard to animals' freedom from pain (Section 8.3.2), and the challenge posed by the creation of chimeras (Section 8.3.3).

### 8.3.1 Animal Experiments

Animal experiments in research have long been a matter of controversy. Under what conditions may animals be exposed to pain and suffering in order for us to gain knowledge? An important aspect concerns the different estimates concerning the necessity of performing research on living models and the possibility of avoiding in vivo experiments in order to reach the goals of toxicological and biomedical research. Fundamental arguments in this debate concern the following assertions (Nuffield Council of Bioethics, 2006):

- The differences between the species mean that transferring the results gained from animals to humans is problematic.
- It is hardly possible to solve the problem of selecting the appropriate species for a particular experiment.
- A reductionist understanding of illnesses is a precondition for animal experiments because at the most only a few physiologically similar phenomena can be represented in animal models.

The advocates of the use of animal models employ primarily two arguments to support their position. The first is that studies on living organisms are an essential part of biological and biomedical research, because of the similarities between species, which as a matter of principle cannot be completely replaced by in vitro studies. The second is that it follows from the necessity of in vivo studies that animal experiments are not only acceptable but even ethically desirable because they make it possible to avoid experiments on humans (DFG, 2004). With regard to the problem of the difference between species, these advocates point out that the level of empirical knowledge has grown substantially and is currently being made more precise by means of biostatistical procedures.

Genetic engineering, molecular biology, and nanobiotechnology make possible new types of research and interventions, such as xenotransplantation or gene pharming, which also lead to a need for further animal experiments. The introduction of new technologies stimulates in this way over and over again anew the debate about the suitability of animal models and their limits. For example, the assumption that rapid and massive progress could be made from the creation of knockout lines of mice has recently become an object of increasing controversy in the scientific community (Qiu, 2006). The criticism that is raised is not a matter of ethics but motivated — in terms of the theory of science — by the question as to the anticipated gain in knowledge. This is supplemented by concerns as to whether the substantial sums for this purpose are being sensibly used. Scientific progress in this way always raises new demands to apply the existing normative framework for animal experiments to more and more cases and, in part, to new ones and to make it more specific. The fundamental issues, however, stay the same.

A new dimension in the evaluation of the procedures of modern research from the perspective of animal ethics is a consequence, however, of the interaction between human enhancement (Chapter 9) and animal enhancement. Animal experiments make up much of the work on animal enhancement (Ferrari *et al.*, 2010), which is frequently in preparation of human enhancement. While it is hardly controversial ethically that animal experiments are in principle legitimate if their results can be used therapeutically for humans, the situation is completely different if the goal changes from human therapy to human enhancement. Controversial forms of human enhancement could — considered from the perspective of animal ethics — make it necessary to have different rules protecting experimental animals than those that currently apply. Even conceptually, human enhancement is by definition not therapy (Chapter 9). If human enhancement is classified as a luxury, then animal experiments for this purpose would be similarly problematic as they are for cosmetics, for example. There is an urgent need for ethical and juridical clarification in particular if — as in the Swiss regulations on animal experiments — there are legally regulated conditions for permitting animal experiments. Exclusion criteria for animal experiments such as using animal experiments for designing new cosmetics, which have already been defined in some countries for genetically engineered changes to animals, could also be applied to research and development for enhancement technologies. This challenge to the existing normative framework must urgently be made the subject of ethical debate at the societal level and, if appropriate, a subsequent legal clarification.

### 8.3.2 Elimination of Animals' Capacity for Suffering

The issue of reducing or entirely eliminating animals' capacity for suffering has been discussed in the context of the evaluation of far-reaching interventions in animals.[5] If animals no longer suffered, then pathocentric ethics would no longer apply to them. To call this enhancement may initially sound cynical. Considering the description of animal enhancement given above as improvement relative to the expectations for human usage, this could in fact be interpreted as enhancement if consequently the human use of animals were simpler because one would not have to take an animal's well-being into consideration, thus making human use more efficient in an economic sense. From an imagined perspective of animals one could say, if animals are going to be used for human purposes anyway, then it is better for them not to suffer. Before continuing this difficult discussion, I would first like to mention a few of the concrete suggestions for illustration.

Animal microencephalic lumps (AML) have been an object of discussion for some time. As a result of genetic engineering, these animals would no longer be in a position to feel any interests or needs in any form and would thus be incapable of suffering. AML

[5] To be quite clear, these considerations are thought experiments and do not describe elements of ongoing research.

chickens, for example, would lack extremities and be featherless, and they could not see or hear or move (Comstock, 2000). From AML pigs and AML cattle it would be possible to regularly remove pieces of meat for consumption since they would grow back assuming that appropriate nutrients are provided. Laying hens could be modified by genetic engineering so that they would not suffer any longer in their tiny cages. This manner of thinking can also be applied to experimental animals: if experiments must inevitably be carried out on animal models, this could lessen the suffering.

AMLs would be reduced to the status of beings incapable of experiencing sensation, an extreme reduction. This could be regarded as a maximal instrumentalization of animals, in which "animal character" would largely or completely disappear, leaving behind a type of "utility object." Such an intervention would basically be so massive that almost all of the qualities of the animals would be changed. Questions of the nature would be raised whether AMLs are still animals and whether they are still subject to animal welfare regulations, or whether they were not rather biofacts (Karafyllis, 2006) in which the created portion so eclipses the rest that the latter no longer plays any role. If we however understand animal welfare holistically as the complex entirety of an animal's physical and mental states, then the loss of the possibility to achieve positive mental status — such as following a massive reduction in cognitive abilities — must be seen as a decrease in integrity (Holtug, 1996).

The creation of such animals does not appear to be absolutely improbable or purely fantastic. It goes back to motifs that today are within the realm of what can be achieved or striven for scientifically. Developments in this direction do not seem completely absurd if we think of the genetically engineered changes in farm animals to make them better adapted to the conditions under which they are kept on farms, to make them disease resistant, to remove the visual ability of chickens, or the creation of so-called zombie mice in the context of research on the neuronal correlates of consciousness (Koch, 2004). At the same time, these are considerations of a rather speculative nature, but ones that an "explorative philosophy" (Chapter 10) can use as a starting point.

The extreme scenarios of AML farm animals and of transgenic mice made incapable of sensation could evoke counterreactions. On the one hand, they could be rejected intuitively in that such animals might be seen as monsters and the repulsive products of technology.

Intuitively, such a manipulation of animals appears to be illegitimate. On the other hand, such rejection might subside precisely because these beings would no longer be viewed as animals in the sense of fellow creatures, but rather as living beings such as plants or even as simple collections of cells and tissue that are incapable of sensations (Schmidt, 2008).

The question as to the maximum reduction in animal attributes and in particular the elimination of their capacity for pain poses a special challenge to animal ethics because the capacity to suffer constitutes a direct source of moral relevance within pathocentric theories and is also an important criterion for other theories of animal ethics (Ferrari, 2008). Although it would be desirable for intensive animal farming to be forbidden because it is associated with pain and suffering for animals, the situation is: if we continue to live in a world in which meat is consumed, it is better for every measure to be undertaken to avoid this suffering or to minimize overall suffering (Shriver, 2009). To afterward replace all the current farm animals by others that have a lower or no capacity for pain would lead to better consequences than the status quo. From this perspective, the elimination of the capacity for pain could be justified not only by human interests — which are usually articulated by recourse to animal welfare anyway — but also by an imagined interest of animals and could be classified as enhancement despite the massive interventions mentioned above (Attfield, 1995). A remarkable counterargument is, however, that the creation of AML animals would be morally reprehensible not because harm would be done to the AML animals but because it is impossible to imagine such animals being created without a larger number of *other* animals experiencing suffering and pain (Comstock, 2000; Ferrari, 2008).

This discussion points to the significance of protecting the welfare and the integrity or dignity of animals, as well as to the limitations to a fixation on animal pain and suffering. Ultimately the issue is the moral status of animals and its consequences. A central question of whether ethical evaluation should be based on the capacity for pain (in pathocentric theories) or on the totality of the species-specific attributes of the animal (partially covered by concepts such as integrity, inherent value, and dignity). The answer will determine the standards establishing which and how many interventions are acceptable in animals and which criteria the evaluation should employ, which in turn will determine the appraisal of far-reaching animal enhancements. In any case, these are complex issues that affect our understanding of an animal as a living being and go beyond individual problems of animal ethics.

### 8.3.3 Transgressing the Boundary Between Humans and Animals

Transgressing the boundary between species, such as between man and animal, is made possible or facilitated by converging technologies. It is the declared goal of xenotransplantation, with reference to the removal of animal organs or tissue for transplantation into humans. Thinking these developments further, out of their narrow therapeutic context, leads to questions concerning the possible ethical problems and limits.

Many of these questions about research on chimeras are similar to those in the discussion of genetically engineered modifications of animals, such as the question whether a species has an intrinsic value and therefore whether transgressing the boundary between species would be morally problematic or unnatural and what this would mean (Robert and Baylis, 2003). The creation of man–animal composite creatures and in particular the possibility of the novel instrumentalization of (quasi-)human living beings lead to the question as to the moral and judicial status of these beings (Chakrabarty, 2003). The fundamental question that would be raised by man–animal composite creatures is: How much animal material is necessary for a human to no longer be a human? In Denmark, for example, animal experiments are forbidden that would lead to the creation of chimeras that strongly resemble humans (Danish Council of Bioethics, 2008).

The rejection of man–animal composite creatures is based on two main arguments. One is that membership in the human race is the criterion for attributing human dignity, which is grounded either in religion or metaphysics (anthropocentric). Consequently, the mingling of the animal genome with that of humans would signify a violation of human dignity (Karpowicz *et al.*, 2005). The other is that man has a special position because of his specific cognitive capacities, in particular his ability to think abstractly, to form complex thoughts, and to use language (see for example Höffe *et al.*, 2002). These arguments are not unchallenged. The recourse to membership in the species of the human race proves to be circular or a mere defense of anthropocentrism if you reject its religious grounding in secular society (Ferrari, 2008). Furthermore, cognitive capacities are to be seen in a continuum between humans and animals as new ethological studies show (e.g., Warneken and Tomasello, 2006).

In confronting the arguments against the creation of man–animal composite creatures, the Danish Council of Bioethics (2008) suggests an argument that is relevant to reflection on the moral status of enhanced animals. It is possible to speak of violating the dignity of creatures if, for example, an animal whose cognitive capacities have been enhanced is not kept in a manner appropriate to its (enhanced) status. Turning this argument around means that enhancement can also result in extended moral obligations in dealing with the enhanced animals.

Overall, with regard to chimera research, animal enhancement does not result in any completely novel ethical issues. There is however a certain ethical connection in that concerns have been raised in the ethical discussion concerning research that has to do with animal enhancement with regard to the creation of chimeras that increasingly could or are supposed to have human capacities. Although this discussion is speculative, it does raise interesting issues about the status of composite creatures that could no longer be unequivocally classified in ethical or judicial categories.

## 8.4 Changing Human–Animal Relationship

Converging technologies aim at interventions in animals that are novel and delve deeper. Genetics refers to itself as a science of knowing how; synthetic biology with its roots in nanobiotechnology aims technically at creating artificial life (Chapter 7); and NBIC convergence is looking for human (Chapter 9) or animal enhancement by means of synergies between scientific disciplines. As "technosciences" (Latour, 1995), they are not primarily interested in an understanding of nature but in *intervention* and in the creation of instruments for the purpose of intervention. From the perspective of nanotechnology, nature (including living beings) is increasingly being understood as a complex mechanism of fundament components at the level of atoms (Chapter 7). At the same time, it is considered something malleable that can be recombined by using the knowledge from nanobiotechnology (e.g., Ferrari, 2010). The capacities of living beings are described in technomorphic terms and conceived of as being malleable in the sense of enhancement. Animals and plants are considered to be inefficient production machines that are to be enhanced, and our natural environment *in toto* is described as a system to be perfected. This suggests a renaissance of ideas about human technology exerting a total control over nature (Section 2.4).

Animal enhancement increases the technical and manipulable part in animals and reduces the natural one (to a maximum in AML animals; Section 8.3.2). It is arguable whether this results in changes in the man–animal relationship. Novel questions have at least already appeared that could signify relevant changes in the human view of animals. Thorough observation of this is indicated because of the strong cultural significance of the man–animal relationship. This, admittedly, is not a genuine task of ethics but rather one of anthropology, hermeneutics, and the philosophy of technology and nature.

A look at current and anticipated future research in the context of animal enhancement shows that new technologies have substantial consequences on the opportunities for using animals. The possibility for transgressing species' boundaries, the extreme modification of animal attributes, the increase in and intensification of animal experiments, and the introduction of artificial items in living beings give us occasion to fundamentally reconsider our relationship to animals. Even in the debate on genetic engineering there was a broad discussion about the yuck factor, i.e., the feeling of outrageousness and aversion to certain scenarios such as those suggested by the keywords "man–animal composite creatures" and "AML hens." One question is whether these adverse intuitions are justified and can be part of rational argumentation.

An example for diverging attitudes and emotions as well as for normative uncertainty in the talk about animal enhancement is the current research in the area of *in vitro* meat. At the end of the 1990s, NASA had experiments conducted to create artificial tissue to provide food for space expeditions. In 1997, the animal protection organization PETA began to cofinance a project between research centers in The Netherlands and in North America with the goal of growing animal-like tissue with the taste, texture, and muscle mass of fish or chicken meat. An international research community now exists whose motivation is not solely animal protection — namely for *in vitro* meat to replace traditional forms of practice that are frequently fraught with suffering for animals — but also environmental protection, since the animal-based food industry is responsible for substantial $CO_2$ emissions, the deforestation of tropical forests, and water pollution (FAO, 2006). Furthermore, economic advantages are expected because the costs of keeping animals and damage to the environment would be eliminated. Finally, health protection is mentioned because the meat produced in a laboratory would be subject to more rigorous testing and free of antibiotics and other animal medications. The production of in vitro meat is therefore tied to positive visions that could lead to sustained consumption, similar to in the visions of green chemistry (Ferrari, 2009a).

The fact that humans compare and distinguish themselves from animals is a decisive element in their self-reassurance and determination of their role. An example demonstrating this is the central role that animals, animal images, animal figures, and animal imitations have played in magical and religious traditions in early human history. In the relatively recent monotheistic religions, many of the regulations for dealing with animals and also animal metaphors are also of great significance. As a result of modern science and technology, a third element gains in influence in these processes of self-determination oriented on animals, namely, the machine. The Cartesian interpretation of man as an animated machine agrees with an interpretation and depreciation of the animal as an inanimate machine. In today's visions of artificial intelligence and autonomous technology, the relationships are changing since advanced machines are getting closer to humans if we define ourselves by our cognitive capacities, which technical systems can imitate to an ever stronger degree.

Work in the converging technologies is today ongoing both to anthropomorphize and to zoomorphize the machine. Bionic robotics approaches animals with the fascination of an engineer for effective technical solutions and produces fascinating artifacts. In other areas of robotics, however, the fashion of automatons similar to humans is experiencing an impressive renaissance. It is expected that some zoomorphic and anthropomorphic machines will serve the emotional needs of people in general need of assistance. The traditional boundaries that have led to clear relationships in the triangle man–animal–machine are becoming increasingly permeable.

It is possible that radical shifts are looming in a small area of the field of converging technologies, namely, in the relationship between animal and machine — the two factors central for human self-understanding — in the form of animal–machine hybrids (Ferrari *et al.*, 2010). The methods and visions in this area could in the future raise the question of whether strongly modified animals — in particular, insects — could or should be regarded animals any more. Without having to recur to these expectations of the future — or even to the visions of an artificial insect brain that seem fantastic — the degree of coalescence between animal and machine can already be interpreted as the sign of a new quality. And robots that can be controlled by the use of animal elements such as rat neurons at least also point in the direction of the de facto disintegration of the boundary between machine and animal.

Technology is increasingly penetrating our cohabitation with other species and modifying the material basis of the relationship in very different ways. Culturally, we can note on the one hand a stronger acknowledgment of animals as being cohabitants of the earth, an increase in animal welfare, and growth in civic involvement for a vegetarian diet and against animal experiments during the last two decades. On the other hand, there is a continuing spiral of increasing exploitation of animals, and animal enhancement is in this tradition. The feeling of being related to animals is increasing just as is the intrumentalization of animals for human ends. Currently, no clear developmental trend can be recognized in these contrasting observations. What is needed are monitoring, hermeneutic questions as to the significance of changes in the relationship between humans and animals, and thorough reflection on these analyses.

## 8.5 Summary and Conclusions

The enhancement of animals has taken place for thousands of years in the context of the human use of animals and is part of the history of human civilization and of our relationship to nature. Depending on the concept of technology that is employed, breeding techniques can also be subsumed under technology. The converging technologies currently draw especially on existing knowledge and experience in genetics, molecular biology, and veterinary medicine, to name the primary disciplines, and involve mostly "evolutionary" processes, not disruptive ones. Thorough observation and study is needed to uncover creeping and very gradual shifts.

In a certain sense, animal enhancement as a goal thus does not appear to be anything revolutionarily new. Yet ethical questions arise at the level of means. This concerns, for example, the degree of the depth of interventions, the question of their reversibility, and the altered conditions regarding the dignity of a creature in a time when the possibilities for accessing the animal, its body including its cognitive capacities, and its environmental conditions have been massively extended. In this context, the fundamental asymmetry between humans as actors (who, e.g., define the goals of enhancement) and animals (as the affected but subordinate partners) has consequences. Insofar as there is a change in the means of animal enhancement by which interventions in animals take place, the question will increasingly arise whether the incapacity of animals to provide their consent will not have to become an object of consideration in a novel manner in human reflection on the legitimacy and illegitimacy of certain interventions. The first issue to be clarified would have to be the relationship (or the desired relationship) of animal enhancement *for* humans and the human interest in the use of animals to animal enhancement in the interest *of* the animals (which, however, has to be determined by humans). The most urgent challenge to ethics in this field may well be in evaluating the legitimacy of animal experiments that are supposed to lead to animal enhancement in order to contribute to human enhancement (Section 8.3.1).

To conduct animal enhancement in a responsible manner is a complex challenge. On the one hand, there are social debates, analyzes from philosophy and animal ethics, and efforts by animal welfare organizations in the form of civil groups. On the other, there are the concrete interests of the users of animals in science, agriculture, sports, the military, and entertainment. This duality leads to a complicated constellation of arguments and actors that takes on many forms. Analysis and debate are needed to clarify what should be understood by "responsible innovation" in the field of animal enhancement. Possibilities of attributing and distributing concrete accountabilities and responsibilities have to be discussed while keeping in mind that attributing responsibilities is a social and political process building on specific rules and arguments (Section 3.2.4).

In dealing with all of these issues, we must always be aware that ethics is only *part of the game*. As in other cases of new and emerging developments in science and technology, the issues in animal enhancement are in many regards not concrete ethical ones at all, but rather the need to reach a social and anthropological self-understanding. Questions such as regarding the development of the relationship of humans to animals and to the tendencies toward technicalization that are at work in this context need first and above all public dialogue and expert discourse. Until now, debates on the relationship between humans and animals have only taken place sporadically, for example, with regard to the conditions under which farm animals are kept. The topic of the technicalization of animals has hardly reached the public or politics.

# Chapter 9

# Human Enhancement

From the very beginning, some of the expectations for technical progress based on nanotechnology have comprised promises concerning the "further development" of man by means of "technical enhancements." In the "converging technologies," the idea of a convergence of biotechnology, nanotechnology, brain research, and information technology was linked with the explicit goal of "improving human performance" (Roco and Bainbridge, 2002; cf. Section 9.1). Advances made in research, but also frequently undifferentiated reporting in the mass media as well as the film industry — in particular, in science fiction films — have attracted attention to this field. The creation of "neuroelectric" interfaces between the nervous system, in particular the brain, and information technology systems such as computers is a prominent example of these visionary expectations. This leads to a whole series of ethical issues, some of them far reaching (Section 9.3), the discussion of which — as in the preceding chapters — is preceded by terminological considerations (Section 9.2) and followed by hermeneutic questions (Section 9.4), here about the relationship between humans and technology.

## 9.1 Improving Human Performance of Converging Technologies

A controversial international debate about human enhancement has been ongoing in the last few years in consequence of a publication

*Responsible Nanobiotechnology: Philosophy and Ethics*
Armin Grunwald

ISBN 978-981-4316-80-4 (Hardcover), 978-981-4363-33-1 (eBook)
www.panstanford.com

based on a workshop held by the National Science Foundation (NSF) (Roco and Bainbridge, 2002). Building on discussions about perfecting humans by means of genetic engineering and about the philosophical criticism of this (e.g., Habermas, 2001), the controversial international debate enriched the debate about the future of man by adding further aspects. In the context of transhumanism (Coenen, 2006), this debate has even taken on ideological qualities.

### 9.1.1 The Vision of Converging Technologies

The recent discussion of human enhancement was initiated by the report of an American research group to the NSF, which has considerable influence on the agenda for the sciences (Roco and Bainbridge, 2002). The title of this report conveys its program: *Converging Technologies for Improving Human Performance*. Nanotechnology and the converging technologies offer, according to this report, far-reaching perspectives for perceiving even the human body and mind to be formable and for improving them through precisely targeted technical measures. Human enhancement is understood in this connection to be *technical* enhancement, initially especially at the level of individual abilities: "The individually desired enhancement of human features and functions — in place of a compulsory state one — is now conceivable, a form of 'liberal eugenics'" (Siep, 2006, p. 309). Only as a result of this mediation at the individual level do we reach considerations of social enhancement:

> Rapid advances in convergent technologies have the potential to enhance both human performance and the nation's productivity. Examples of payoff will include improving work efficiency and learning, enhancing individual sensory and cognitive capacities, revolutionary changes in healthcare, improving both individual and group efficiency, highly effective communication techniques including brain to brain interaction, perfecting human–machine interfaces including neuro-morphic engineering for industrial and personal use, enhancing human capabilities for defence purposes, reaching sustainable development using NBIC tools, and ameliorating the physical and cognitive decline that is common to the aging mind. (Roco and Bainbridge, 2002, p. 1)

Behind these thoughts is the idea of "NBIC convergence," according to which nanotechnology, biotechnology and genetic engineering, information and communication technology, and cognitive science and brain research are converging and, as a consequence of this convergence, creating radically new opportunities:

> Nanotechnology is the means for manipulating the environment at the molecular level. Through modern biotechnology, humans endeavor to direct their own evolution. Information technology and cognitive science are both exploring ways to increase the speed and range of information a person can assess, whether by artificial computing or expanding cognitive capacity. (Williams and Frankel, 2006, p. 8)

We must first ask what is meant by "convergence." Several alternative interpretations are (for more details, see Coenen, 2008a; see also Wolbring, 2008b) as follows:

- An evolutionary convergence of the technical fields named above in the sense that an overarching line of technology is formed that replaces the previously separate fields, and in which the individual fields largely merge.
- The convergence of the disciplines of the natural sciences with those of the technical sciences in the sense of stronger interdisciplinarity and transdisciplinarity, such as is the foundation of the European answer to the American position represented by Roco and Bainbridge (Nordmann, 2004).
- The development toward a new unity of the sciences on the basis of nanotechnology, such as is suggested by Roco and Bainbridge's reference to Leonardo da Vinci as a polymath and in the expectation of a "new Renaissance" (Roco and Bainbridge, 2002). This idea of convergence links together, on the one hand, with sociobiological programs (Wilson, 1999) and on the other with visions of a new "unity of science" (Bunge, 2003).
- A concept of research policy that permits normative positions to be formulated for the further development of the scientific disciplines involved and for their agenda.

This book is based on the interpretation that "atomic reductionism" is the guiding motif behind many of the visions and expectations related to nanotechnology, as is expressed in the slogan "shaping the world atom by atom" (NNI, 1999; see Chapter 2). From this perspective, it is logical to look in the same direction to find the epistemological program of the "emerging technologies." Convergence would, in this line of thought, mean restoring the common roots of the scientific disciplines involved, namely that they operate and work using atoms and molecules. The scientific disciplines would converge in these roots, which, in analogy to plants, can be viewed as different shoots that have developed out of common roots. Nanotechnology, as the science of technically working with atoms and molecules, would provide the necessary key competence.

Operating with atoms and molecules would in the perspective of atomic reductionism thus become an engineering version of a "theory of everything." This would not be a theory of physical fundamentals that provides explanations at an abstract level. It would be a practical theory of operation and manipulation in order to construct matter in a targeted fashion. Ultimately, nanotechnology — as such a theory of everything in this line of thought — would be the instructions for building and operating a molecular assembler (Drexler, 1986) or a similar construction. This assembler would signify the triumph of atomic reductionism. At the same time, it would show man to be the one who could build such an amazing machine and who could ultimately create himself anew technologically — which would lead us back to "enhancement" — and thus modify himself according to his ends: "When God fashioned man and woman he called his creation very good. Transhumanists say that, by manipulating our bodies with microscopic tools, we can do better" (Hook, 2004; see also Dupuy, 2007). In a radicalized version of nineteenth century physical reductionism, the spheres of the living and of the social are supposed in this perspective to be explained and made the object of technical manipulation starting from their atomic basis:

> Science can now understand the ways in which atoms form complex molecules, and these in turn aggregate according to common fundamental principles to form both organic and inorganic structures. [...] The same principles will allow us to understand and when desirable to control the behavior both of complex microsystems [...] and macrosystems such as human metabolism and transportation vehicles. (Roco and Bainbridge, 2002, p. 2)

If this program could be implemented, then — according to the conviction of its promoters — even the categorical difference between the living and the nonliving would disappear, at least with regard to the capacity for matter to be manipulated at the atomic level (see also Chapter 7). The following theses are examples of the optimistic belief in progress that is expressed in this program. They are supposed to be milestones on the road to such a new society and to a new man; they also provide an impression of what is understood under technical enhancement of man:

> Fast, broad-bandwidth interfaces directly between the human brain and machines will transform work in factories, control of automobiles, ensure superiority of military vehicles, and enable new sports, art forms and modes of interaction between people.
>
> The human body will be more durable, healthy, energetic, easier to repair, and resistant to many kinds of stress, biological threat and aging process. (Roco and Bainbridge, 2002, pp. 4ff.)
>
> Convergences of many technologies will enhance the performance of human warfighters and defenders. (Roco and Bainbridge, 2002, p. 287)

These passages claim to do more than simply demonstrate future *possibilities*. As can be recognized by the fact that most of them are formulated in a constative future tense (Grunwald, 2006a), they want to anticipate the future development as something that is already fixed today and thus simply to be constated. Roadmaps and milestones are supposed to provide the link between today's research and the visionary futures (Roco and Bainbridge, 2002, pp. 4ff.). In this sense, the visions claim to formulate not *simply possible* futures but to be anchored in today's scientific activity by being part of the current agenda and milestones to be attained, and, by going down this path, to lead to making the far-reaching visions come true (Grunwald, 2006a).

### 9.1.2 Improving Human Performance: The Cultural Background

Man's dissatisfaction with himself is presumably as old as mankind. Dissatisfaction with one's physical endowment, physical and mental capacity, dependence on external events such as illnesses, the inevitability of aging and ultimately death, or — and this is presumably particularly frequent — his or her appearance are well-known and virtually daily examples of an individual's self-experience and represent furthermore a general self-experience of man throughout all of history. The experience of one's own inadequacy when confronted by one's own or society's expectations as well as one's limited capacities in dealing with challenges and blows of fate — which is frequently perceived as inadequate — can be found in a wide spectrum of human self-experience. This self-experience extends from annoying forgetfulness concerning both harmless as well as important items of knowledge to a collective experience of inadequate morality, such as in conflicts and wars, and of moral failure in view of the temptations of power or wealth or simply for the sake of convenience.

Stories, fairy tales, and sagas process this experience, such as the story of the fountain of youth or the legends in which humans acquire superhuman powers. Spider-Man and his colleagues of both sexes are modern expressions of such dreams. Cultural practices were developed to compensate for perceived deficits, such as beauty flaws, for example by means of cosmetics or clothing, and forgetfulness, by means of crib sheets (whether paper or electronic). Superhuman capacities — i.e., those clearly surpassing the average capacities of humans — have been and continue to be achieved by means of training, such as in sports or the ability to play a musical instrument. Even a cultural achievement such as our judicial system can be interpreted as compensation for man's experience of deficits and as an aid in stabilizing civilization regarding its fragile nature. Considering man to be a "deficient being" (Gehlen, 1986) who builds his own civilization, including technical opportunities, to compensate for his deficiencies is a prominent anthropological interpretation in view of these experiences — although one to be criticized (Gutmann, 2004). A new manner of experiencing deficits is itself linked with technical progress. This is the experience of not being able to keep up with technical progress, of feeling inferior in view of the possibilities of technical systems, and of experiencing oneself to be antiquated even relative to one's own technical creations (Anders, 1964).

As a branch of the economy with substantial and still growing revenue, cosmetic surgery is today probably the most commonly used form of compensation for features perceived by an individual to be deficits. What does not please and does not correspond to one's own expectations or to external ones is adapted by technical means. Doping in sports, which has led to public scandals over and over again for decades, represents the attempt to enhance (certain) humans. Such doping is, however, not only a cause of concern from the perspective of sports ethics, but it can sometimes even lead to serious damage to the health of those involved, even to their premature death. Such doping continues and is being further developed despite its health impact, despite the social ostracism, and despite the threats of relevant sanctions. The enhancement of certain individual abilities is even on the advance in everyday matters. Pharmaceuticals are an example, specifically those that are supposed to enhance the performance of individual, such as in testing situations (which is occasionally referred to by the concept of "everyday doping," linking it conceptually with sports).

While the forms of compensating for or overcoming deficits in specified properties or capacities refer to *individuals* (athletic performance, personal beauty, good results on examinations), the *collective* enhancement of humans is certainly not a new topic either. Even in the early modern period, there were breeding programs for enhancing humans, some referring to Plato's Republic (after Siep, 2006, p. 309). The frequently lamented deficits of humans from the perspective of morality and civilization (e.g., in the expression *Homo homini lupus*) led in the European enlightenment, which embodied a belief in progress even in a moral sense, to approaches that attempt to employ *education* in a dedicated manner in order to enhance man *as a whole*, i.e., ultimately enhance man's and society's constitution in general. Beginning with the individual, above all in school education, a far-reaching higher development of human culture, civilization, and morality was to be stimulated and supported. For example, the critical theory of the Frankfurt school set on the emancipatory function of education, a few aspects of which — along with many others — can be found in ideas on education for sustainable development (Kastenhofer *et al.*, 2010).

In totalitarian regimes, "human enhancement" was at the service of the respective ideology. In Nazi Germany and in the context of its biologically racist ideology, for example, "enhancement" was understood to refer to breeding to strengthen the allegedly Aryan ideals. The ideal was represented by physical features (blond, blue-eyed, athletic) in connection with an unconditional subordination under the Nazi regime. "Human breeding" was its declared program with regard to the features that were accessible biologically, while with regard to social qualities the multiple possibilities of indoctrination and propaganda were utilized for what was understood to be "enhancement." Stalin also employed propaganda and indoctrination in order to "enhance" individuals according to the orthodox ideology.

On the one hand, the idea of human enhancement is in this way largely linked with individuals' experience of their own individual and collective imperfection. Attempts to implement the idea stand in what is at least an ambivalent tradition. Seen in the context of man's negative experience with enhancement fantasies and the measures employed by totalitarian regimes in the twentieth century to make them come true, one cannot deny that approaches to human enhancement can exhibit totalitarian traits. Human enhancement in favor of some propagated form of *future* society can be instrumentalized to suppress *contemporary* humans. For this reason, caution is due as soon as there is talk of a dedicated "enhancement" of humans and society.

Indeed, even the simple use of the word "enhancement" must be examined carefully. "Enhancement" is not a one-place predicate because "enhancement" is not undertaken in itself but always only relative to certain criteria (Section 8.2). Enhancement relative to certain criteria (e.g., practicing Stalinist orthodoxy) can mean deterioration relative to other criteria, for example, with regard to the ideal of an emancipated citizen. Caution is thus due in order not to fall into the obvious rhetorical trap of everyday language, namely of considering an "enhancement" to be positive in itself. As a matter of principle, we must inquire as to the criteria according to which a simple change is evaluated as an "enhancement." This must suffice as a critical introduction to the ethical issues concerning technical utopias for human enhancement (see on this also Section 9.3).

### 9.1.3 Enhancement Utopia 1: Neuroenhancement

"Neuroenhancement" refers to an enhancement resulting from an implant or medication that are connected either to the nervous system or to the brain or that act on them. If man is technomorphically modeled as an information-processing being, then it is possible to distinguish technical enhancements for taking up data from the environment (sensors), technical enhancements of the brain's capacity to process and store data, and technical enhancements for controlling external motor systems.

The expression "converging technologies" applies in the full sense of the phrase in this context (Roco and Bainbridge, 2002).

Biotechnology, brain science, neurophysiology, and information and communication technologies meet within nanotechnology. One key question among others is how to transform electrical impulses at contacts or wires into impulses that are "understood" and processed by the nervous system. A successful interconnection between an artificial limb (e.g., hand or leg prosthesis) and the nervous system would make it possible to compensate for the loss of a natural limb to an extent which is presently unimaginable. This sort of technical limb reconstruction would open up possibilities to expand conventional body functions.

Part of neuroenhancement extends to neuroimplants. As a rule, the starting point is the motivation to compensate technically for a *loss* of sensory function (e.g., of an eye or an ear). As a result of advances in nanoinformatics, such as miniaturization or the increased capacity of implants to take up and process data, the spatial dimensions and performance of the neuroimplants now approach those of natural systems. Not subject to human physiological limitations are the detection, recognition, and interpretation of external signals (e.g., acoustic or optical ones). A visual implant can in principle be extended technically so that it can receive data beyond each end of the visible spectrum of electromagnetic waves. This makes it possible to imagine that a visual prosthesis could be implanted in a human that would make it possible for them to see in the dark as if using a night vision device. By using an artificial accommodation system, an optical zoom could even be integrated into the implant. The capacity to enlarge sections of a perceived image at one's discretion would presumably be quite attractive in many professions (e.g., soldiers, pilots, locomotive engineers, surgeons). These ideas about technical enhancement of human sensory organs are oriented on technical possibilities, such as they have been employed outside the human body, as in cameras, microscopes, and telescopes. They thus have technical models (Siep, 2006, p. 308).

Cognitive enhancement (Sarewitz and Karas, 2006) spans the cognitive functions of the brain. If the brain is modeled in the sense of information technology, i.e., as a machine that stores and processes data, these functions would on the one hand include expanding the *storage function* of the human brain and creating the possibility to back up the information stored in the brain on a brain chip. By connecting a chip directly to the visual nerve, it might be possible to record and externally store all visual impressions in real time.

In this way, all the visual impressions acquired in the course of a human life could be called up at any time. Via a wireless connection, the impressions could even be coupled with external bodies of knowledge or stored externally.

On the other hand, we have to consider the *processing* of data in the brain. A chip implanted in the head and connected to an external network could also be expanded as a new interface to the brain. It could be used, for example, to "load" the content of books directly into the brain, or a device could be created that could load different language modules onto this chip and activate them as needed, making the tedious work of learning a foreign language superfluous. I cannot emphasize enough that such thoughts are pure speculation, both now and far into the future. Yet they indicate a direction of thought that is certainly not irrelevant to, for example, a modification of the image of man or for the relationship between man and technology.

If man were to succeed in connecting an artificial limb or extremity (e.g., a hand or leg prosthesis) to the nervous system, loss of motor capacity (such as from an amputation or accident) could be compensated in a manner that is hardly imaginable today. It would namely be possible for the brain to exert control over artificial limbs in the same manner as it does over natural ones. Such a technical reproduction of control functions or entire limbs would then constitute the starting point for possible enhancements (on the conceivable consequences for sports, see Wolbring, 2008a).

Suitable neuroelectric interfaces would make it possible to connect still-to-be-invented motorized terminal devices to the nervous system. This device would be in addition to our arms and legs and would be controlled directly by the brain. This could provide attractive supplementary competence for training purposes and certain occupational groups: "Cognitive enhancements are envisioned by many as a future component of education [...] and by some as a future component of jobs" (Wolbring, 2008b, p. 28). It is however unclear how the human brain would process the information about novel organs and what effects this extension of our senses would have mentally if, for example, in addition to our natural limbs a third arm were to be controlled in the form of a robot by neuronal signals from the brain. The military especially exhibits great interest in such speculative possibilities. Relevant projects of DARPA, the advanced research arm of the American Department of Defense, are supposed to contribute in the short term to the following (after Coenen, 2008a):

- Revolutionizing prosthetics by utilizing brain activity for controlling assistive technologies and by developing limbs whose sensory and motor functionality is fully functional
- Developing systems with which computers would significantly enhance a soldier's performance, in particular in stress and while maintaining the necessary control of numerous devices
- Enhancing the performance of soldiers affected by sleep deprivation

Even further reaching possibilities belong to the medium-term and long-term expectations on technical human enhancement, such as controlling jet fighters solely by means of brain activity (projected for 2045), the development of exoskeletons that make normal movement possible despite heavy loads, or bionic equipment with which soldiers are supposed to be able to climb walls like a gecko, i.e., without the usual climbing equipment.

### 9.1.4 Enhancement Utopia 2: Antiaging and Immortality

Attempts have been undertaken for a very long time to extend man's life span, and some have met the requirements of modern scientific standards at least since the beginning of the twentieth century. The hopes to achieve longer life and reduce the negatively felt effects of aging also have spurred such diverse developments as the increase in sport activity among the elderly, the creation of a market for new dietary supplements, the popularization of Botox medications as antiwrinkle applications, or the upsurge in cosmetic surgery. It can be noted that new products from science and technology that are based on ideas from the converging technologies and nanotechnology and that in part even played a role in Drexler's (1986) visions currently exert a strong fascination.

Ideas and expectations that aging might be slowed markedly or stopped entirely play a central role in the discussion of human enhancement. Such hopes are being nurtured by several developments in nanomedicine (Section 2.2), which however is supplemented by rather speculative assumptions. Among the relevant developments is a new diagnostic procedure that is supposed to permit continuous monitoring of a person's health status in great detail (Freitas, 1999). Also being developed are procedures based on nanotechnology that enable carefully targeted drug delivery to the affected parts of the body. These procedures alone would in many cases make a substantially earlier and more efficient treatment of degradation processes possible. There are hopes that these procedures would produce far fewer side effects than the classical therapies, which in comparison appear decidedly coarse. If aging is a degradation process at the cellular level — an understanding that is a matter of real controversy in medicine — then aging could be retarded by immediately discovering and repairing any manifestation of a degradation process.

Even further reaching ideas are in circulation, for example, that nanotechnological resources could be used to install a second immune system, one that is technical in nature. Ultimately, such ideas are based technically on Drexler's assembler (Chapter 2) and the associated idea of being able to build nanorobots. Such intelligent nanomachines could move in the bloodstream, serving as a technical immune system monitoring the human body in order to constantly maintain an optimal health status (Drexler, 1986). According to these visions, any degradation and every sign of physical decline should be recognized immediately at the atomic level and be stopped or repaired. In this manner, the machines might succeed in healing injuries perfectly within a short period of time and ultimately in stopping aging.

Whether such visions can be realized, whether they are possible in principle, and how long a period of time is expected to be necessary for visible advances to be made are all issues that are highly uncertain. Yet some scientists consider it possible that the human life span could be extended from the currently medically accepted limit of 120 years to 250 years within the next few decades. Regardless of their speculative nature, it is important to pay attention to the effect that such expectations might have on public opinion, especially considering the antiaging boom. Even if, behind the speculation, there were no possibility for such ideas to become reality (which currently cannot be estimated), they could have real consequences on expectations, man's image of man, and the agenda of science.

Despite the speculative nature of such visions, ethical reflection began early. Moor and Weckert (2004) considered the question of what quality of life could mean in a society where humans could expect a life span of 500, 5000, or 50,000 years. Another speculative vision in this context is that it might be possible in the future to store the contents of a brain's consciousness on a computer by means of a neuroelectric interface and thus to transfer human consciousness to a machine. Since this machine can be repaired over and over again in case there is malfunctioning or technical aging, it would in a certain sense achieve a state of immortality.

## 9.2 Semantics of Technical Enhancement

Building on these general considerations of the concept of enhancement (Section 8.2) it is now necessary, on the one hand, to conceptually specify the distinctively *technical* aspect of enhancement, and on the other to contrast enhancement and medical healing.

### 9.2.1 Enhancement Beyond Healing

Although part of the conceptual field situated between healing and enhancing, improving performance, doping, and modifying human nature is the object of philosophical debates (e.g., Jotterand, 2008b), the conceptual problems are often uncritically disregarded. These conceptual problems must however be analyzed in advance in order to identify the need for ethical reflection since, depending on the concepts, different normative contexts are involved. Differently situated normative uncertainties could result.

Large segments of the enhancement debate attempt solely to clarify the border between *healing* and *enhancing*, or they argue that they cannot be clearly differentiated. "The accent in the discussion of bioethics today is usually on the question of whether it is at all possible to draw a line separating therapy and enhancement" (Siep, 2006, p. 306). Yet terminologically and in everyday understanding, human enhancement is categorically different from healing an illness or compensating for the consequence of an accident. Healing is oriented toward the regulative idea of a healthy individual. Healing reaches it goal when the patient is healthy, while enhancement does not contain any criterion within itself (Section 8.2).

This does not mean in any way, however, that ontologically — i.e., on the basis of objective data — healing can be fundamentally and unequivocally distinguished from enhancement. Medical judgment about this distinction is based not only on data, but also on normative criteria for "healthy" that have grown in cultural practice. The concepts of "healthy" and "ill" are thus the subject of positive and negative valuations of conditions that are dependent on sociocultural context and normative criteria (Engelhardt, 1976, 1982; Margolis, 1981; Nordenfelt, 1993). They are thus defined *between* a reference to nature and a value judgment, and contain both descriptive as well as normative components. Consequently, the attributes that a healthy human is supposed to possess cannot be determined by natural science although scientific data are relevant. Ontologically, i.e., by means of concrete attributes that are precisely specified, it is thus impossible to draw a clear distinction between healing and enhancement.

What is understood to characterize a healthy human — i.e., when healing reaches its goal — can thus not be determined objectively and naturalistically. The answer differs depending on the time point in history and on the civilization. Contextually, however, it is usually clear. The ophthalmologist who gives a patient an eye test has an understanding of what a healthy human eye is capable of performing. He or she will only suggest technical compensation (e.g., glasses) if the results deviate from their own understanding and if the deviation reaches a certain magnitude. The goal of this measure is to reach the statistically determined normal state. Healing ends when this state has been reached. It is impossible to imagine traditional medical conduct without the constellation that a presumed normal state functions as a normative evaluative criterion for recognizing deviations and diagnosing the necessity of interventions. Even though the border between healing and enhancement at the object level is controversial and may be unclear, in discourse it represents a border. The discourse of healing is a different one from that of enhancing, and the normative frameworks of both discourses, for example, differ distinctly (see below). In the realm of healing, the medical oath of healing is relevant and is supported in reflection and analysis by medical ethics. Therapy is subject to a sense of obligation, advisability, or expectation — an *Ought* — inasmuch as its purpose is the (re)creation of species-specific biological functioning in an understanding of being healthy that is adapted to a specific culture. Talk about human enhancement forsakes this discourse, however, and (still) has no regular place in the traditional system of applied ethics. In the case of enhancement, there is no *should* (Siep, 2006), but only or at best a *wanting* (Hormberg-Schwetzel, 2008).

It cannot be denied that healing itself — i.e., the restoration of certain body functions — frequently contains an element of enhancement, in two different senses. On the one hand, an unintended side effect of healing can be an enhancement in a different respect. The replacement of limbs by artificial replacements can lead not only to a restoration of lost body functions but also to enhancements, such as the higher mechanical stability of artificial limbs or a longer operational lifetime. Pistorius, the athlete amputated below the knee, had precisely this held against him. This was the reason given for his exclusion from competition (at least initially) (Wolbring, 2008a). In this way, i.e., as the unintended consequence of acts intended to heal, enhancements can enter routine medical practice and thus the lifeworld of anyone affected. In this sense, technical human enhancements could enter societal practices and would be normalized in an insidious manner.

On the other hand, something that is a healing or restorative intervention in one situation can be used in other situations as a technical enhancement. The technical requirements involved in technically restoring the lost body functions of someone wounded in war are not very different from the requirements for equipping a soldier technically. In this way,

> even the focus on the social roles "handicapped" and "soldier" in the relevant debates reveals a concrete strategy for organizing acceptance for performance-enhancing modifications of human corporeality and the human psyche. The promoters of radical human enhancement might find these test groups for a performance-enhancing society attractive in that in a military context, with its structures characterized by commands and obedience, the rejection of strongly interventionist techniques by an individual is relatively difficult to maintain and that in both groups there is often a high motivation and willingness to take a risk with regard to such procedures. (Coenen, 2008b)

The healing of those wounded in war, the public acceptance of which is guaranteed, can be employed — in the sense of a dual use — for enhancement in normal soldiers. Viewed in this way, society does not have a fundamental choice as to whether it wants human enhancement or not since technical enhancement will gradually and inevitably enter the practice of healing.

### 9.2.2 Healing, Doping, Enhancement, and Alteration

In regarding enhancement it is crucial that the *point of departure* of the enhancement be clarified (Section 8.2). Possible definitions are as follows:

1. The physical or mental endowment of a specific human *individual*
2. The standard of an average healthy person, as measured for instance according to statistical surveys of human performance potential
3. The achievement potential of humans under optimal conditions, that is to say, the achievement potential at the upper end of the statistical distribution

In the first case, a pair of glasses would even be an enhancement in some sense of the word for an individual whose eyes did not meet the expectations for healthy human eyes. The second case would only constitute an enhancement if a particular measure meant that the standard of a healthy human were exceeded. In the third case, one would ultimately only speak of enhancement if the abilities achieved exceeded what is typical at that range of the performance spectrum.

I will consider only those measures as human enhancement that take the condition of a healthy human being under optimal conditions as the starting point and go beyond this. Corresponding criteria are — according to the examples above — for example, the perceptibility of electromagnetic radiation on the basis of its wavelength, the storage capacity of the brain, or the rate at which knowledge is forgotten. Thus, I will only regard those alterations as human enhancements that in some manner make humans more efficient than is normally expected under optimal conditions. The term "normally" in this context also encompasses the training methods used in sports, coaching techniques, etc. This means that normal cosmetic surgery, in contrast to what is widely maintained, does not serve to enhance humans, at least as long as its aim is to achieve accepted ideals of beauty, i.e., those which are not out of the ordinary within cultural development and history.

This usage creates a semantic gap between healing and enhancement when the outcome of an enhancement does not exceed a customary human level. It is precisely here that there is a gap related to doping in sports. Doping is neither healing nor enhancement going beyond abilities that are humanly possible. To make the linguistic usage more precise and at the same time draw attention to each of the different normative regimes that regulate these options, I will employ the following terminology on the basis of the distinctions made above:

1. *Healing*: The elimination of an individual's deficits relative to the accepted standards of an average healthy human being.
2. *Doping*: An increase in an individual's performance potential without there being a deficit in terms of the accepted standards of an average healthy human being and without the individual's performance exceeding what still appears as conceivably normal, i.e., within the spectrum of usual human performance, whether in sports or in normal life.
3. *Enhancement*: An increase in performance going beyond abilities that are regarded to be "normally" achievable by humans who are healthy, capable, and ready to perform under optimal conditions.
4. *Alteration* of the human composition: For example, the invention of new organs or bodily functions.

In cases of doping — unlike "enhancement" as defined here — the topics that dominate the enhancement debate, such as man's self-image and the future of human nature, are not touched upon. While for doping, even for new forms, normative frameworks exist, which can be taken into account, they do not exist at all for human enhancement, i.e., in terms of the production of "superhuman" or "transhuman" capabilities and features. The normative uncertainties involved thus differ in terms of category. For this reason it makes sense to draw this distinction and work with it, for example, in ethical reflection. There is hardly any overlap between considerations of how we should handle the consumption of drugs to improve brain performance, which is currently strongly increasing, and questions about the direction a society would take in which the ability to discriminate between humans and technology would in the future be largely removed. The conceptual means we employ to speak about these topics should reflect such differences.

The distinction between doping as performance improving in the sense of option 2 and enhancement in the sense of option 3 is essential in the interest of a clear use of language and with regard to the respective normative framework, and thus with regard to the possible normative uncertainties and the necessities of ethical reflection. The reason is that doping is characterized above all by two normative demands on human action or by the problems resulting from the fact that these demands are insufficiently met: the demand for fairness and that one act responsibly toward one's own body.

- *Fairness* is the essential demand placed on athletic activity that is violated by doping. Individuals that dope gain unpermitted advantages over those who follow the rules.
- Doping poses *risks* to the human body (initially unknown ones, as a rule), on the one hand short-term threats to one's health (from a legal point of view, sports doping is in large measure drug abuse) or even to one's life, and on the other completely unknown long-term risks.

These challenges are clearly associated with normative uncertainty, because the criterion of adhering to the underlying accepted normative conditions has not been met, and this criterion must be met if there is to be a standard situation in a moral sense (Chapter 3). Ethical reflection, for example, about sports, coupled with considerations of prevention, education, and sanctions are therefore appropriate (for gene doping, see Gerlinger *et al.*, 2008).

If the distinction between doping and enhancement is taken seriously, it leads to several very clear classifications. First and trivially, sports doping as we know it today would not fall under enhancement. Everyday doping, such as taking stimulating pharmaceuticals like Ritalin (Farah *et al.*, 2004) prior to examinations would not be enhancement, unless it would lead the individual to exceed the customary abilities of test candidates to an inordinate degree. Even steps in cosmetic surgery would therefore not be enhancement, but partly healing — if suffering from beauty deficits made one ill — and partly doping — beauty doping — if the goal is to win beauty competition. The goal of the latter, analogous to the goal in sports doping, is to gain advantages in the respective competition. Both are linked to the known ethical problems of a lack of fairness and of unknown risks.

An example that illustrates the dependence of an individual measure on the context and the interpretation is Viagra. If it is taken for an erectile disorder, then this is a measure in the context of eliminating deficits, as measured by the customary attributes of humans, and thus belongs in the realm of healing. If it is taken by a healthy person and leads to an inordinate increase in the duration of erection, then this is performance enhancement. Between these two there is a region in which the increase in duration of erection is still in a customary range but nonetheless constitutes an improvement for the individual. This would then be a type of doping for improving one's competitive ability in this special sphere. This illustrates that the attribution of the terms "healing," "doping," and "enhancement" depends on interpretations of the respective situation and in particular on what "customary" is taken to mean.

The fourth category — the modification of human properties — goes beyond enhancing properties that are already present, in the sense of a performance-enhancing society (Coenen, 2008b; see also Section 9.7.2):

> The fundamentally novel quality of these thoughts about human enhancement is the fact that it no longer concerns an improvement of the body's own functions and performances but the imitation of performances by nonhuman beings and devices. (Siep, 2006, p. 311)

There is no model in the natural human organism for *alterations*: "This means adding new features to brain functions (brain to brain interface, web access, etc.) through technological means unknown so far" (Jotterand, 2008b, p. 18). Even though such "alteration" frequently stands at the center of thoughts about enhancement and is an ideal candidate for this role because it promises the farthest reaching challenges — or threatens them — this case will not be considered further in this chapter because the considerations are completely speculative and therefore nearly arbitrary. Many of the ethical (or pre-ethical, e.g., anthropological) aspects of alteration are likely to correspond to those of enhancement.

The boundaries between these four categories are visibly dependent on interpretation and may be a matter of controversy. Even though these boundaries may not be ontologized at the object level, they exert a decisive influence on the ethical debate because each is involved in a different normative context: in the case of healing, medical ethics; in the case of doping, the fairness imperative and considerations of risk; in the case of enhancement, largely normatively uncharted area (Section 9.3), exactly as in the case of alteration.

## 9.2.3 Technical Enhancement

Engineers know very well what a technical enhancement is. Every technology is described by certain parameters, which include the performance characteristics. Enhancement would mean bettering one or more of the already attained performance characteristics according to the accepted standards of technology, such as motor performance, efficiency, life span, or price. This description assumes that it is possible to carry out the necessary determination of the *starting point* of the enhancement itself in a technical manner. The criteria and the direction of the enhancement result from this definition relative to the technomorphically determined starting point. As a rule, the magnitude of the enhancement is determined directly by comparing, usually quantitatively, the values of the corresponding parameters before and after the enhancement. The first result of this terminological consideration is thus that technical modeling of humans is a prerequisite for the technical enhancement of humans that is more than merely metaphorical. The technical enhancement of man implies a technicalization in the sense of technical modeling *before* an enhancement measure is initiated.

From a practical perspective, as shown for example by cochlear and retinal implants, the technical compensation for deficits, such as organs lost as a result of illness or accident, comes before technical enhancement. The standard to be reached and goal of the technical compensation is to employ the technical implant to reach the customary human performance. The goal — fully corresponding to the medical oath — is to re-establish the patient's full level of performance by technically reproducing the natural functioning of the body, which in turn produces an *identical result*. This does not mean that the technical processes employed for this purpose have to reproduce the natural ones. For example, a retinal implant certainly does not have to use the same mechanism to detect incident light as the natural retina. Yet the outcome must be an identical result in the sense that the data received by the sensors and passed on to the brain produce the same impressions that the natural organ did. The retinal implant must produce a visual impression that is equivalent to that produced by a natural eye. In this case, a criterion for reaching the goal of an identical result could be, for example, that an ophthalmologist has the implanted patient take an eye test, objectively examines the visual performance, and compares this result with the standard set by a natural eye. In addition to this demand that the results be identical, further technical measures in the organism have to be adapted. At least the size, the surface biocompatibility, the supply of energy, and the provision of the necessary interfaces have to be taken into account.

Now, let us assume that one succeeds in technically recreating, for instance, a sensory organ such as an eye with equally good results. An artificial eye of this kind would — as is customary in technical development and production — be given a version number by its manufacturer. This would be "eye 1.0." Version 1.0 will not be the last one, however, because engineers and physicians will be thinking of the next version as soon as version 1.0 has been developed and tested. Continuous enhancement is a technological imperative in modern technology. The following enhancement might take an entirely different direction than the previous one, for example, a reduction in costs or in the service interval for eye version 1.0. Although the human sensory capabilities of version 2.0 consequently do not have to be improved (e.g., by providing night-vision ability or zoom options), enhancement nonetheless forms part of the spectrum of the technological imperative, as applied to eye 1.0. Human technical enhancement is thus revealed to be a step that is a logical consequence of the technical restoration of failed or deficient body functions. The transition from *restorative* to *enhancing* interventions is a gradual one *from a technical perspective*. The technological imperative leads *of necessity* from healing to enhancing if normative arguments do not restrict it.

It should not be forgotten, however, that technical enhancements in one sense frequently represent deterioration from *other* perspectives, something that engineers are very familiar with. Specialization based on one specific criterion of performance is often deterioration according to other criteria. For example, more safety often means higher costs. In human enhancement, this ambivalence concerns first the increased dependence on functioning technology, in this case on functioning enhancement technology. We must furthermore remember that technical enhancements, by modifying certain human performance characteristics that can be described technomorphically, lead to a specialization that is diametrically opposed to the multifunctionality of natural human limbs and sensory organs. An athlete, for example, who has artificial legs implanted to improve his performance in athletics, i.e., track and field events, will very probably have problems swimming that he would not have had without the technical enhancement. The fundamentally ambivalent impact of technical progress (Grunwald, 2009a) is thus transmitted by human technical enhancement into the immediately human realm.

Enhancement, then, has no intrinsic limits or measures but opens up an infinite space of the possible, just as technical progress does in general. The process of human enhancement does not stop once a given status has been reached, i.e., in the sense that a target has been reached. It serves on the contrary as the starting point for the next enhancement, and so on. Technical enhancement, similar to technological advance, would be an infinite process if there were no external societal measures for "containment" or for setting limitations.

## 9.3 Human Enhancement: Ethical Analysis

Technical enhancements of humans leave the normative framework of medical ethics. A field of activity is opened for which normative conventions have not been established, a classic case for ethics (Section 3.3). At any rate, it is easy to find normative uncertainties and ethical issues (Section 9.3.1) as well as patterns of ethical argumentation (Section 9.3.2).

### 9.3.1 Normative Uncertainties

Technical enhancement or alteration of humans constitutes a step into a new sphere in a normative sense, too. It does away with the idea of a healthy individual whose performance is customary, which can be identified statistically, and of which medical communities have an explicit or implicit picture. This means that the word "healing" loses part of its meaning, since terminologically it necessarily refers to something healthy as being the state to be achieved. If the framework of healing is abandoned, the normative context of medical ethics is no longer responsible, at least initially. The fact that no normative context currently exists for enhancement is a clear challenge to ethics (Ach and Pollmann, 2006; Schöne-Seifert *et al.*, 2009).

To understand these challenges more precisely, the initial task is to localize the normative uncertainties that have arisen (Section 3.3.). Expressed abstractly, this is simple. If the body and mind could be formed technically, the question would be how far people *may*, *should*, or *want* to go in (re)forming the human body and mind for the purpose of enhancing them. Controversies surrounding the answers to this abstract question have already taken place. While transhumanistic positions see not only a moral acceptability but even a duty to achieve enhancement as rapidly as possible, other authors warn against this, giving ethical reasons for their position (e.g., Siep, 2006). To structure the field of possible answers in the sense of policy options for cognitive enhancement, the following classification has been suggested, in which different moral positions can be recognized (Sarewitz and Karas, 2006, p. i):

- Laissez-faire: emphasizes the freedom of individuals to seek and employ enhancement technologies based on their own judgment
- Managed technological optimism: believes that while these technologies promise great benefits, such benefits cannot emerge without government playing an active role
- Managed technological skepticism: believes that quality of life arises more out of society's institutions than its technologies
- Human essentialism: starts from the notion of the human essence (whether God given or evolutionary in origin) that should not be modified

Although this classification is very generic and surely does not do justice to the arguments of the respective protagonists, it nonetheless provides a good impression of the range of possible answers and of the normative uncertainty that can be observed concerning technical enhancement. In the other field of enhancement, i.e., the fight against aging, it appears more difficult to take a position since the extension of man's life expectancy is certainly an object of medical ethos. The measures being discussed in this regard exhibit however an entirely different degree of intervention in man (e.g., the idea of a technical immune system, as discussed above).

At least there are new options available regarding decisions concerning the future of man. Society *must* form an opinion about these options despite all the normative uncertainties, although this does not have to take place immediately. Against the background of the fundamental issues in the ethics of technology, especially as to man's autonomy, distributive justice, and dealing with the uncertainty of knowledge, the abstract question of what we should, want to, or may do in this field can be broken down into the following more concrete questions. These questions have in the meantime come to represent something like a canon of ethical questions directed at human technical enhancement (see, e.g., Williams and Frankel, 2006; Jömann and Ach, 2006; Siep, 2006; Jotterand, 2008b):

- What are the *criteria* according to which the enhancement will or is supposed to take place? Are all of the criteria ethically justified? How are decisions about criteria made, and who decides? Is it possible to indicate an ethically grounded ranking of the priorities?
- Are enhancements in one sense linked with deterioration in another? How could the outcomes be weighed against one another? How can a measure of enhancements be determined?
- Will the rights of those affected be touched on without them having been asked for permission or without this being possible?
- Will human technical enhancement push open an infinite space of further and further enhancements, as is suggested by the technological imperative, or are there *limits* to enhancement? How can limits be explained and how resilient are they argumentatively? What does their justification depend on and what are their premises?
- Which risks to the individuals affected have to be taken into account and which measures can be taken prophylactically to prevent the misuse of enhancement technology? Is informed consent sufficient, or can we imagine situations in which people seeking enhancement must be protected against themselves?
- What are the consequences of human technical enhancement from the perspective of *just distribution*? What are the prospects for a deeper division of society into technically enhanced superhumans and unenhanced people who thus are in a position of inferiority?
- Will any enhancement options that are accepted devalue currently accepted and practiced forms of life such as that of the handicapped? Or even make them impossible?
- What do we think of the heightened performance of individuals that has already become part of our society: cosmetic surgery, sports doping, and performance-enhancing pharmaceuticals?
- Can a spiral be triggered that creates a compulsion for individuals to seek more and more enhancement in order, for example, to remain competitive on the labor market?

- Should public funds for research be made available to scientifically develop enhancement proposals and to put them into practice? Which role in this will economic arguments play (this takes up large space in Williams and Frankel, 2006)?

These questions label certain points in a confusing debate that is still developing. In the following section, the issue will be to structure this material, to reconstruct the patterns of ethical argumentation that are already in use, and in this way to gain an overview of the confusing state of the debate.

## 9.3.2 Patterns of Ethical Argumentation

The topic of human technical enhancement resulting from converging technologies was quickly taken up by philosophical ethics (e.g., Kushf, 2004b; Siep, 2006; Jotterand, 2008b; Schöne-Seifert *et al.*, 2009). Its discussion was prepared by the debate about the genetic enhancement of man (Habermas, 2001). Various patterns of argumentation are employed, not only from ethics but also with regard to issues from anthropology and the philosophy of technology. The debate is characterized on the one hand by the large uncertainties that are involved and by the speculative expectations regarding technical possibilities and social consequences, and on the other by great normative uncertainties. In methodological terms, most of the reflection can be understood as mental experiments in which assumptions are made about the empirical manifestations and consequences of enhancement technologies and then examined from an ethical perspective, for example, with regard to the infringement of the rights of those affected.

The patterns of argumentation are incommensurable. Anthropological thoughts about man's naturalness or about the man–technology relationship cannot simply be balanced by ethical considerations of the consequences or by statements about human rights. The various patterns of argumentation are presented and discussed critically in the following sections before a diagnosis of the current situation is given in Section 9.3.3.

### 9.3.2.1 *Ethical Consideration of the Consequences*

If we assume that human technical enhancement will be possible in the future, ethical analysis conducted from a consequentialist perspective will ask about the expectable consequences and the (unintended) side effects (Section 3.2).

First to be named is *distributive justice* (see on this in general Section 3.2 and on nanotechnology Section 5.3). The enhancement of an individual's human capacities would probably be tied to a high expenditure of resources, knowledge, and capital, which would strongly restrict the circle of beneficiaries. The demand that there be an equal distribution of accessibility therefore appears naïve: "It is likely that neurocognitive enhancement, like most other things, will not be fairly distributed" (Farah *et al.*, 2004, p. 423). This raises the question of who could afford to undergo enhancement, and which consequences the associated unequal distribution of opportunities for enhancement would have for those who cannot afford to undergo enhancement. The range in the distribution of individual abilities that has always been present in society would be greater, as would the difference between the high performers and ordinary people. Technically enhanced individuals could have the idea of securing their advantage or increasing it by further limiting the opportunities for others to obtain technical enhancement. The result might be a societal fault line between the superhumans and the normal humans. Siep (2006) introduced arguments of this type as his strongest ethical arguments against the planned technical enhancement of humans: "Securing such an evolutionary advantage over those who are previous conspecifics definitely represents a harm" (Siep, 2006, p. 318). The ethical reason would be that this development would violate the rights of the nonenhanced (i.e., those without access to enhancement technology).

This would shake the existing system of providing health care in many Western countries to its foundation. The determination of reimbursement for medical services by a health insurance provider is currently oriented to the largest degree (but not wholly) on healing. For example, ophthalmological measures are paid by the insurance organization as long as the eye's visual acuity is sufficiently below what is expected statistically. In a normative context, the meaning of a normal state of health has largely been operationally established. If the idea of a normal state of health is eliminated by technical enhancement, much work will be needed to reorient the healthcare system on other normative ideas. "Above all, this would eliminate the scale for measuring health and normality, which until now has been used to measure claims for help and compensation" (Siep, 2006, p. 320).

Another aspect of ethical reflection on the consequences concerns *human autonomy*. Human technical enhancement according to the models described above would be performed on responsible adults who have provided their informed consent. Yet compulsion could develop out of the presumed autonomy of those affected in the course of the spread of enhancement technologies. "If neurocognitive enhancement becomes widespread, there will inevitably be situations in which people are pressured to enhance their cognitive capabilities" (Farah *et al.*, 2004, p. 423). The reason would be in the fact that refusal to undergo technical enhancement might be tied to disadvantages with regard to participation in social processes, such as in working life. Technical enhancements could become the standard in certain occupational groups, just as today the possession of a driver's license is a prerequisite that is taken for granted among those seeking employment in many occupations. Someone who disregards the standard will have to reckon with disadvantages.

> With the advent of widespread neurocognitive enhancement, employers and educators will also face new challenges in the management and evaluation of people who might be unenhanced or enhanced (for example, decisions to recommend enhancement, to prefer natural over enhanced performance or vice versa, and to request disclosure of enhancement). (Farah *et al.*, 2004, p. 422)

While in principle the autonomy of affected persons is maintained in this case, de facto it is counteracted by a compulsion on individuals to conform. A refusal to conform would have serious consequences.

The creation of new ways of life could devalue established forms or ruin their survival. For example, the way of life of the handicapped could lose its social acceptance if it were possible to compensate technically for the handicap (Wolbring, 2006). This is a fear that has already been discussed with reference to prenatal diagnostics in the context of permission for late abortions, possible genetic interventions to avoid disabilities, and the discussion of eugenics. In the context of human technical enhancement, the fear would pose itself in intensified form.

Even at the level of the individual, it is necessary to reflect on the unintended consequences and their acceptability. For example, human enhancement does not have to be a success; it might fail, leading in a concrete case to a worsening of health problems or to death. This raises issues of accountability and liability as they are known from the field of medicine. Yet even in the case of a success, undesired effects might appear such as dependence on the enhancement technology that could lead to helplessness in the case of failure or psychic changes. It is true of the latter, however, that it is difficult to identify *specific* concerns about technical enhancements: "And if we are not the same person on Ritalin as off, neither are we the same person after a glass of wine as before, or on vacation as before an exam" (Farah *et al.*, 2004, p. 424). Thoughts about the possible undesired individual consequences of technical enhancements result on the one hand in postulates to inform those affected as fully as possible before implementation of the measure, and on the other hand in the demand that the implementation of the technical enhancement should be executed in a manner to make it as reversible as possible.

Finally, one also has to think of the social consequences of a dramatically longer life expectancy of humans (Moor and Weckert, 2004; Sethe, 2007). The factors that are customarily named are concerns about a loss of creativity, increasing thoughts about security since the terror spread by death gets greater and greater the more seldom it happens, the view of the world's population increasing beyond all measure, or even the question about how reproductive behavior will change or has to change.

It is evident that a broad diffusion of enhancement technologies in society will have far-reaching consequences. What they will be is, however, uncertain. Prospective knowledge of consequences is epistemologically fragile (Chapter 10) and occasionally strays in the vicinity of sheer speculation (Nordmann, 2007a). An indicator is the frequency with which the word "could" is used in the relevant texts, the present chapter included. In this sense, the following ethical considerations about consequences do not constitute a search for orientation that is relevant to practical action or policy (such as on nanoparticles, see Chapter 6). They are rather the tentative preparation of a new field of ethical reflection, entirely in the sense of explorative philosophy (Chapter 10).

#### 9.3.2.2 *The Naturalness of Man*

In view of the new possibilities for human technical enhancement, there are frequent references to man's naturalness as being a factor that limits or even hinders any technical enhancement. May man —

as he is, whether created by a creator god or the result of evolution — liberate himself from this connection and give up his naturalness in favor of an artificial future that he himself has created? It is precisely at this point in public debates that much anxiety is created.

The situation is complex, however, in an anthropological as well as an ethical regard. Man is not simply "natural," but at least also "cultural." Plessner (1928) speaks, for example, of man's "natural artificiality" and describes man as being a natural being and a cultural being at the same time. No unambiguous argument regarding man's technical enhancement can thus be gained from his naturalness. On the one hand, technical enhancement is unnatural because it transcends the present natural state. On the other, however, it is also natural for man not to stop at any level he reaches but to focus on his next step.

> While some authors [...] declare human "nature" to be sacrosanct, others are of the opinion that human striving for perfection and self-transcendence are part of being human and are therefore to a certain extent actually "natural." (Jömann and Ach, 2006, p. 35)

The appeal to man's naturalness can thus be employed in any manner desired, i.e., simultaneously both for and against technical enhancement. It therefore cannot contribute anything to this issue. In a nutshell, "Man's nature is thus contingent" (Clausen, 2009).

Another argument for rejecting the justification for employing the argument of man's naturalness operates with the naturalistic fallacy (Frankena, 1939; Quante, 1994; Engels, 2008). It would, according to these authors, be a naïve naturalistic fallacy to argue that we may not technically enhance man's evolutionarily acquired abilities to see, hear, think, etc., solely because they have arisen naturally and developed through evolution. The fact that we humans happen, for example, to have eyes that only work in a certain segment of the electromagnetic spectrum does not have any immediate normative significance: "Behavioral norms do not follow immediately from evolutionary facts" (Siep, 2006, p. 312). Some authors claim that limiting man's capacities to the naturally provided organic attributes would be an arbitrary museification of humans and would mask the cultural side of being human, part of which is to transcend oneself, i.e., to think and develop beyond the what already exists (see above; Clausen, 2008).

In one sense it is possible to use man's naturalness as an argument, if not a very specific one. The species *Homo sapiens* has developed by evolution in the course of millions of years. It appears per se suspiciously risky to intervene in this slowly growing "nature" to any significant degree and on — compared to evolutionary periods of time — a very short time scale. If technical enhancement were to take place very significantly and very quickly, we might lose the matter of course in a manner in which we recognize each other to be *humans*, a scenario that would surely cause anxiety. The result of these considerations is, however, only some general obligations to be cautious and to give preference to small and reversible steps instead of immediately undertaking a radical transformation of man's body and mind (which, however, is far beyond the possibilities of science and technology).

The concern that technical enhancement might endanger or even destroy man's naturalness thus overall does not pose a strong argument. Human nature "functions as an entity reflecting anthropological processes of self-clarification" (Clausen, 2009). In other words, it is worthwhile to discuss and quarrel about this concern, but we cannot expect it to provide any clear answers to the ethical questions raised by human enhancement (Section 9.3.1).

#### 9.3.2.3 *The Question as to Ought*

The primary issue in the previous sections was *may,* as in the question whether human technical enhancement is permitted morally. The ethical question can also be posed in a more intensified form, namely regarding *ought to* or *should*: Isn't it possible that we are obligated to pursue human enhancement? Might enhancement perhaps even be ethically imperative because, for example, it would alleviate human suffering, because it might help solve great problems, and because it would make it possible to achieve a substantially greater quality of life? Is human enhancement not imperative for reasons very similar to those for providing medical aid when someone is ill? Might we not have to pursue technical enhancement in order to secure the survival of human civilization in view of our exploitation of the natural foundations of life and of the continued growth of the earth's population?

This question as to the *should* of human enhancement has been answered negatively until now (Siep, 2006). In particular, arguments concerning uncertainty have been given. They assert, on the one hand, that it is highly uncertain whether the goals associated with *should* will be reached, and on the other that there might be significant problems from unintended consequences, such as with regard to distributive justice (see above). *Should* cannot be justified by uncertainty about reaching the goal in combination with the possibility that there will be unintended and undesired consequences.

For the moment, this line of argument is certainly convincing. In particular, no one has indicated clear and reasonable purposes for which people should be technically enhanced. Considering the uncertainty that is involved, however, this can only be a *preliminary* answer to the question about *should*. For example, specific requirements of certain professions could provide a normative background that technical enhancements should be performed in narrowly delineated areas. Yet if it were the case "that we want our firefighters stronger, aircraft pilots with clearer sight, and our researchers smarter, yet we want our athletes to play fair" (Williams and Frankel, 2006, p. 10), then questions would be raised regarding the criteria employed for such decisions, and regarding who should and should not be enhanced. This would also constitute normative uncertainty and the need for ethical consultation.

### 9.3.3 Assessment of the Current Status of the Ethical Debate

In the ethical debate (meant here is not a public debate but a professional one in philosophical ethics, see Chapter 3), a lack of agitation tends to predominate regarding the perspectives of human technical enhancement despite intensive work on the subject. This is entirely different from the situation several years ago in the debate about reproductive cloning.

So far no *strong* arguments have been raised against technical enhancement. Under "strong arguments" I refer to those that work on the consequences that will necessarily follow the introduction of enhancement technology and that are therefore not dependent on uncertain developments in the future. One argument whose claim is to be strong is Habermas' against the use of genetic interventions in the germ line (Habermas, 2001). His line of thought is that a person subject to such an intervention during the embryonic stage of development would have been instrumentalized, could not have defended himself or even freed himself of this instrumentalization in later life, and thus could not be the full author of his own biography, which would affect the *conditions of morality*. If such human technical enhancements are not carried out on people unable to give their consent, an informed consent could and would have to be created. If only responsible persons, after receiving information about the risks and providing their informed consent, could undergo technical enhancement, there would be no danger that technically enhanced individuals were no longer fully the authors of their own biography, in contrast to the case of genetic intervention in the germ line to influence future individuals. Responsible individuals, on the contrary, would possibly — at least according to the opinion of the promoters of the converging technologies — gain autonomy by being able, according to their intentions, to liberate themselves from naturally given conditions. Informed consent takes on a central role in ethical argumentation.

The ethical debate has instead focused in a consequentialist manner on the unintended side effects, in particular on questions of distributive justice (see above). Arguments of this type are, in terms of validity, rather weak because they have to make a series of more or less uncertain assumptions about the future (for a systematic discussion, see Chapter 10). Furthermore, arguments based on the possible manifestation of unintended side effects do not have to be understood as arguments against the new technology per se. They can be understood as arguments that something must be done politically and socially, and perhaps also technically, to limit, prevent, or compensate in some suitable manner for the manifestation of these unintended consequences. These partially speculative reflections on the consequences thus serve less to help decide whether enhancement technologies are socially desirable or at least acceptable. Their purpose is rather to think about problems at an early stage so that it is possible to think about avoiding or solving problems just as early. This will keep society from being surprised by the consequences.

Viewed from this perspective, it does not seem implausible for enhancement technologies to be introduced according to a market model in the absence of strong ethical arguments. It is not only possible but even probable that there might be an emerging demand for enhancement technologies caused by increasing competition, as has been explained by various developments:

> [...] several market pressures leading to rapid development of HE [human enhancement] technologies: 1) global competitiveness; 2) brain drain/depopulation economics; 3) national security concerns; and 4) quality of life/consumer life-style demands. (Williams and Frankel, 2006, p. 3)

This is even clearly shown by the use of doping in various areas (e.g., sports, beauty, and sexuality). Precisely a society that has instituted the idea of competition as its central motor at nearly every level, from the economy to the military to lifestyle, will be confronted by ongoing efforts to achieve enhancement. Phrased differently, competition and enhancement belong inextricably together. Since, from a technical point of view, doping and enhancement are only different in degree and because the technological imperative leads directly from the former to the latter (Section 9.2.3), the pressure of competition will result in measures that go beyond doping and lead in the direction of enhancement. In a liberal market model, regulation would be limited to compensation for the side effects of a market failure (e.g., from the clarification of liability issues, such as what would happen if an enhancement does not succeed) and to ensuring distributive justice and access:

> Governments may also need to protect the level playing field for consumers, so that the already-enhanced do not act, through non-market means, to protect their status by preventing others from becoming enhanced. (Sarewitz and Karas, 2006, p. 10)

To many people, this scenario may seem unattractive or even scandalous in a cultural, anthropological, or ethical sense, especially considering the situation that the possible human technical enhancements often meet with uneasiness or even substantial anxiety. This anxiety certainly plays a role as a *fact* to be taken seriously. In consequentialism, this fact is taken into consideration ethically by including moral anxiety as a cost in the overall evaluation of a technology, possibly resulting in its negative balance. From the perspective of deontology or discourse ethics, however, the fact there is anxiety does not have any normative consequences. For example, it has not been clarified whether it is not simply the lack of familiarity or of familiarization with the thought of human technical enhancement that is the cause of the anxiety. In this case it would be evident that anxiety could not be attributed any ethical relevance, such as an argumentative role in justification debates. Yet the empirically observable anxiety poses a challenge to ethics, namely for it to search for patterns of argumentation that are masked by this anxiety and to make them explicit.

No decisions that would require ethical reflection are currently pending with regard to human technical enhancement. The issue is instead to prepare the field in which the respective discourses over ethical justification will have to be conducted in the future (Chapter 10). The debate is thus still in the run-up to the formation of an ethical judgment and considers, for example, issues from anthropology or the philosophy of technology or the issue of the naturalness of man.

## 9.4 Changing Relations Between Humans and Technology

The ethical debate over human enhancement is superimposed by another discourse that is frequently not distinguished from it, namely that about the attempt to reach an understanding about the possibly changing relationships between man and technology and about the hopes and anxieties about technicalization. In the following sections, I will first demonstrate with reference to neuroelectric interfaces that ethical issues are less dominant that those from anthropology and the philosophy of technology (Section 9.4.1). I will then examine in a more general sense the possible technicalization of man by nanotechnology and enhancement (Section 9.4.2).

### 9.4.1 Neuroelectric Interfaces

Neuroelectric interfaces are often considered a field of new technical developments for which there is per se a high need for ethical reflection. Precisely the descriptions of the field in popular science usually do not lack references to possible moral misgivings, possibilities for misuse, and ethical problems. Neuroelectric brain–computer interfaces (Stieglitz, 2006) can work in both directions. On the one hand, they can capture, interpret, and use the electrical activity of nerves or sections of the brain in order, for example, to control an external device such as a computer mouse. On the other, external electrical signals can be applied in order to stimulate certain nerves, such as causing a muscle to contract, for example to restitute lost or defective body functions (Hildt and Engels, 2009).

> So, using neuro-prosthetic technologies should actually shape and modify brain processing to improve control of each prosthetic technology and facilitate the ability of paralyzed individuals to convert their motor intentions into purposeful communication and interaction with the world. (Scott, 2006, p. 141)

The principle consists in using volitional exertion of the brain to produce electrical brain potentials that can be recorded by the appropriate sensors (e.g., EEG) and analyzed. The result of the analysis can be used to control an external device. Via feedback, the patient can train his own brain waves and optimize them to improve the functionality of the technical system. Invasive systems make it possible to measure brain activity in certain regions in a more targeted fashion, to interpret them more specifically, and to translate them into action more purposefully. They enable significantly more flexibility and precision of movement than the use of surface electrodes. Such systems are currently in an empirical experimental state and must be optimized for each patient according to the principle of trial and error. Analogous to the smooth transition between healing and enhancement (Section 9.2), this makes it possible to imagine new or enhanced interfaces between man and technology being achieved technologically (Farah *et al.*, 2004).

Among the visionary ideas about the future of neuroelectric interfaces are playing piano with an artificial hand, a paraplegic playing football again, or implanting a chip in the brain that contains foreign languages (Hennen *et al.*, 2007). Some scenarios of the future consider the possibility of being able to use technical chips to compensate for lost brain functions in degenerative illnesses such as Alzheimer's disease. An artificial hippocampus — the natural one receives and processes different sensory impressions, forwards information, and provides the transition from short-term to long-term memory — in the form of a technical implant (brain chip) would replace the nonfunctioning region of the brain. As long as the goal is to heal the illnesses described above or to restitute lost memory, these developments take place within the context of classic medical ethos despite their futuristic sounding possibilities. Albeit, ideas on technical enhancement can be seamlessly attached to these developments.

Few references to neuroelectric interfaces fail to mention comments about possible moral misgivings, possibilities for misuse, and ethical problems. A frequent theme is that technical measures near the brain are considered to touch on a sensitive issue regarding the traits of an individual's personality (Clausen, 2006). Initially, estimates of this type only express *anxiety*, namely that technical operations near or on the brain pose a threat to personality rights, that the traits of one's personality could be endangered, that a person's individuality could be damaged, and that the possible or probable consequences could include violation of human dignity, the instrumentalization of man, a threat to the private sphere, and discrimination in providing medical care (Hildt, 2009). In this sense, the frequently printed statements that call for ethical limits or at least for a procedure that is as safe and judicious as possible are psychologically very understandable.

In addition to the positive expectations, for example, that neurological illnesses could be healed, the new possibilities created by neuroelectric interfaces have also led to insecurity, which is associated with the lack of an established routine in dealing with this progress. This uncertainty is however partly due to the media reporting about this field, in which frequently no clear distinction is made between visionary speculation, sober descriptions of the measures being conducted at the frontline of clinical experiments, and realistic expectations for the coming years. The public is thus occasionally given the impression that all types of neuroelectric interfaces and neuroimplants are already in use or about to be used. This impression does not correspond in any way to today's practice, as a glance at the current state of research shows (Fiedeler, 2008).

This insecurity is also, however, not yet per se an expression of the necessity of comprehensive ethical reflection. Emotional anxiety in itself does not represent an ethical argument against neuroelectric interfaces in humans. It is also not a compelling argument that ethical reflection is needed. Initially, this anxiety regarding neuroelectric interfaces—although appropriate—is simply intuition. Although this intuition does exist, its existence *alone* does not yet have any ethical dignity in the sense that normative uncertainty is manifested here that requires professional ethical consultation. Finally, the anxiety could result from the simple fact that people are unaccustomed to neuroelectric interfaces, and it could vanish as society accepts the procedures and products that might follow.

Without a doubt, it is the task of ethics to trace this de facto anxiety. Although this intuitive anxiety does not have to mask any normative uncertainty, it might. Analysis of the normative background might be able to uncover the normative uncertainty that might arise. This would make it possible to determine the need for ethical reflection in a concrete and problem-oriented manner, instead of simply calling placatingly for ethics in response to the anxiety that results from uncertainty. In this sense, the objective is to clarify which normative framework is present for the development and application of neuroelectric interfaces, and where the existing framework is being challenged by new developments in science and technology to such a degree that the criteria for the existence of a standard situation in a moral sense (listed in Section 3.1) are no longer satisfied. Failure to meet the criteria might result because of controversies about moral judgments and about the consequences required for the further procedure (analogous to the case of nanoparticles, see Chapter 6).

As long as the goal is to heal an illness or to restore the attributes of a dysfunctional or destroyed organ, the normative context is provided by the established normative components of healing and of the environment needed for it (Müller, 2006): the medical ethos, the doctor–patient relationship, patient autonomy in the form of informational self-determination, ethical standards of research and clinical test procedures, providing fair access to medical care services, etc. It is in medical ethics that this normative framework is reflected, further developed, and related to new situations, therapeutic procedures, and technologies. In the background are questions as to which rights of those affected could be impaired, or whether and how conflicting goals or means can or may be balanced.

Those directly affected are first of all the individuals who use the neuroelectric interfaces, i.e., as understood in this chapter this is primarily patients who hope for an improved quality of life. The question here is whether the limitations, risks, and burdens that might arise are in an appropriate relationship to the expected advantages, or whether the dignity of the affected individuals could be violated. A decisive role in this connection is played by informed consent since decisions of this kind have to be made individually. Indirectly affected are persons associated with the patient and possibly also future persons who via inheritance could be confronted with the consequences of the use of neuroelectric interfaces (although it is difficult to image such a scenario). In addition to thinking of the individuals who are affected, the social effects must also be taken into consideration. In this connection, we can think of those indirectly affected who do not have access to the possibilities for analogous (expensive) treatment despite possibly having an identical indication (EGE, 2005, pp. 97f.).

Considering the state of research and development (after Fiedeler, 2008), I cannot see that neuroelectric interfaces *currently* pose a challenge to this normative framework. Since the incremental progress from current research and clinical tests continue the existing medium-term and long-term developments, it will not cause any abrupt, new challenges to forming an ethical judgment. The anxiety that can sometimes be seen in parts of the public does not correspond to any normative uncertainty inasmuch as neuroelectric interfaces are taken into consideration that either currently exist or are possible in the foreseeable future and that are employed for medical purposes.

This finding corresponds to the manner in which the ethical aspects of neuroelectric implants are communicated in the public and in science. It is not the *current* developments that are considered problematic (entirely in contrast to the current situation with regard to nanoparticles, where there are already marketable products), but those that might be possible in the future. Previous ethical debates were rather a discussion well ahead of a real need, which fits the analysis provided above. It is not the normative uncertainties of today's neuroimplants that is at issue, but the *possible* normative uncertainties of *possible* future implants and interfaces. Independent of the possible future applications that have been presented in the manner of mental experiments (on the problematic nature of speculative nanoethics, see Chapter 10), some of the issues that are definitely already involved today lead to tasks for explorative philosophy (in part after Hennen *et al.*, 2007; EGE, 2005):

- Because of their location in the body, implants have a greater closeness to their carrier, and this is true not only of implants incorporating neuroelectric interfaces. The question is still open whether and to what degree an implant could thus become an element of the carrier's identity and of his person.
- It can be unclear what the effect of certain interventions via brain–computer interfaces or stimulations will be on the autonomy of actors. A significant feature of brain–

computer interfaces is that their functional integration into brain processes could make the autonomy of the actor and his independence of the technical system questionable (Clausen, 2006).

- The question as to the relationship between a neuroimplant and its carrier is posed in heightened form when the attribution of actions and their consequences becomes ambiguous. The situation could arise that a person is not entirely responsible for actions that he or she carried out because a neuroimplant did not provide the expected performance. Issues of liability could be posed if, by providing certain data, a retinal implant induces its carrier to perform an act that leads to an accident or some other damage. The clarification of such issues of liability could be markedly complicated.
- A very different, morally sensitive topic is associated with neurotechnologies that could compensate for disabilities, such as cochlear implants that restore some of a deaf person's hearing. Many people were surprised that the response of the deaf community was not exclusively positive (Reuzel, 2001). Parts of this community felt that such technical aids threatened the identity of the community, which has its own language (sign language) and culture, especially regarding technical aids for children. The question raised here is whether the disabled are viewed or consider themselves to be people with a deficit that must be overcome or as members of a special culture. This debate is connected with the question about the status and the social recognition of the lifestyle of the disabled in general. For example, is life with a disability acknowledged to be a special lifestyle? Or is it society's task to liberate the life of the disabled from their disabilities by utilizing technical and other means?

These questions and open items cannot be attributed directly to applied ethics. They are rather issues from anthropology, philosophy of consciousness, and theories of identity, which can or even must, depending on the circumstances, form the basis of ethical considerations. The debates about neuroethics are thus less concerned with ethics than with reflection on the changing relationships between humans and technology.

## 9.4.2 Technicalization of Man by Nanotechnology?[1]

The tendency toward "subsuming humanity under the realm of technology" is a word collocation that is widespread in the debate about technology and the "future of human nature" (Habermas, 2001). This wording contains generalizations in two respects simultaneously:

A general view toward "increasing technological stakes in humanity," in which it is no longer a question of individual, particular technologies, but of technology in general

A general view of humanity and not of individual human beings and, with it, a generalizing view of the genus *Homo*, or in (philosophical) anthropological respect, of "human nature" in general

These perspectives on generalization can imply various semantic connotations, especially regarding the distinction between the individual and social sides of human beings. Accordingly, subsuming humanity under technology can result in a greater regulation of human beings, both individually and collectively. Although people's apprehensions concerning a possible technicalization are similar — loss of human autonomy and control, possibilities for external manipulation, and instrumentalization of human beings — they might express themselves quite differently. The following semantic aspects of the phrase "humans as technology" (see below) can therefore be distinguished that concern nanotechnology and converging technologies:

- Implantation of technical artifacts into the human body (e.g., pills, artificial replacement parts, prostheses, monitoring devices) as a "move towards technical humans" on an individual level. In this context, ethical aspects such as informed consent or the right to informational self-determination are the focus of the debate, whereas the consequences that this technical humanity might have on a massive scale would take a back seat.
- Technical organization of society with the corresponding effects on the individual (bureaucratization, militarization, monitoring, etc.): technicalization "of humanity" as a collective

[1] This section refers to arguments employed in Grunwald and Julliard (2007). I am indebted to Dr Yannick Julliard for his permission for me to refer to our common work.

entity, which can lead to a transformation of human society into a "technical" society, including the reproduction of society, its need and ability to repair itself, and the organization of the latter.

- Technical acquisition of knowledge about the human body and the mind (e.g., about genetic dispositions or about physiological processes in the brain), with the consequence that the human body and mind increasingly become the object of technical interpretation.

The radicalization and the ultimate triumph of *Homo faber*, who, nanotechnologically equipped, sets out to manipulate the world atom by atom (NNI, 1999; see Section 2.4), is an idea characteristic of nanotechnology. In a revival and a radicalization of the physical, even mechanistic, reductionism of the nineteenth century, this new *Homo faber* believes to have everything under control, including the realms of life and society, because this control starts at the atomic level. The *Homo faber* story seems to be common to nanotechnological visions, whether they are visions of salvation or disaster. Dupuy (2005) thinks in the tradition of Anders and Sartre about a metaphysical program underlying the converging technologies, the technical basis of which would be a molecular assembler that could be universally employed: "The aim of this metaphysical program is to turn man into a demiurge or, scarcely more modestly, the 'engineer of evolutionary processes.' [...] This puts him in the position of being the divine maker of the world" (Dupuy, 2005).

An unbroken optimistic belief in progress based on science and technology stands at the core of all of these visions. Finally, the solution to all of the world's problems (even including questions of development, poverty, and hunger) are expected from nanotechnological progress. In the circles of so-called transhumanism, these optimistic estimations of progress are ideologically exaggerated. According to them, it would be humanity's responsibility to overcome its own deficiencies by creating a technical civilization that, in the final analysis, makes human beings, with their sufficiently well-known deficits, superfluous. A man-made and then independently further evolving civilization is to take the place of human civilization (Coenen, 2006). This perspective — here, the anthropological consideration that technology serves to overcome the deficits of the "imperfect being, man" (Gehlen, 1986) is radicalized in the direction that humanity should abolish itself in favor of technology — would self-evidently be a move to include human beings into technology in the sense of a self-abolition of human civilization in favor of a technical civilization. This "abolition of humanity" — giving up the defects of humans in the dialectical sense of the word — is here no menace, but is understood as a normative objective. This specific kind of technicalization would be welcomed by transhumanists.

The presumably ultimate triumph of *Homo faber* could, however, turn out to be a final Pyrrhic victory. As a result of the renowned contribution by Bill Joy (2000), an apocalyptic dimension was added to the numerous positive expectations for the future as a result of nanotechnology. The point of departure was the apprehension that self-duplicating nanorobots based on Drexler's assembler could one day get out of control and make humans superfluous (Joy, 2000), or that humans could fall prey to their own technical achievements. In this "prey scenario," it is a matter of a loss of control by humans. A seizure of power by robots stands at the center of interest. Instead of serving ends set by human beings, the nanorobots could make themselves autonomous and take over control of the planet Earth. A technical civilization would then cease to be dependent on mankind (see Section 2.6 on these issues).

The question of whether humanity moves into the realm of technologies is, in both types of debate, to be understood as an end to human civilization, in favor of a technical civilization — provided, however, with different connotations between expectations of salvation and of the Apocalypse. Increasing the importance of technology in this context does not refer to artifacts or to implantation of technology into human beings, but to our self-perception of humanity's development as a genus. This self-perception is frequently associated with tendencies to reduce human beings to the level of what is scientifically observable (and with it, to what can be influenced technically; Habermas, 2001), and to the continuing "naturalization" of human beings. Technical interpretations of the human being as a complex machine (e.g., as a cybernetic machine) can be observed. If there were a purely technical description of humanity serving all needs, then human beings would be seen as technology.

In my view, the semantics of "moving humanity in the realm of technology" is the generalization that at present most probably justifies the discussion of technology, in the form of nanotechnology, playing an increasing role for humanity. Scientific, technical progress in neurophysiology, biotechnology, and genetic engineering precisely by means of nanotechnology and especially nanobiotechnology have, in many circles, allowed the idea of a technical human to come much closer, i.e., the description of human beings as a machine, on the basis of an atomic reductionism, is made more plausible by nanotechnology.

This reflection can be recognized by the interpretation of humans as, ultimately, a technical system. Enhancement is then quite logical, as it is for every technical system. It is semantically something completely different from the implantation of technical artifacts in the human body. "Enriching" humans with technical artifacts would not prevent us from speaking of human beings as "transtechnical" entities, i.e., as beings which may profit from the technology in their bodies, but not beings in which the technology was central to them.

## 9.5 Conclusions for Responsible Action

The analyses show, on the one hand, that enhancement ethics is not covered by the spectrum of the subdisciplines of applied ethics, but that it rather raises new issues. The practical consequences (Section 9.5.1) are related on the one hand to the attribution of responsibility in the course of further actions (Section 9.5.2), and on the other to reflections on the structure of present society (Section 9.5.3).

### 9.5.1 Need for Orientation on Human Enhancement

The first question is whether and on which factors there is currently a problem of orientation at all on which ethical reflection could make a contribution. Human technical enhancement in the sense of a terminologically intended enhancement or modification is by no means already on the threshold to the lifeworld. It is still deep in the laboratory. Their technical feasibility has often not been proven or is a matter of controversy. There is no reliable knowledge about the consequences and side effects of technical enhancements. It could be concluded in this situation that there is no need for orientation. Ethical reflection and debates about the consequences of technology could be surmised to be fundamentally irrelevant in this field. This is true with some justification for, for example, ethical reflection that is concerned with what the social consequences might be if human life expectancy reached 5,000 or 50,000 years and how these consequences might be judged (Moor and Weckert, 2004; see Chapter 10).

A different reaction is for us to perceive the prospects for human technical enhancement either as an apocalyptic danger or as the promise of paradise, such as if we imagined that cyborgs were to appear in our society and we had to answer the question of whether we would grant them human rights and give them identification cards. This idea seems to be behind some expressions of anxiety or even disgust at technical enhancements. At this point it is important, however, to warn against a widespread fallacy. In debates about the consequences of technology, extrapolation into the future is often carried out very differently in different areas:

- For technical developments, very substantial advances and far-reaching social consequences are anticipated.
- For the possibilities of society, in contrast, to deal with these technical advances and their consequences, many people proceed from society's current *capacity* for coping with problems.

This unequal treatment leads to an anachronistic situation with a reputed potential for catastrophe (Grunwald, 2006a). Namely, technology continues to develop very dynamically, while society is imagined to remain static at today's level. Today's society, in this line of thought, would obviously hardly have the capacity to deal with cyborgs or with people who are technically completely transformed or extremely old if they were to appear suddenly in society in great numbers and without any advance warning. A situation is created here by the unequal handling of future developments that produces at least anxiety but that may even exhibit apocalyptic features: the possibilities of technology *as extrapolated into the future* are juxtaposed with the processing capacities of *today's* society. The inevitable consequence of this is a dystopian vision of dramatically excessive demands being placed on society by rapid advances in science and technology promoted by enhanced humans.

This line of argumentation is, however, a simple fallacy. The time journey into the future and the dramatic changes that it imputes are only undertaken for technical development, not however for the individual's or society's mechanisms for overcoming problems. This arbitrary unequal treatment is not justified either methodologically or empirically. At the most it is possible that it is easier to predictively describe technical progress than it is societal. Yet even if this were the case, this would not justify a selective manner of drawing conclusions. For this reason, apocalyptic conclusions based on such a procedure are to be rejected as misleading. There are demands and possibilities for shaping precisely the further development of society instead of letting it simply be at the mercy of the dynamics of technical development. The question is thus in which fields something can be done today to shape society without any work having already been conducted on concrete technical enhancements (in the specific terminological sense described in Section 9.2.2).

- *First* is *research funding*, which is important for determining the topics and priorities of science and thus exerts influence on future presents. It provides opportunities for shaping long before possible concrete offers for human technical enhancement reach the market by supporting certain directions of research instead of others.
- *Second* is the sensibilization of society for the relationship between man and technology, including in controversies with ideological movements such a transhumanism (Coenen, 2006). The debate about human technical enhancement and the relevant ethical issues is one about man's self-conception via, for example, his image of man, the relationship between man and technology, the relationship between man and nature, and the future of man in light of new technical opportunities.
- *Third*, it thus also plays a role in providing orientation for dealing with issues that are far from technical enhancement, such as the everyday handling of the human body (e.g., whether it is considered a machine that requires maintenance or part of the human identity). It can contribute to a better reflected handling of certain *present* practices.
- *Fourth*, such debates take place in a situation in which orientation regarding normative uncertainties takes place in advance and then held in reserve. Orientation is created in anticipation of situations in which real decisions may have to be made even faster than anticipated (see also Chapter 10).

These answers to the question as to the actual need for reflection make it obvious that the task today is by no means to decide about the future of man and his nature. Customarily, technical enhancements come gradually. Although the possibility cannot be excluded that there might be disruptive leaps from today's "normal" humans to technically completely modified humans and cyborgs, this is only true in the fundamental sense that almost nothing can be excluded with regard to the future. Such disruptive leaps are profoundly implausible, especially considering the complexity of the measures in technical enhancement, the difficulty to control all of the relevant processes, and the related high vulnerability to risk.

It is not implausible, however, that there will be enhancements and modifications of humans in the future. The ethical debate has not yet produced any strong arguments against enhancement procedures (Section 9.3.3). Human technical enhancement will however be a gradual process consisting of many small steps. Technically enhanced, optimized, or perfected humans or human–machine composite beings will not appear suddenly in droves. The gradual nature of the emergence of technical enhancements reduces the challenge of dealing with the normative uncertainties, which initially appeared to be so large as to be hardly manageable. These uncertainties appear to reach maximal size here since there is no normative context for enhancement at all that anyone could use for at least rough orientation, even if in opposition to it. By considering these challenges both in the context of a time line and with regard to the presumably long periods of time that are involved, they lose their fright, or at least some of it. We have time to reflect on the possibilities of human technical enhancement, in particular regarding their consequences and ethical issues, and to develop a normative framework.

Many of the presumed ethical problems in this field are simply pseudoproblems. A beautiful example of this is the fact that Ken Warwick, a known advocate of human technical enhancement and himself someone who wears neuroprostheses, regards himself to be the first cyborg and is also classified by others as a cyborg: "He is the world's first cyborg, part human and part robot. In 2002, he breached the barrier between man and machine (Jotterand, 2008b, p. 23). This diagnosis, made in the cited article, is supposed to make it obvious that the talk of human technical enhancement and of the emergence of man–machine composite beings is not simply speculation but existing social reality in the form of at least one existing cyborg. The perception created in this way is supposed to demonstrate to us the urgency of ethical reflection, namely that it would be high time for us to concern ourselves with the moral status of composite beings.

This example proves just the opposite, however, and disavows the excitement. One reason: I can only assume, but it is more than plausible that Ken Warwick is not kept in a zoo or a high-security tract, that he has an identity card, and that he is covered by general human rights and civil rights. If he were truly a cyborg (after Jotterand, 2008b), then the ethical question of the moral status of such composite beings would at least be clarified: they would be humans. Many of the visions associated with enhancement might be normalized in this way when concrete cases and people are involved.

### 9.5.2 Responsible Action

In contrast to ethical questions that can be posed abstractly, i.e., independent of concrete actors, the attribution of responsibility and the act of taking responsibility always concretely entail some actor accepting some responsibility for some situation (Durbin and Lenk, 1987; Grunwald, 1999). It is time to speak about responsibility when answers to questions about responsibility (e.g., for distribution, attribution, guilt, or achievement) become unclear or a matter of controversy in a specific field of action. The purpose of speaking about responsibility is to reach agreement about the structure of responsibility in the respective field in order to make reflected and responsible action possible. The purpose of discourse on responsibility is clarification of specific areas of responsibility for the purpose of practical action.

Responsibility is not a natural object in the sense that someone has or does not have responsibility, but rather the result of an *act of attribution*. The phrase, "Who has which responsibility?" is abbreviated and hides the central process of attribution. The attribution of responsibility constitutes an action that takes place given certain objectives and relative to rules of attribution (on this, see also Jonas, 1979, p. 173). The rules of attribution themselves require justification by, for example, limiting the group of individuals who are capable of being responsible and providing criteria as to which conditions individuals have to satisfy in order for it to be possible to hold them responsible. The attribution of responsibility is a prescriptive act.

The concept of responsibility must be reconstructed as at least a three-part concept: *someone* (a responsible subject) is responsible for *something* (the results of acts forming the object of responsibility) before an *authority* (e.g., a person, an institution, a body of rules, or morality). The three-part reconstruction leads to the *causal responsibility* for an act (Lenk, 1992, pp. 27f., 82). The moral dimension does not become accessible until, in a four-part reconstruction, the question is whether responsibility *should be accepted relative to which body of rules. Moral responsibility* is thus understood as the attribution of responsibility relative to a body of rules (Grunwald, 1999).

In view of this conceptual situation and based on the ethical challenges posed by human enhancement, it is possible to establish the following theses on responsibility in this field, which require further reflection and concretization, structured according to the groups of different actors involved:

- *Sciences in the vicinity of NBIC technologies*: The responsibility of the participating scientists (e.g., in nanotechnology) comprises the usual facets of scientific responsibility (Lenk, 1992) with regard to the goals, research practice, and early reflection on possible consequences. In addition to providing information at an early point in time, participation in the public dialogue belongs to responsibility.
- *Ethics*: Ethical reflection is desired in this field both as an academic profession and also to provide orientation for the formation of opinion in the discipline and in public. The work of philosophical ethics must be carried into the public debate.
- *Medicine*: Medicine has been molded for thousands of year by the ethos of healing. Fields such as cosmetic surgery and doping have contributed to raising doubts as to whether even now this ethos alone should provide orientation for medical acts. Technical enhancements would lead to further going modifications in the self-understanding of a physician: the physician as body designer? Reflection on the profession is needed here in order to be able to help shape this process in a responsible manner.
- *Scientific policy consultation*: Since enabling technical enhancements would bring extensive social changes with it, policy consultation has the responsibility not simply to consult

political principals, but to contribute to the social dialogue by spreading their results on a wider scale.

- *Research funding*: The funding of research influences the course of science. In determining the topics for funding programs, attention must be paid that — inasmuch as technical enhancement is addressed — there is ethically reflected debate on the topics selected and that the ethical reflection is conducted in an adequate manner.
- *Social sciences and social theory*: Beyond its individual context, the political and social aspects of technical enhancement must be examined. Examples to be examined are references to a performance-enhancing society (Coenen, 2008b) and to the idea of ever-present competition that is inherent in the capitalist form of economy (see below).
- *Media*: Differentiated presentations can be expected of the media, both with regard to ethical issues and to the object of research. (This is frequently not the case because the technical possibilities are strongly exaggerated; on neuroimplants; see Fiedeler, 2008.)
- *Society*: Finally, all of society is confronted by new questions or by new inquiries regarding existing lifestyles. The debate about technical enhancement could be the occasion for an individual to reflect on the previous handling of his or her own body and on his or her perspective toward it.

These thoughts about human technical enhancement extend man's options for thinking and acting. They raise doubts about what until now went unquestioned, for example, that man has no alternative but to rely on the customary functions of his body. The transformation of something that had to be accepted as given into something that could be manipulated is the hallmark of technical progress. In the degree to which the human power to exert control is increased, new rooms are opened for visions and shaping, but at the same time, as side effects so to speak, so were the challenges to compensate for the loss of traditions by finding new forms of orientation. Considering the visionary nature of the prospects of human technical enhancement and the long to very long periods of time envisaged for the realization of certain partial steps, then in all probability there is sufficient time for us to grapple with the issues that are raised. It generally holds that this reflexive debate should take place in the early phases of development. It is then that the greatest possibilities exist for exerting influence on the process of scientific development. The chances in the case of human enhancement are good that ethical reflection and social debate will not come too late but that they can critically accompany the development of science and technology.

### 9.5.3 Approaching an "Enhancement Society?"

We can continue to ask what it means for human technical enhancement to produce anxiety on the one hand, but on the other for it to also encounter a remarkable positive resonance. This is true not only of the anthropological fact that suffering from one's own inadequacies has accompanied man for a long time (Section 9.2.1), but presumably also of a current factor in the background of social policy. In the United States, an item that has already been the object of scattered discussion as a possible political goal is the creation of an enhancement society (Coenen, 2008b). The appearance of an attractive world market for enhancement technologies and procedures has already been predicted. In particular in view of the fact that societies are aging, enhancement could be an appropriate means for companies and economies to achieve competitive advantages. The question of social diagnosis is whether we stand at a transition from capitalist achievement society to an enhancement society (Coenen, 2008b).

The changes in the world of employment and the disruptions in the last decades have presumably also been accompanied by changes in our understanding of achievement. It often no longer appears to be sufficient for an individual to produce an expected performance in a context defined by salaried employment. More and more people feel compelled to constantly improve the preconditions of their performance and possible actions in an increasingly flexible world of employment. Lifelong learning comprises one element in this ongoing challenge. Such concepts oriented on education could be supplemented or replaced in the future by technical enhancements. An above average and constantly improving performance at work, a beautiful and strong body, high resistance to stress: these abilities (Wolbring, 2008b) are moving to the top of the agenda of many people. They are both expression of an atmosphere that constant enhancement is necessary just to maintain one's quality of life, not to mention to improve it, as well as a motor for maintaining and strengthening this atmosphere.

Individual actors are now embedded in a given form of society and in a capitalist economy. Part of a society in which competition is the prime driver of development is the related competition between institutions and between individuals. Rankings and ratings, constant evaluations, the necessity of presenting oneself, and beating the competition, whether at work, in love life, or in the evening game show, whether in the attempt to become the next top model or to use some crazy idea to get onto the news or into the Guinness Book of Records — competition is ubiquitous, and competitiveness is measured by these abilities. In this way, culture becomes an aspect of competitiveness or a location factor, education becomes an article of competition, and soft skills are required to increase one own competitiveness. Competition and abilities belong inseparably together and thus, in the dynamics of development, so does the improvement of abilities. This is because every success in competition is only for a finite period of time and is constantly threatened by the possibility that others might catch up or pull ahead. In this manner, the idea of competition and human technical enhancement also belong inseparably together:

The report (Roco and Bainbridge, 2002) referred to the term "productivity" over 60 times and to "efficiency" 54 times. These terms are closely linked to the term "competitiveness." (Wolbring, 2008b, p. 29)

The trend in the shift from an achievement society to an enhancement society is shown in efficacious social and cultural developments such as very widespread everyday doping. A thought that has hardly been expressed is that the possible developments toward an enhancement society — in which enhancement technologies develop according to a market model and then spread — are associated with the overarching social system (not even Wolbring, 2008b goes this far). If this connection could be substantiated — and there are some arguments for it — then this would say something about the location and direction of the social debate that would have to be conducted about this situation. It would then not simply be about ethical issues, which could be answered in the one or other direction, but also about the form of society in which we live and about its implications. The politically explosive nature of this question is obvious.

# Chapter 10

# Explorative Nanophilosophy: More Than Applied Ethics

At various points in the preceding chapters I have demonstrated that ethical judgments — in fact, even determining that normative uncertainties are the starting points of ethical reflection — must rely on knowledge about the consequences of developments and about the future that is uncertain to a greater or lesser extent. This fundamental problem affecting prospective ethical reflection — dealing with uncertain knowledge is a regular feature of the ethics of technology (Section 3.2) — is manifested in some fields of nanotechnology in particularly pointed form, such as in synthetic biology or the "technical enhancement" of humans. In this chapter, I will therefore systematically pursue the dependence of ethics on prospective assumptions and its reliance on epistemological reflection about these assumptions. The starting point is the recent discussion about "speculative nanoethics" (Section 10.1). Critics pointed out first of all the dangers and problems of nanoethics, which would rely too much or even exclusively on speculation about future developments, thus raising suspicions of being arbitrary (Nordmann, 2007a). This criticism has led to differentiated considerations of the relationship between ethical reflection, the knowledge available about the future, and temporal relations (Roache, 2008; Grunwald, 2010a), which I will describe and further develop. The outcome — based on fundamental consideration of the character of knowledge about the future and of its function in decision-making processes

*Responsible Nanobiotechnology: Philosophy and Ethics*
Armin Grunwald

ISBN 978-981-4316-80-4 (Hardcover), 978-981-4363-33-1 (eBook)
www.panstanford.com

(Section 10.2) — is that speculative elements in fact have a place in reflection on future technologies, but that (a) they must be clarified epistemologically and (b) reflecting on these speculative elements can never be classified as applied ethics. A field of explorative philosophy has rather been opened, which encompasses hermeneutics, anthropology, and epistemology (Section 10.3).[1]

## 10.1 The Debate on "Speculative Nanoethics"

In the wake of the emergence (Mnyusiwalla *et al.*, 2003) and rapid thematic development of nanoethics (e.g., Khushf, 2004a; Grunwald, 2005), fundamental criticism followed swiftly (Nordmann, 2007a; Keiper, 2007; Nordmann and Rip, 2009). Although this criticism acknowledged that the gap diagnosed at the start of nanoethics had been more or less closed — i.e., the gap between the rapid progress of nanotechnology and hesitant ethical reflection (Mnyusiwalla *et al.*, 2003) — it claimed that a new gap had instead opened in its place. Nanoethics was said to have become much too concerned with speculative developments and to be concerning itself too little with the questions of nanotechnology design and applications that were actually pending:

> but a new gap has opened up because most nanoethics is too futuristic, focusing on nano-enabled devices that can read our thoughts, for example, at the expense of ongoing incremental developments that are more ethically significant. (Nordmann and Rip, 2009, p. 273)

The DEEPEN project put it into the phrase: "Move away from speculative debate! Return 'ethical concerns' to the sphere of politics!" (DEEPEN, 2009, p. 7). If this diagnosis is true, then large parts of nanoethics are misguided and concern themselves with irrelevant and purely speculative ideas, while the really important developments are not taken into consideration. Such fundamental criticism must surely either result in a radical reorientation of nanoethics or be refuted for good reasons. In order to identify this fundamental alternative, in the following four subsections I will first reconstruct and critically analyze the four central arguments raised by the critics of speculative nanoethics.

[1] Parts of the argumentation presented in this chapter, in particular Sections 10.1 and 10.4, were initially published in Grunwald (2010a).

### 10.1.1 The Main Diagnosis: "Most Nanoethics Is Too Futuristic"

My criticism starts from the critics' diagnosis that, namely, "most nanoethics is too futuristic" and that "There is now a market for ethics of nanoscience and technology, and ethicists and others have responded to the demand for nanoethics with an over-supply of nanoethics" (Nordmann and Rip, 2009, p. 273). This diagnosis criticizes two facets of quantitative relations ("most" and "oversupply"):

- "Most" nanoethics are too futuristic. This criticism applies to the largest part of nanoethics (i.e., "most"). It takes a critical view of the portion of speculative nanoethics in the entirety of nanoethics.
- There is an "oversupply." In this criticism, the "supply" of speculative nanoethics is compared with a "demand." This results in a completely different argument.

Critical examination of these assertions is difficult because the diagnosis is not sharp.

- No clear criteria are provided to determine the border at which a nanoethical study is "too futuristic," and no quantitative analysis is attempted to determine how many studies are above this threshold and how many below it.
- The authors do not describe how an "adequate" unit could be determined to measure the supply of nanoethics or where it should be, something that has to be known in order to diagnose an "oversupply" between supply and demand.

Nordmann and Rip's initial diagnosis is presumably meant in a much more elementary sense, namely that there is simply "too much" speculative nanoethics *per se*. They create the impression that there is hardly anything else to nanoethics since they do not mention nonspeculative ethics in any way. This itself produces, however, a false impression. By glancing in the relevant anthologies (e.g., Baird *et al.*, 2004; Allhoff *et al.*, 2007; ten Have, 2007; Jotterand, 2008a), and journal articles (see the issues of *Nanoethics*), it is easily possible to determine that the portion of nanoethics that is not speculative is quite considerable. It includes above all the ethical analyses of the handling of uncertain environmental dangers and health risks from nanomaterials, which are frequently based on environmental, health, and safety (EHS) studies (e.g., Grunwald, 2008c or Weckert and Moor, 2007), and the consideration of questions of fairness, which are now present in numerous varieties such as the significance of nanotechnology for developing countries (e.g., Schummer, 2007; Invernizzi, 2008; Foladori, 2008).

At this point I concede that there is much speculative ethics. What is not clear, however, is that there is *too much*, and if so according to which criteria. This requires precise analysis. Such a diagnosis would at least have to acknowledge that nanoethics does not consist *exclusively* of speculative considerations.

### 10.1.2 How Speculative Is "Speculative Nanoethics"?

The central criticism leveled by Nordmann and Rip (2009, see also Nordmann, 2007a; Keiper, 2007) at current nanoethics is that it is "too speculative." They explain what "too speculative" means by referring to the implausible nature of the developments that nanoethics assumes to come from nanotechnology. According to Nordmann and Rip (2009, p. 273), nanoethics even goes so far as apparently taking developments into consideration that contradict the laws of nature.

These critics show convincingly how what was originally a conditional if–then chain gradually and unnoticeably becomes presumed expectations or even certainties: "As the hypothetical gets displaced by a supposed actual, the imagined future overwhelms the present" (Nordmann and Rip, 2009, p. 273). Indeed, it is not unusual for a discussion of the consequences of nanotechnology to include second- or third-level conditionality, namely when these consequences might occur as a consequence of the use of nanotechnical products that themselves only *might* or *could* become reality, and then only if the respective technical development *would* take place in the direction envisaged. As a rule, in multilevel conditional sentences of this type it is also possible for the result to be just the opposite of what is assumed. It would then be impossible to decide which of the contradictory alternatives should be given preference and for which reasons.

The methodological criticism of speculative nanoethics is clear: Nanoethics is based on mere speculation that cannot be validated and employs "mere possibility arguments."

> I will use the term mere possibility argument (MPA) to denote an argument in which a conclusion is drawn from the mere possibility that the choice of an option, behaviour, or course of action may lead to, or be followed by, certain consequences. (Hansson, 2006, pp. 317f.)

According to this methodological criticism, nothing in nanoethics is firm enough to deserve consideration. Nanoethics reflects, in this view, on objects that are purely speculative bubbles: "nanotechnology takes as its subject a science still aborning; many of the ethical and social ills it seeks to address are mere speculations" (Keiper, 2007, p. 55). It is said that this, in part, takes place without any consideration of whether the objects concerned are even theoretically possible according to the laws of nature. Nanoethics itself is therefore said to be before its time: "Nanoethics [...] bears all the signs of prematurity. Its time may come some day but it is too soon to say just when and how" (Keiper, 2007, p. 67).

The philosophical debate that followed the publication of the *Imperative of Responsibility* (Jonas, 1984) demonstrated, however, that the *mere possibility* of thinking something cannot be employed as an argument or part of a valid chain of arguments, but only leads to aporias. If the future conditions that ethics reflects on were in a radical sense purely speculative, then the ethical advice that might follow from this reflection could not lead to anything of consequence. If the consequences and side effects of technology could not be evaluated prospectively *in any manner whatsoever*, the question of responsibility would be superfluous: that which is not known cannot be subjected to any ex ante ethical reflection, and no one can be held responsible for it. The object of ethics would get lost in the speculativeness of the considerations of consequences, making ethics obsolete: ethics would not set limits on technology, but technology — as a result of its uncertain future — would set limits on ethics. This was diagnosed as early as 1993 (Bechmann, 1993), long before the criticism of speculative nanoethics (Nordmann, 2007a).

The question is, however, what follows from this statement. To believe that ethics is only supposed to concern itself with developments that are certain to materialize would definitely be a hasty and inappropriate conclusion — which by the way Nordmann and Rip do not make. In that case, ethics would fall into the opposite and frequently mentioned trap of always providing advice too late and missing opportunities for influencing those developments that are precisely in their early phases, when developments are still uncertain (see the criticized "ethics last" model in Moor and Weckert, 2004). Thus the only viable result of the accurate diagnosis that there is a high degree of speculation in parts of nanoethics is the demand that procedures be developed in order for us to be better able to judge the degree of speculation (e.g., as vision assessment, see Grin and Grunwald, 2000 and Grunwald, 2007a). The simple diagnosis that some nanoethics is speculative in nature does not (yet) allow conclusions for the future orientation of nanoethics. To this end, two additional arguments are needed: (a) concern that some objections to nanotechnology might be artificially created (Section 10.1.3) and (b) the opportunity-cost argument (Section 10.1.4).

### 10.1.3 The Anxiety that Unjustified and Artificial Concerns Might Emerge

It is of course possible to ask what the consequences are of the speculative nature that has been diagnosed for many nanoethical works. The answer provided by Nordmann and Rip includes an argument that stood at the beginning of nanoethics, namely the anxiety that unanswered ethical questions could cause nanotechnology to lose acceptance. At the beginning of the debate about the risks of nanotechnology it was noted that nanotechnology and the ethical reflection on it have developed at two very different speeds (Mnyusiwalla *et al.*, 2003). The concern, which was referred to as one over a "policy of innovation," is that the grave and growing gap between rapid advances in nanotechnology and its inadequate ethical appraisal could lead to a development in which the expected advantages and opportunities promised by nanotechnology might not be achieved, for example, because of public rejection (see Section 4.1 for more details). The conviction behind this diagnosis is the following: the earlier the possible ethical problems resulting from nanotechnology innovations are recognized, the greater the possibility to deal with them constructively.

Nordmann and Rip (2009) turn this argument around and fear that the supposed arbitrariness of speculative nanoethics could lead to completely unfounded and unjustified ethical problems which then could become issues for nanotechnology in the real societal debate. These issues, even if they were completely artificial, could inflict serious damage on the further development of nanotechnology:

"worries about the most futuristic visions of nanotechnology can cast a shadow on all ongoing work in nanoscience and technology" (Nordmann and Rip, 2009, p. 274). A premature and thus inevitably, at least in part, speculative handling of the ethical questions of nanotechnology would produce artificial problems, whose communication to the public might cause damage to innovation policy in the field of nanotechnology and could prevent exploiting its chances.

Nordmann and Rip (2009) conclude that nanoethics should deal with "realistic" concerns rather than with speculative ones. The problem, however, is that it is hard — perhaps impossible — to decide *in advance* which concerns are justified and which are not. Decisions of this type are laden with numerous uncertainties, and as history shows, full of surprises and unexpected developments. The result of discontinuing speculative nanoethics and of shifting the focus to "truly pressing" problems (leaving open the question of who determines which problems are truly pressing) might be that we overlook significant problematic developments and are unprepared for, for example, an emergency or an unexpected problem.

It is, however, quite probable that nanoethics — in its speculative version — will (also) discuss (presumed) problems that will prove to be unnecessary from a future perspective. But who is to know *in advance* which problems will really be relevant in some future and which will not? This decision — here mere speculation and there justified concern — cannot be made in a simple manner. However, this shows the need for developing criteria and procedures for being better able to distinguish between mere speculation and more plausible futures.

Within the framework of a "reflective science" (Siune *et al.*, 2009), the argument that artificial problems might be raised and could even inflict damage on innovation policy in nanotechnology is not an argument against speculative nanoethics even if such damage were to actually occur. In other words, this is not an argument even if, for example, speculative discussions of cyborgs were to endanger the prospects for nanomaterials being used in lacquer (although this case does not seem to be very probable). Viewed the other way round, to push this argument could be regarded as giving implicit preference to a "wait and see" strategy instead of dealing with the consequences (which would of course be uncertain) in a precautionary manner.

### 10.1.4 The Opportunity–Costs Argument

To turn to a different argument raised by Nordmann and Rip against speculative nanoethics, is it not the case that if (too) large a portion of nanoethics is concerned with speculative questions, then there will be insufficient resources (people, time, money, creativity, attention, etc.) for the nonspeculative and nonetheless ethically relevant aspects of nanotechnology? Nordmann and Rip formulate the opportunity–costs argument as follows:

> There are good reasons to think that the opportunity costs of speculative ethics are too high, with less spectacular but more "here and now" ethical issues not getting the attention they deserve. (Nordmann and Rip, 2009, p. 273)

Ethical reflection on speculative visions of nanotechnology would, according to this argument, tie resources, making them unavailable for other issues (see also Nordmann, 2007a).

This argument, despite its apparent intuitive plausibility, is not as strong as is asserted. It is based on two premises of doubtful validity:

1. Apparently taking the term "nanoethics" literally, Nordmann and Rip consider the entire speculative debate about nanotechnology to belong to *applied ethics*. They draw a good portion of the intended strength of their argument precisely from this classification: if applied ethics focuses too much on nano speculations, then there might not be enough resources left for fields closer to real application, which is the actual topic of applied ethics. Yet many of the studies about the speculative aspects of nanotechnology can in no way be attributed to applied ethics if the participating persons, the arguments employed, and the literature used are taken into consideration. Many of these studies stem, to give several examples, from the philosophy of technology, anthropology, philosophy of the mind, and the theory of artificial intelligence.
2. The persuasive power of the opportunity–cost argument depends fundamentally on the finiteness and constancy of the variable "available resources." Metaphorically, a model based on a pie diagram is being used, and if a larger part of the "nanoethics" pie is used for speculative topics, then there will be correspondingly less for the nonspeculative topics. But this

model is not adequate in this case, which is related to point 1. It is precisely the speculative topics in nanotechnology that have motivated philosophers and science, technology, and society studies (STS) researchers from outside applied ethics to concern themselves with the questions of nanotechnology. The new visions of nanotechnology have attracted *additional* resources from parties interested in such issues. The pie has grown considerably. The opportunity–costs argument thus does not hold up.

### 10.1.5 Resume

According to the preceding analysis, Nordmann and Rip's criticism is valid for one but *only* for one sore point. The valid criticism points to a problem of a possible "pure speculativeness" of speculative nanoethics (Section 10.1.2), in which statements might run the danger of being purely arbitrary. The other arguments, however, cannot be accepted:

- The diagnosis that "most nanoethics is too futuristic" (Nordmann and Rip, 2009, p. 273) is one sided. It conceals the fact that a substantial part of current nanoethics is indeed concerned with concrete questions concerning current developments (Section 10.1.1).
- The argument of generating "artificial arguments" and thereby possibly causing damage fails because there are no instruments available to decide *in advance* which ethical problems of future technologies will really be relevant in some future and which will not (Section 10.1.3).
- The critics' central argument, which is based on a consideration of possible opportunity costs — i.e., that the attention given to (irrelevant) speculative nanoethics takes resources away from dealing with the relevant ethical questions in nanotechnology — cannot be sustained (Section 10.1.4).

The general situation is that there is, on the one hand, a clear need for orientation concerning future technologies and that satisfying this need unavoidably requires some knowledge about future applications and their impacts, but that, on the other, this knowledge is highly problematic and is in danger of leading to arbitrary results. Facing this situation, we must look deeper into the structure of such precarious prospective knowledge in order to derive some orientation.

## 10.2 Searching for Orientation by Investigating Futures

In the last few years, it has been possible to observe a considerable increase in communication about technically induced "new futures," especially in the fields of nanotechnology and the converging technologies (Selin, 2006). This increase in visionary communication can be interpreted as an indication of a new occasion for regaining orientation as well as an attempt to do so (Grunwald, 2007a). It therefore seems appropriate to investigate the role of visionary communication in order to arrive at a better understanding of the problems and challenges involved.

Information gained from reflection on the future is a commonly used means for facilitating the decision-making process in modern societies (e.g., Luhmann, 1997). Prospective knowledge of consequences, prognoses of technical progress, expectations and fears, as well as aims are bundled together as "futures" (e.g., in the form of scenarios), which serve to provide orientation *today* for pending decisions (cf. Fig. 10.1). Proceeding from *present-day* problematic situations and their diagnoses, orientation *for today* is sought via the roundabout route of debates about the *future*. Communication about possible or expected future developments is often an *attempt to regain orientation* when facing choices and the uncertainty that can accompany scientific and technological advancement. A situation of heightened contingency could in fact be confronted constructively if orientation for decision making were successfully re-established, ideally by social agreement on future scenarios (planning for them, and determining what is desirable and what is to be prevented).

This includes a *normative* expectation concerning the role of communication about the future (Giddens, 1984; Luhmann, 1997). Whether it really would be possible for debates on possible future scenarios to re-establish orientation is not self-evident because in modern societies the plurality of the population can directly affect any judgment about future developments and make it far from easy for a consensus to be reached (Brown *et al.*, 2000). The well-known social conflicts will also enter into the consideration and assessment of futures. Even worse, as can easily be seen in the field of nanotechnology, attempts to bring about orientation by means of communicating about the future frequently cause new problems of orientation instead of providing solutions.

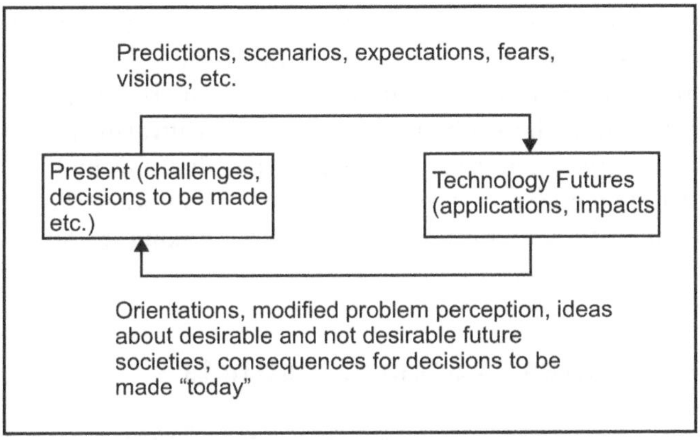


**Figure 10.1** The decision-making cycle via future thinking.

A fundamental problem of visions or scenarios of the future is the inevitably high degree to which material other than knowledge is involved. In many cases, entire conceptions of the future, or aspects of them, are simply "given" due to a lack of knowledge. Nanotechnology futures are also uncertain, normatively shaped to some extent, and often controversial. Consider, for example, the different views on converging technologies expressed by Dupuy and Grinbaum (2004) and Roco and Bainbridge (2002). The future prospects of the converging technologies show the *maximally imaginable disorientation*: they oscillate between expectations of paradise and of catastrophe (see Section 2.5). If no approaches were available for assessing and scrutinizing diverging futures in a "rational" sense, then the arbitrariness of futures would destroy any hope of gaining orientation by reflecting on future developments.

Additionally, images of the future also seem to be exposed to the influence of ideology. Participants who are advocates of social issues, substantial values, and particular interests seem to simply produce the future visions that fit them in order to then use these visions to promote their particular positions in debates (Brown *et al.*, 2000).

To provide orientation by reflecting on futures is therefore a highly ambitious undertaking. For this to succeed, the loop of decision making (see Fig. 10.1) may not be a vicious circle, i.e., an idle state of knowledge, but must demonstrate added value that provides orientation compared to the situation *prior to* entering the cycle. To satisfy this requirement, however, several problems must be overcome. The epistemological problem of the "objectifiability" or the "argumentative quality" of futures will be addressed in the following section because this was the primary issue resulting from the examination of the debate on "speculative nanoethics." It is essential that the problem of the feared arbitrariness of futures be sufficiently resolved. Otherwise the decision-making cycle (Fig. 10.1) would be no more than self-deception. The solution to this aporetic situation can be found by looking in-depth at the character of "futures."

## 10.3 Futures as Social Constructs

We make statements and prognoses about the future, simulate temporal developments and create scenarios, formulate expectations and fears, set goals, and consider plans for their realization. All this occurs in the medium of language (Kamlah, 1973) and is thus in the *present*. Forecasters and visionary writers cannot break out of the present either, always making their predictions on the basis of *present* knowledge and *present* assessments. Future facts or processes can be neither *logically deduced* (Goodman, 1954) nor *empirically investigated*. The only things that are empirically accessible are the *images* which we make of the future, but not the future itself that will at some time become the present.[2] For this reason, we can talk about *possible* futures, about alternative possibilities for imagining the future, and about the justification with which we can expect something in the future. These are always *present futures* and not future presents (Picht, 1971).

If we talk about cyborgs or far-ranging human enhancements being possible in the future, we are not talking about how these developments will "really" occur but how we *imagine it today* — and such visions differ greatly. Futures are thus something always contemporary and change with the changes in each present. "A future

[2] For this reason, we refer to "futures" in the plural in this chapter. The plural expresses that "the" future is not accessible to us but that we can only form many and different visions of it, i.e., futures.

is thus not something separate from the present, but a specific part of each present" (Grunwald, 2006a).[3]

Futures do not exist per se, and they do not arise of their own accord. On the contrary, futures are "made" and linguistically constructed in a more or less complex manner. The designing of futures is purposive action, intended especially to provide orientation in the sense of the decision-theory cycle (Section 10.2). Futures, regardless of whether they are forecasts, scenarios, plans, programs, visions, or speculative fears or expectations, are "produced" using a whole range of ingredients such as the available knowledge, value judgments, and suppositions. This construct character of a future (i.e., that its character is the result of a construction process) is essential for scrutinizing them with regard to content and quality (see below).

Future constructs are created in accordance with available knowledge, but also with reference to assessments of relevance, value judgments, and interests, often in the context of commissions awarded by decision makers in politics and the economy. The construct character of futures can thus be exploited by those representing specific positions on social issues, substantial values, and particular interests, in order to produce future visions corresponding to their interests and to employ these to assert their particular positions in debates (Brown *et al.*, 2000). This leads to the question of whether and to what extent one can work against the usurpation and instrumentalization of nanotechnology (or other) futures by evaluating pictures of the future in a comparative fashion according to objectifiable standards.

A process of evaluating futures is required in which their "rationality," i.e., their intersubjective and transsubjective argumentative quality can be analyzed and ultimately tested. Only such a procedure could help to overcome the fear of relying on completely arbitrary futures in making ethical assessments of future technologies, as is expressed by the critics of "speculative nanoethics" (see Section 10.1). The decisions about the quality of scientific results and hypotheses, i.e., the argumentative validity of statements and the justification for demands, are made *discursively* (e.g., Habermas, 1988). A discourse that takes place between opponents and proponents of different futures is the philosophical method in which the argumentative quality of futures must be

[3] For this reason, it is also possible to speak of "past futures," for instance if historians concern themselves with future projections in earlier periods of time.

judged. Yet greater "argumentative strength" is not synonymous with the probability of later occurrence, as is often thought. What can be said *with validity* is not whether claims about futures will come true but only whether their coming true can be *expected* on the basis of present knowledge and a present assessment of their relevance (Lorenzen, 1987).

Because of the social construct character of futures, their argumentative quality will strongly depend on what was put inside their construction. The question arises as to which ingredients are invested in the shaping of futures and in which way these ingredients have been assembled and composed in arriving at the respective statements about the future. As far as their knowledge structure is concerned, futures are initially opaque constructs consisting of highly diverse elements: scientific knowledge of diverse origin and validity, knowledge from the realm of experience, ad hoc suppositions, estimates of relevance, *ceteris paribus* conditions, etc. In analyzing what argumentative quality can be assigned to statements about the future, how the quality of predictions should be evaluated, and how quality is to be determined, it is important for proponents of a given position (e.g., those promoting a particular nano future) to be confronted with their own claims and with the arguments of their opponents (e.g., representatives of competing nano futures). To do this, one must rank the knowledge components that have contributed to each statement about the future and the respective premises according to their validity. This amounts to an epistemological *deconstruction* of the futures under analysis, dismantling them into their linguistic and epistemologically relevant components. In a rough approximation, the following gradation of knowledge and nonknowledge components can initially be made:

- *Present knowledge*, which is proven according to accepted criteria (e.g., of the respective disciplines) to be knowledge (e.g., according to the issue at stake from the field of nanotechnology, engineering, economics)
- *Estimates* of future developments that do not represent current knowledge but that can be substantiated by current knowledge (e.g., demographic change, energy needs)
- *Ceteris paribus (All other things being equal) conditions*, whereby certain continuities — business as usual in some sense or a lack of disruptive changes — can be assumed as a framework for the prospective statements

- *Ad hoc suppositions*, which are not substantiated by knowledge, but "given" (e.g., the future validity of a German phase-out of nuclear energy, or the nonoccurrence of a catastrophic impact of a comet on the Earth)

Whoever raises claims to validity and to be avoiding arbitrariness when speaking about future developments has to name the preconditions that have to be presumed as the condition for making a grounded statement about the future. A discourse about questions concerning the argumentative quality of future statements is thus turned into a discourse about the knowledge components and preconditions that are present in the respective positions, and about the methodological integration, such as in the underlying model that formed the basis of specific future statements. A dispute about the quality of technology and nano futures does not refer, therefore, to the events predicted to come about in a future present, but to the reasons for the respective future that can be given on the basis of current knowledge and current judgments of relevance. These reasons must then be weighed in discourse.

A discourse about the argumentative quality of statements about the future thus becomes a discourse about the preconditions that are present and that have led to the statement about the future. This at least demonstrates programmatically a way in which to use scientific tools to confront the feared danger of arbitrariness or the often presumed ideological substance of energy futures. The objective of an epistemological deconstruction of these futures is to discover something about their epistemological status and about how tenable their arguments are. In order to make this program work, more operable approaches have to be developed (Section 10.4.2). The application of such procedures creates more transparency, namely concerning the knowledge components and their limits, concerning the uncertainties that are involved and that must be explicated in such a deconstruction (although they otherwise can stay in the dark), and concerning the exposure of the values, norms, and even interests that are involved in promoting particular nanotechnology futures.

## 10.4 Explorative Philosophy of Nanotechnology

Precisely in the context of nanotechnology and new biotechnologies, the concept of ethics has been broadly diffused in recent years. The criticisms raised against "speculative nanoethics" (Section

10.1) are based on the notion that applied ethics should contribute to solving decision-making problems in the public and political sphere, which is close to my notion of "problem-oriented ethics" (Chapter 3). If this is taken seriously, parts of the debate on nanotechnology can indeed not be classified as applied ethics. The notion of "explorative philosophy" suggested in this chapter might help to differentiate between several modes of philosophical thinking about nanotechnology and possibly to prevent some misunderstandings.

### 10.4.1 Explorative Philosophy Beyond Applied Nanoethics

Since the very beginning of ethical reflection on science and technology a discussion has been ongoing about what the appropriate relation in time is between scientific-technological advances and ethics. Ethics often seems to pant helplessly behind technical progress and to fall short of the occasionally great expectations. The rapid pace of innovation in technology has had the effect that ethical deliberations have often come too late, namely when all of the relevant decisions have already been made, i.e., when it is far too late to shape technology. In response to this diagnosis, ethics joined the move toward "upstream engagement" in the field of STS. The hope was and still is that addressing the earlier stages of development more directly would allow for a greater contribution to shaping technology. At this point, however, we must be aware of the Collingridge dilemma from technology assessment (Collingridge, 1980), according to which it is in principle easier in the early phases to influence the course of events, but in the early stage of development the required knowledge is absent that would enable one to intervene in a constructive manner. Nordmann and Rip's criticism can be read as a diagnosis that nanoethics, in avoiding the Scylla of the Collingridge dilemma (namely, coming too late), must fall victim to the Charybdis (i.e., inevitably being too speculative).

The Collingridge dilemma is exaggerated. The question of whether ethics should start early or late, whether it should be prospective or only start when reliable statements about consequences are available, poses a false alternative. The issue here is not an either-or one but the differentiation of ethical reflection in line with the problem at hand and with the validity of the available knowledge of the consequences. Ethical reflection differs conceptionally and methodologically depending on whether it is concerned with measureable consequences of technology or with just imagined or even merely speculative ones. Above all, it also serves *different purposes*. The question, for example, as to whether it is responsible to permit nanoparticles to be included in food is a concrete question to be considered in the context of regulation, labeling requirements, corporate commitments, and individual responsibility. In contrast, early thoughts about synthetic biology or human enhancement serve rather to promote conceptional understanding and clarification of the issues from a normative perspective or to facilitate the development of clear terminology and ethical alternatives, but without there being anything immediate to be regulated.

The initial question is about the purposes that ethical reflection on nanotechnology is supposed to serve. Only then would it be possible to determine whether and under which conditions "speculative nanoethics" might help to satisfy at least some of these purposes, or to find that it is obsolete. There is no difficulty in determining sensible purposes of ethical reflection, even for a rather speculative form of ethical reflection. The character of the latter would rather provide orientation but not any direct guidance for action:

- Speculative reflection on nanotechnology that operates with experimental ideas is confronted with the danger of being arbitrary. However, if its purpose is not to provide concrete orientation for action, but rather to provide clarification in preparation for a possible future emergency or provide a preliminary conceptual and substantive structure for a possible future field of applied ethics, then the argument of pure speculation misses the point. Such explorative reflection would certainly not be applied ethics but perhaps the preparation for future applied ethics.
- Early philosophical reflection on a new technology can draw attention to aspects that can only be judged after better knowledge of the consequences becomes available. Or it could point to critical questions that require more precise examination in the future. In this manner, it would focus the attention on specific although speculative elements of the future.

- Despite its speculative nature, early reflection could even have concrete consequences for today's decisions, such as with regard to research funding or the scientific agenda. The debate about stem cells was conducted, for example, in part as speculative ethics, namely inasmuch as interventions in the rights of embryos were being weighed, in part, against very uncertain expectations regarding the possibility of being able to use the results of research on these embryos to heal severe illnesses or to lessen their impact.
- By means of early (and in part speculative) ethics we can learn something "about and for us" today: "What do these visions tell us about the present, what is their implicit criticism of it, how and why do they require us to change?" (Nordmann, 2007b, p. 41). A good example is "human enhancement," much of the debate about which can truly be called pure speculation. The debate has however beyond a doubt already led to a more intense consideration and better understanding of ongoing developments "beyond healing" (such as for doping in sports or everyday life).

The purpose of the reflection changes with the degree of speculativeness of the object being considered. The more speculative the considerations of the consequences, the less they can serve as orientation for concrete (political) action and decisions. Conceptual, pre-ethical, heuristic, and hermeneutic issues then become, in contrast, more significant. The primary issue is then to clarify what is going on in the speculative developments being considered, what is at issue, which rights might possibly be compromised, which images of man, nature, and technology are formed and how they change, which anthropological issues are involved, and which designs for society are implied in the projects for the future: "Epistemology, moral philosophy, and natural sciences have always benefited from thought experiments on future technology" (Sethe, 2007, p. 362).

This makes it clear that it is certainly not sensible to classify such very early reflection (which Nordmann and Rip diagnose as speculative nanoethics) as applied ethics. *Ethics,* in the sense of the discipline, constitutes a rather small portion of such reflection, while considerations that are conceptual, hermeneutic, epistemological, philosophical (philosophy of technology, philosophy of science, philosophy of culture, etc.), or anthropological in nature play a large role. Ultimately this leads to the recommendation not to speak of "speculative nanoethics" but to consider these forms of reflection as elements of an explorative philosophy of nanotechnology.

### 10.4.2 Elements of an Explorative Philosophy of Nanotechnology

Explorative philosophy is concerned with very early and in part unavoidably speculative evaluations of new developments in science and technology and of their consequences for man and society. As such, it is in no way supposed to provide orientation for action in areas of concrete development. It is neither competition to nor part of applied ethics (see above). Its task is rather to *prepare* for possible coming debates in a conceptual, theoretical, and methodological sense as well as with a view to basic distinctions and relationships from the philosophy of technology and anthropology such as man-technology, life-technology, or nature-technology. Existing work — which is sometimes subsumed under "nanoethics" and can then be criticized as speculative — might be better classified against this background as explorative philosophy of nanotechnology in the following fields.

#### 10.4.2.1 *Nano Epistemology*

The emergence of nanotechnology has raised expectations of a new unity in science (Roco and Bainbridge, 2002). Following these expectations, nanoscience — as the science of "shaping the world atom by atom" — should integrate the former classical disciplines of physics, chemistry, biology, and the engineering sciences into a new type of science. Obviously, these projections challenged the philosophy of science and philosophical epistemology in order to uncover the underlying assumptions of those expectations and to question the conditions of their validity. The assumptions can be traced back to a naturalistic understanding of atoms as something similar to bricks or stones in the macroscopic world. "Shaping the world atom by atom" would then mean building new structures from atoms in complete analogy to building systems from macroscopic elements such as bricks. Epistemological reasoning has questioned the validity of this simple analogy and the "atomic reductionism" behind it (Schmid *et al.*, 2006; Chapter. 2).

### 10.4.2.2 *Nano Anthropology: The Relationship Between Humans and Technology*

Nanobiotechnology is emerging at the interface between biotechnology and nanotechnology. It bridges the gap between inanimate and animate nature, and aims at combining biological modules at the molecular level as well as producing functional building blocks on a nanoscale that might include technical materials, interfaces, and bounding surfaces (Schmid *et al.*, 2006). The point of departure is the fundamental belief that life processes take place on the nanoscale because the essential building blocks of life are just this size (e.g., proteins and DNA). The processes in a cell can be analyzed with nanotechnological methods and possibly rendered technologically utilizable. In nanobiotechnology, the language of mechanical engineering is employed to describe the mechanisms and parts of cells: cells and their organelles are interpreted as micro- or nanomachines. Literally speaking, we see nanotechnology infiltrate molecular biology, genetics, and neurophysiology, all of which are integrated under technical points of view. The nanotechnical (and possibly feasible) duplication of fundamental life processes is the essential prerequisite for crossing the borderline between technical and living systems.

In transferring this process to the relationship between humans and technology, speculation emerged about the convergence of humanity and technology. The concept of "cyborgs" as technically enhanced humans or as humanly enhanced technology can be raised. An aspect that frequently turns up in visions of nanotechnology is related to the borders between humans and technology and to the question of whether information or material can be passed from the one to the other. This aspect deeply affects our self-understanding and, consequently, our distinction between what a human being is and what he/she creates with the aid of technical achievements and applications. These developments — currently no more than a concept in spite of their widespread presence in the ethical discussion — obviously pose questions of the type as to whether and to which extent this would increasingly place human beings in the realm of technology (Grunwald and Julliard, 2007). The presence of such questions also indicates, however, the necessity of renewed reflection on our self-understanding as human beings and our relationship to technology.

#### 10.4.2.3 *Nanotechnology Hermeneutics: Philosophical Interpretations of Nanotechnology*

The fascinating new opportunities of manipulating matter provided by nanotechnology but also the many visions about nanotechnology that are related to social issues have motivated thinking about the deeper changes in human civilization and its relationships to nature and technology. Roughly, three different positions have been expressed so far (see Section 2.5). First, there are assumptions that a new Baconian approach is arising. Nanotechnology — as "shaping the world atom by atom," to use a phrase from the National Nanotechnology Initiative — could be interpreted as a new manifestation of optimism by making everything appear to be technically possible. Second, the opposite story starts from the "enabling" character of nanotechnology and assumes creation of the greatest uncertainty imaginable: everything could be possible, and probably nothing could be controlled (Schmidt, 2008). The third story regards nanotechnology as a "cipher of the future" that serves as a catalyst for societal, philosophical, and scientific debates on issues such as the future relationship between humans and technology and the future of human nature in avoiding strong substantial claims about controllability or other issues.

All of these fields — and there might be others, too — are speculative in a sense. Thinking about these issues nonetheless seems worthwhile even though no direct policy actions depend on their results. It is more for understanding what is at stake and at issue, a contribution to a "hermeneutics" of possibly changing elements of the *human condition*. In this way, explorative philosophy can prepare the groundwork for applied ethics and for the technology assessment of the developments when they become more concrete. Ultimately this promotes a democratic debate on scientific-technical progress by investigating alternative approaches to the future of humans and human society with or without different nanotechnology developments.

### 10.4.3 Epistemological Grounding

What remains is the justified point raised by the critics of speculative nanoethics that no legitimate conclusions could be drawn if the reflection addresses merely speculative and arbitrary futures ("mere possibility arguments," cf. Hansson, 2006; see Section 10.1.2). In order to circumvent this criticism, explorative philosophy may not be epistemologically blind but must develop the methods and procedures for assessing the degree of rationality behind images of uncertain futures. It must clarify the cognitive and normative content of the partially speculative future conditions and consequences of nanotechnology:

> Instead of welcoming without scrutiny anyone who cares to add to the stock of promises and concerns about nanotechnology, we need to encourage discussions about quality of promises. (Nordmann and Rip, 2009, 274)

The epistemological task is to examine both the cognitive and evaluative content and the consequences of technology-based visions. An epistemological "deconstruction" of prospective statements is necessary in order to be able to qualify the object of subsequent ethical reflection with regard to its applicability and validity.

Making a success of epistemological analysis of uncertain nano futures requires that we think carefully about the structure and the grounding of any nano futures, such as visions (Grunwald, 2007a), scenarios, and projections (Dupuy, 2007). One of the anthropological characteristics of human beings (an element of the *unique status* of humans) is that humans are able to think about futures: to make plans, to predict future developments, to create scenarios, to develop visions, and to express hopes and fears. These futures are imagined as pictures of a future which we cannot access empirically or with our senses, such as by direct observation.

Epistemological analysis of future conditions would initially have to uncover the cognitive content of the visions, i.e., the portions of knowledge and lack of knowledge that are involved, their respective premises, and the way they are combined to form coherent images of the future, such as scenarios. An important aspect would then be to examine the conditions needed for such futures to become reality and the periods of times that are involved. Furthermore, the *normative content* of the visions would have to be reconstructed analytically: the images of a future society or of the development of man, and the possible diagnoses of current problems, the solutions to which are supposed to be facilitated by the visionary developments of nanotechnology.

The *de facto* importance of futuristic visions in the nano debates is the main argument for postulating an early vision assessment in order to allow for more rationality, reflexivity, and transparency (Grunwald, 2009b). Even a vision without any facticity at all might have real impact on debates, on the formation of opinion, on acceptance, and even on decision making. Futuristic visions are able to change the perception of the world and to increase the contingency of the human condition (Grunwald, 2007a). This importance also emphasizes the need for early assessment of such visions *in an ethical regard*, in order to allow for more transparent discussions about the premises involved, about the normative content (Kushf, 2004a) and its justification, as well as about potentials for moral conflict.

The vision assessment approach has been proposed in order to uncover the epistemological and ethical grounding of (nano) futures (Grin and Grunwald, 2000; Grunwald, 2009b). Vision assessment can be analytically divided into several steps which are not sharply separated and not linearly ordered but which serve different subobjectives and involve different methods: (a) vision *analysis* — which itself can be separated into a *substantial* part (what is the content of the respective vision?) and a pragmatic part (in which way is it used in concrete communication?) — (b) vision *evaluation* (in which way could the content of the vision be evaluated and judged?), and (c) vision *management* (in which way should the persons and groups affected deal with the visions?). In particular, the task of vision assessment in an ethical regard is to confront the various and, in part, completely divergent normative aspects of the nano futures directly with one another in order to identify the roots of possible moral conflicts and normative uncertainties. This can, on the one hand, be done by ethical analysis and desk research; on the other, however, the representatives of the various positions should discuss their differing judgments in workshops directly with and against one another, in order to lay open their respective premises and assumptions.

Based on the observation that visions and other types of futures are strongly influencing the scientific agenda, political decision making, public attitudes, and the structure of risk–chance debates in nanotechnology (and also in other fields of science and technology), it is necessary to provide tools for allowing transparent democratic debate about the different futures put forward by different actors. Vision assessment aims at uncovering the epistemological and normative ingredients of future statements in order to permit better informed and more rational formation of opinion, assessment and decision making. In this way, vision assessment can contribute to ethical reasoning, which must avoid running into the problem of dealing with "mere possible arguments" (Hansson, 2006).

# Chapter 11

# Conclusions and Perspectives

In this concluding chapter, I would like to undertake several integrative analyses in order to reflect the status of the ethical discussion of nano(bio)technology. This reflection includes the attempt to summarize the previous effects of nanoethical activity (Section 11.1), a consideration of the frequently feared moral resistance to nanotechnology (Section 11.2), and several tentative remarks about the anticipated prospects of the ethics of nanotechnology (Section 11.3).

## 11.1 Ten Years of Nanoethics: What Has Been Achieved?

It is difficult to identify the starting point of the debate over nanotechnology and society. There are, however, some arguments for assuming that everything began around the year 2000: the Bill Joy debate, the first consideration of the possible health or environmental risks caused by nanoparticles, and the first activities related to nanotechnology and sustainable development. We can thus look back on about 10 years of debate over nanotechnology and society, including ethical reflection, whose start was in fragmented and isolated activities but which soon developed fulminant momentum (Sections 2.6 and 5.1). This is one motivation for assessing the situation: What has been reached so far? What has

*Responsible Nanobiotechnology: Philosophy and Ethics*
Armin Grunwald

ISBN 978-981-4316-80-4 (Hardcover), 978-981-4363-33-1 (eBook)
www.panstanford.com

not been reached? And what can be said about the outcome of these 10 years in relation to the expectations raised at an early stage?

Obviously, a manifold of research and reflection in the social sciences, in ethics, and in the humanities has been performed, which now constitutes a considerable body of knowledge that previously was not available. This is a remarkable step toward understanding the cultural, social, and ethical dimensions of nano(bio)technology in particular and new and emerging technologies in general (see Chapters 5–9). However, my question is more directed toward the impact of ethical reflection outside these fields (cf. in more detail Grunwald, 2011b).

One aim of ethical inquiry into nanotechnology is to influence the further course of development or, in short, *to have an impact* rather than to be a mere ornament. The early stage of many developments in nanotechnology provided and provides an advantageous opportunity: the chance and also the time for concomitant reflection, as well as the opportunity to integrate the results of reflection into the process of technology design, thereby contributing to the further development of nanotechnology (Moor and Weckert, 2004; Fleischer, 2003). Ethics and other types of prospective analysis and assessment, such as science, technology, and society (STS) studies, have largely recognized this opportunity: the hype surrounding nanotechnology has been followed by hype around reflection on the social implications and impact of nanotechnology. It seems appropriate to question which impact the second form of hype has had up to now. In a "strong" understanding of the social shaping of technology approach (SST; Yoshinaka *et al.*, 2003; Grunwald and Hocke-Bergler, 2010) we should expect an impact on the "shaping" of nanotechnology, and this "strong understanding" was dominant in the early phase of nanoethics: the objective was to influence nanotechnology development. The central message was that a "better" nanotechnology could be constructed by using SST: "to achieve better technology in a better society" (Schot and Rip, 1997). The social construction of technology has even been extended to the social construction of the *consequences* of technology. In order to achieve an environmentally and socially friendlier technology, network-oriented approaches to the sociology of technology have tried to control the problem of the unintended side effects of technology by applying adequate strategies for shaping technology during its genesis. Behind these approaches are highly ambitious models of social construction and *constructability* of technology.

The main diagnosis today is, however, that there is only weak evidence for the strong understanding, i.e., weak evidence that ethical contributions played a substantial role in the formation of nanotechnology itself (following Grunwald and Hocke-Bergler, 2010). Study of the literature and observation of the public discourse have detected hardly any traces of ethical deliberations on the pathways and roadmaps of nanotechnology and, thus, of an influence of ethics on the nanoscientific agenda itself.

Ethical deliberations, however, clearly had and still have a concrete impact in other fields. They have not directly affected the nanoscientific agenda, but they have complemented the view on what should urgently be done in other fields of research (such as nanotoxicology). And by motivating public debate, they have contributed to nanotechnology becoming a public phenomenon. There is more evidence to support a weaker interpretation of the contribution that ethical debate has made to the formation of nanotechnologies. This weak understanding — which does not imply a weak influence — could be described as follows:

> Weak understanding: "formation of nanotechnologies" means "formation of the perception of nanotechnologies," where the "perception" could be the public perception, the perception of the scientists, of stakeholders, etc. — with possible impacts on the embedding of technology into society, and with a more indirect influence on the research agenda. (Grunwald and Hocke-Bergler, 2010)

Ethical issues, positions, and judgments have become an accepted part of society's debate about nanotechnology. They can be found in the academic sphere, but especially in public, such as in events at academies and educational institutions, at discussion forums organized by foundations, in the features section of larger newspapers, in policy consultation, and in the political debate itself. In the meantime, ethical topics have come to belong to the standard repertoire of scientific policy consultation. The example of parliamentary technology assessment demonstrates that ethical topics have, taken together, increased in significance in parliament. Policy consultation on nanotechnology has in the meantime become a broad subject, in which ethical issues are taken into consideration as a matter of principle.

Social issues in nanotechnology receive broad treatment in public dialogs in Germany (at events such as the federal government's nanoforum, which is coordinated by the Federal Ministry for the Environment, and at similar dialogs at the state level), in the activities of nongovernmental organizations (e.g., BUND, 2008), in conferences held by church academies and by foundations belonging to political parties, as well as in the relevant workshops of economic groups. Workshops and consumer conferences take place on specific fields of application, particularly in the food sector, where the issues covered are primarily related to risk acceptance, risk acceptability, and labeling requirements. Ethical issues even frequently play a central role at these conferences. The two issues at the forefront are the risks posed by synthetic nanoparticles and the rather futuristic topic of the relationship between humans and technology.

The imprint of ethical arguments accordingly permeates the entire public debate on nanotechnology. It is therefore possible to speak of the *de facto* presence of ethics in the debate about nanotechnology. Moral arguments are raised by scientists and spread by the media, and they are present in the realms of politics and economics. This ethical reflection has contributed to *sensitizing* social groups with regard to nanotechnology. The fact that nanotechnology is a challenge to science, to reflection, and to the societal organization of technology is not something that was made clear by the older economic estimates of the market potential of nanotechnology (e.g., Bachmann, 1998); it was made clear by the pattern of ethical argumentation that was introduced into the debate, as it was everywhere following the Bill Joy debate.

Professional ethical reflection has deconstructed and reconstructed the patterns of the moral arguments that have been raised by Eric Drexler, Bill Joy, the transhumanists, Jean-Pierre Dupuy, and other representatives of striking positions. It has also linked these patterns to the relevant discussions and arguments in subdisciplines of applied ethics. This has, as it were, demystified nanotechnology by "normalizing" it. This normalization may have contributed to the fact that the social dialog on nanotechnology has until now proceeded in civilized channels. The revolutionary pathos of nanotechnology's early phase was constantly in danger of turning into a rejection that was just as radical. "Tremendous transformative potential comes with tremendous anxieties" (Nordmann, 2004, p. 4; see also Grunwald, 2006a).

No one can assert that ethical reflection has changed the course of nanotechnology (as asserted by a strong understanding of "shaping technology"). Yet it has — in concert with other forms of research into and reflection on the social aspects of nanotechnology such as STS studies, technology assessment, and risk studies — contributed to the fact that nanotechnology has been embedded in society in a manner that is different to how it would otherwise have been. In other words, ethical reflection was not a force shaping nanotechnology itself — this technology is still at a stage that might be too early for this to happen — but it was a force shaping the interface between nanotechnology and society. This occurred, for example, by means of the early demand that research be conducted into the possibly toxic properties of nanoparticles, by opening the debate on the relationship between humans and technology in light of the possibilities for technical human "enhancement," and by insisting on opportunities for participation and on dialog.

Ethical reflection on nanotechnology, as has been shown, has not met the high expectations on it that it contribute to the formation of nanotechnology itself in the sense of influencing the research agenda. A strong interpretation of the position of social construction cannot be verified. The ethical debate on nanotechnology in its different forms has however had a remarkable impact on the social and scientific environment of nanotechnology. This means that ethical reflection — although it has not been able to contribute directly to "shaping technology" — is not merely ornamental. This is, in my eyes, a positive result. Ethical analysis and STS studies have really contributed to normalizing nanotechnology, which means to a successful piecewise embedding of nanotechnology into society despite the risks and even while taking these risks seriously.

## 11.2 Moral Arguments Feeding a Broad Antinano Movement?

Nanoscientists, policymakers, and funding agencies have frequently been concerned about the public's perception of nanotechnology. Several years ago (ca. 2003–2005), the idea became popular that nanotechnology — following nuclear power and genetics — might be the next field of technology to run into a communication disaster at the interface between technology and society (Section 2.6). Postulates of a moratorium concerning the use and release of nanoparticles, brought up by nongovernmental organizations fed expectations and fears in this direction to a considerable extent. However, up to now, nothing in the direction of such a communication disaster has happened even in the situation of severe controversies around nano(bio)technology such as in France. The interface between nano(bio)technology and the public is still more or less relaxed. The debate on the possible risks posed by nanotechnologies (and of nanoparticles in particular) is still going on (cf. Chapter 6), but it has not led to broad rejection or protest. The question to be considered in this section is whether analysis of ethical issues surrounding nanotechnology can identify any indications that a *morally motivated* front rejecting nanotechnology might develop. I will attempt to ground the hypothesis that this is not to be expected with a high degree of probability according to our present knowledge. This diagnosis is supported by numerous reasons of varying significance.

*First*, nanotechnology consists of many nanotechnologies, each with their own entirely different ethical aspects that taken together comprise nearly all of the relevant issues of the ethics of technology (Chapter 6). This makes it appear difficult to imagine that a coherent front rejecting nanotechnology in *general* could develop. Nanotechnology does not exhibit any "unifying aspect," such as the technical modification of the genetic features of living beings by genetic engineering. The guiding motif of nanotechnology — i.e., technical shaping at the atomic or molecular level — is too abstract. The diversity and heterogeneity of nanotechnology are themselves in a certain sense a protection against focused and coherent counterpositions that would affect nanotechnology *as a whole*.

*Second*, there is no specific argument that is also plausible to larger sections of the population that could form the motivating core of a general moral rejection of nanotechnology. In the debate about nuclear energy, for example, there were (and are) such central topics, such as the possibility of catastrophic accidents, illustrated by the names Three Miles Island and Chernobyl, and the problem of a permanent repository for radioactive waste. Both of these problems have a moral core if there is supposed to be an answer to the questions of whether it is morally acceptable (1) to incur the risk of catastrophic accidents if the probability of their occurring is sufficiently small, or (2) to burden following generations for many thousands of years with the problem of wastes that require special handling. Although scenarios with catastrophic developments can also be constructed for nanotechnology in the context of what is conceivable (e.g., the gray goo scenario; see Section 2.6), their plausibility in our lifeworld and likelihood of acceptance are small, however, compared to that of nonfunctioning technology as the cause of catastrophic accidents, for example.

*Third*, general rejection cannot be constructed out of really new concerns regarding nanotechnology applications. The topics of ethical reflection stem to a large extent from previous discussions of technology (e.g., data privacy, fair distribution, and health risks, see Chapter 5). In most cases they concern intensification in degree or the accentuation of known problems, such as the question as to the possible toxicity of nanomaterials and nanoparticles (Chapter 6). Such types of questions are familiar in social debates. As a rule, their consideration leads to greater attention being paid to certain aspects in shaping technology (e.g., protection again misuse) in order to avoid undesirable developments and the corresponding problems of acceptance. The concerns in this case are hardly suited to be the basis of a front rejecting nanotechnology.

*Fourth*, the moral status of objects of research can entail anxiety in the public and the corresponding moral conflicts, such as currently is the case in the debate on the moral acceptability and legal regulation of stem cell research. In that case, the moral status of the embryo is the central item of discussion, linked with the question of whether and to what extent the simple prospect of successful cures made possible by successes in stem cell research justify treating them as being expendable in research. The guiding motif of nanotechnology — manipulating material at the level of atoms in a targeted fashion (NNI, 1999; cf. Section 2.1) — poses in contrast prima facie an ethically unsuspicious matter. Atoms and molecules are not the objects of any special moral protection.

Essential prerequisites that have been typical for previous communication disasters and problems of acceptance at the interface between technology and society are thus not given in the case of nanotechnology.

In addition, there are indications that the socially dominant attitude has changed compared to earlier times. It appears to be currently widely accepted (a) that a postulate of "zero risk" is inadequate, and (b) that technological advance is certainly related to risk and uncertainty. On the one hand, it might be that the social perception of technology in general has changed or that society today has other concerns than technological risk. On the other hand, however, it seems plausible that ethical deliberation on nanotechnology and related activities have contributed to this "relaxed" development in the following way. In the field of nanotechnology, there have been a lot of activities in recent years that make it clear that dealing responsibly with risk is now seen as part of scientific advance and its political shaping, funding, and regulation. Major examples are activities related to the ethical, legal, and social implications, toxicological research, and debates over regulation. Risk is not ignored or denied (especially not by nano scientists and managers), but is being actively dealt with. In this way, *trust* has been generated, and trust is a major issue in avoiding communication disasters.

The result — namely that a risk of a communications disaster for nanotechnology is currently not to be feared on the basis of moral positions — does not provide reason for us to let up in our efforts to conduct a constructive debate over the possible risks and other consequences of nanotechnology and how to deal with them. The opposite is rather the case because the "constructive" character of the current situation is fragile. Ongoing efforts by all parties are needed to maintain this situation.

Currently, there are thus no signs for a morally motivated rejection of nanotechnology *in general*. The situation can, however, be different in *certain nanotechnological fields* of action or application. If synthetic biology could at some time create artificial life using the means of nanobiotechnology (Chapter 7), it is probable that a heated social debate with moral conflicts and an uncertain outcome would entail, i.e., including the possibility of fundamental rejection. Psychological factors in combination with the dynamics of media society could also lead to widespread public aversion to nanotechnology. For example, the fact that nanoparticles cannot be noticed with the usual human senses, such as vision, hearing, and smell, could be used to create diffuse anxiety. This would, however, be more a problem of media responsibility and thus of media ethics than an ethical problem of nanotechnology and its consequences.

## 11.3 The Future of Nanoethics

It is always difficult to say something convincing about future developments. People may think, "What nonsense!" On the other hand, reflections about the future are tentative and yet might help to provide some orientation about *possible* developments.

### 11.3.1 Nanoethics as Concomitant Reflection on Nanotechnologies

Since the very beginning of ethical reflection on science and technology, a discussion has been ongoing about what the appropriate relation in time is between scientific-technological advance and ethics. The rapid pace of innovation in technology has led to concerns that ethical deliberations often come too late. Ethics in this perspective, could, at best, act as a repair service for problems that are already out in the open. In contrast, the "ethics first" model postulates comprehensive ethical reflection on the possible impact *in advance* of the technological development. Ethics actually can provide orientation in the early phases of innovation, for example, because future projections and visions emerging on the basis of scientific and technical advances may be subject to ethical inquiry. It is possible for ethics to reflect and discuss the normative implications of items long before their entry into the market because scientific and technical knowledge will make early ideas available about the items, their capabilities, and their societal impacts (both risks as well as chances). For example, Jonas (1987) worked on the ethical aspects of human cloning long before cloning technology was even available in the field of animals, and animal ethics today deals with the more or less speculative possibility of having AML animals (i.e., animal microencephalic lumps, see Chapter 8). Obviously, ethical reflection in the ethics first model has to deal with the situation that the relevant knowledge about technology and its consequences is uncertain and preliminary.

This does not necessarily mean that ethical deliberations have to be conducted for absolutely every speculative scientific or technical idea. The problems associated with considering new technologies in a timely manner appear most vividly in the diverse questions raised by the visions of salvation and horror in connection with nanotechnology and human enhancement. What sense is there in concerning oneself hypothetically with the ethical aspects of an extreme lengthening of the human life span or with self-replicating nanorobots (Moor and Weckert, 2004)? The ethics first perspective is exaggerated in these cases to such an extent that all relevance is lost: "If discussions about the ethics and dangers of nanotechnology become fixated on a worry that exists only in science fiction, we have a serious problem" (Ball, 2003). We should not forget that ethical reflection binds resources and that there should therefore be some evidence for the "validity" of these visions if resources are to be invested in them, which could then be lacking elsewhere (Nordmann and Rip, 2009). Consequently, methods of assessing visions of human enhancement are required, which allow for epistemological investigation of the visions under consideration (see Chapter 10 for a detailed analysis of this situation and regarding the criticisms concerning "speculative nanoethics").

An ethical judgment made during the very early stages of a development could provide orientation for shaping the relevant *process* of scientific advance and technological development (for example, with regard to the question of equity or of the risk of misuse). In the course of the continuing concretization of the possibilities for applying a particular nanotechnology, it is then possible to continuously concretize the — initially abstract — estimates and orientations on the basis of newly acquired knowledge, and finally, to carry out an ethically reflected technology assessment. In this sense, ethical analysis is an ongoing process accompanying scientific and technological advances (as was described in Chapter 10).

Due to the fact that nanotechnology and the converging technologies are at an early stage of development, we have the rare case of an advantageous opportunity: there is the chance and also the time for concomitant reflection, as well as the opportunity to integrate the results of our reflection into the scientific agenda and design of technology, and thereby to contribute to the further development of science and technology (Moor and Weckert, 2004). In view of the still visionary nature of many the prospects in nano(bio)technology, and of the long and very long time spans within which the realization of certain milestones can be expected, there is in all probability enough time for us to analyze the questions posed. In the field of nanotechnology, the chances are good that ethical reflection and the societal discussion do not come too late, but can accompany scientific–technical progress critically and, in particular, can help influence science's agenda by providing ethically reflected advice (even if this does not seem to have happened yet, see Section 11.1) and help improve the communication between science and society.

### 11.3.2 Nanoethics as Interdisciplinary Research

Common to ethical reflection of any of the links in the chain of nanotechnological development is the necessity of interdisciplinary dialog. As experts for the structures of justification and for patterns of normative argumentation (Grunwald, 2003), ethicists are of necessity dependent on factual knowledge from the natural sciences and engineering, but also from the social sciences, about their respective fields. They are similarly dependent on metaknowledge about the epistemological status of this factual knowledge. Varying contexts, partners, and recipients of this ethical advice make different forms of organization necessary in the interdisciplinary and transdisciplinary dialog. Two models have been practiced in pointed form in the context of the developing ethics of nanotechnology. They are the model of the interdisciplinary expert group (above all in the context of the ethical, legal, and social aspects) and the model of ethical parallel research.

The field of activities concerning the ethical, legal, and social aspects and implications of nanotechnology is not characterized by a specific approach to science organization. It shows signs, on the contrary, of exploring these aspects without any compulsion or even the claim to want to set a unifying or collective perspective (Section 5.1). It has consequently manifested itself in different models, which range from expert groups (Royal Society, 2004; Schmid *et al.*, 2006) to the collaborative work of many authors, coordinated above all by means of electronic media (Nanoforum, 2004). A variety of methodological and organizational forms have developed, corresponding to the lack of conceptual inhibitions, if not to say the haphazardness, that prevails.

A further model for conducting interdisciplinary dialog on nanotechnology's ethical issues consists in getting ethical expertise directly involved in projects whose focus is primarily on science and engineering. Such projects have been initiated and conducted especially in research supported by the European Union in its Sixth Framework Program. Among them is the Nano2Life project, in which different forms of organizing ethical parallel research and of intensifying communication between nanotechnology and ethics (University of Münster) have been practiced (Ach and Siep, 2006). These include the establishment of an ethics advisory council, joint workshops, and visits by ethicists to laboratories. Experience shows that a factor that poses a special barrier to communication is the difference in the levels of aggregation between concrete research on nanoscale processes in a lab and the consideration of greatly differing fields of application. In this "integrated" sense, the tasks of ethical parallel research are at several different levels, namely to

1. *Identify* the ethically relevant issues in the context of the project and thus also in the context of technology and application. The focus is on reconstructing the respective underlying normative conditions and evaluating them with regard to the presence of a "standard situation in a moral sense" or to existing normative uncertainties (Chapter 3). In carrying this out, consideration must be given to knowledge from science and engineering and to the estimates about future applications and their consequences made on the basis of this knowledge. Even this circumscribed task may lead to the involvement of a wide spectrum of actors and of people who are potentially affected (e.g., patients in cases of medically relevant developments).
2. *Clarify* the moral uncertainties that are identified. The focus here is on the specific nature of the ethical challenge, determination of those spots at which normative uncertainty is manifested in the structures of justification, clarification of the rights of persons and groups who might be affected, creation of connections to established fields of ethical reflection and to other philosophical debates, clarification of responsibility for the next steps, and the conditions for attributing responsibility. This comprises primarily analytic work.
3. *Tackle* the ethical challenges that have been identified and clarified. At this point, the results of the reconstruction and analytic work are to be converted into answers or orientations to *practical* issues. The issue is to prepare suggestions as to how or according to which procedures (e.g., regulatory, participatory, or in the context of self-obligations) the normative framework can be further developed in order to transform the normatively "modernized" situation into a standard situation (Chapter 3). Another aspect of this is

to submit the points of orientation that are worked out to discussion in the field and to public debate.

This helps us to recognize that ethics can play specific roles in its capacity as parallel research. These roles include the evaluation of emerging technologies from an ethical perspective, the uncovering of possible normative uncertainties, and possibly the search for morally less problematic alternatives, which of course is only possible in close cooperation with engineering and the natural sciences.

All the same, it is important to point out that the model of parallel research also harbors problems. The independence of its ethical judgment can be threatened especially if the necessary distance to the technical developments and those working on them is lost. Inasmuch as ethics becomes part of the development process and would identify itself with the technical success, there might be an accusation that its acceptance was "purchased" or that it was nothing but grease in the process of innovation. Yet this critical argument demands a critical answer: If ethics succeeds by means of ethical arguments — for example, regarding data protection and privacy — to help shape the design of the relevant product to preventively avoid potential resistance that might have resulted from insufficient consideration of data protection and privacy, this is not a case of acceptance being purchased but of work being successful in helping to shape technology. If this is to be judged, the parallel research must maintain methodological transparency, conceptual clarity, and be monitored critically by the academic profession of ethics.

Ethics in and for nanotechnology must take place in an interdisciplinary dialog, which can be organized in different ways. We should furthermore not forget, however, that academic consideration retains its significance even in this ethical reflection that has a very strong practical orientation. It continues to play an indispensible role as a provider of ideas, for ensuring the connection of the argumentation with other fields of ethics, and in relation to the maintenance of quality standards in ethical arguments.

### 11.3.3 Disentanglement of Nanoethics

The ethics of nanotechnology, just like some other forms of work for investigating and evaluating the social consequences of nanotechnology, has undergone major development within a period of just a few years. However, there was (and to some extent, there still is) some uncertainty as to whether nanotechnology has created *new* ethical questions, whether it has created ethical questions at all, and if it has, what the appropriate scientific and systematic framework would be for adequately dealing with these questions and providing answers to them (Nordmann, 2007a; Allhoff, 2007). Today, no one doubts any more that nanotechnology involves *relevant* ethical issues. The question of whether these challenges are new or to what extent they are new is only of secondary importance considering the practical necessity of ethical reflection.

We have argued, however, that "nanoethics" will probably not and should not become an autonomous new subfield within applied ethics, analogous to, for example, medical ethics, information ethics, or bioethics, as was expected until recently (Lin and Allhoff, 2007; see Chapter 4). Nanotechnology leads to some partially new ethical questions in these fields by providing options for new applications, but it does not add an entirely new field to the range of applied ethics.

The concept of nanotechnology will instead probably disintegrate in the future. Looking back at the development of nanotechnology, it is notable that the term "nanotechnology" was used almost exclusively in the singular until around 2004, but since then "nanotechnologies," in the plural, has had increasingly wider repercussions (prominently, e.g., in Nordmann *et al.*, 2006; on reactions to the difficulties of defining nanotechnology, see Section 2.3). The term "nanosciences" is also usually used in the plural form. There is much emphasis on the fact that the term "nanotechnology" is used as a blanket covering many forms of research and development that are very heterogeneous in detail. The term has, however, had a great impact. It succeeded in giving the field public perception, created political awareness and a readiness to provide sponsorship, instilled a certain fascination, and for quite some time cast a favorable light on technical progress. One probable development is that the prefix "nano" will remain and may even become more widespread. We will continue to talk of nanomaterials, nanobiotechnology, nanomedicine, or nanoelectronics, but as differentiation in science and engineering progresses there will probably be less talk of nanotechnology being an entity as such.

Ethical reflection will probably follow this trend, leading to a disentanglement (Nordmann, 2007b) of the different spheres of nanoethics, each of which will return to its established subfield of applied ethics. The right place for professional ethical reflection on nanotechnology to be conducted is where central ethical issues are an object of discussion in the context of normative uncertainties, where the relevant traditions exist, and the necessary competence is present. This is precisely the case in the subfields of applied ethics. Disentanglement here means that questions of nanomedicine are dealt with in medical ethics, questions of data protection and privacy in information ethics, questions of dealing with nanoparticles in risk ethics, and questions of technically modified life in bioethics. Nanoethics will consequently not become a new branch of applied ethics. There will, however. be a need for ethicists to continue to demonstrate their readiness to openly address ethical issues related to nanotechnology (or nanotechnologies, as the case may be) across the borders of the classical subdivisions of ethics and conduct a dialog with scientists in the natural sciences and in engineering.

There is one field that, however, is not properly covered by the existing subdisciplines of applied ethics. It is concerned with issues related to the technical "enhancement" of humans (Chapter 9). Medical ethics in the traditional sense is hardly responsible for this field because the objective of this enhancement is not healing. The ethics of technology only takes these issues into consideration to a small extent. Bioethics simply disregards the social aspects of technically enhanced humans. For these reasons, it is possible to follow the reasoning of those who now occasionally call for enhancement ethics becoming a field of its own. This field could feed on issues from anthropology, technical philosophy, and the philosophy of the neurosciences that precede consideration of an ethics of technical enhancement, and refer to the relevant areas of practical philosophy, such as the theory of justice and risk ethics. Enhancement ethics might develop on the basis of considerations of doping in sports ethics, yet it would have to generalize the issues beyond the specific context of sports. In view of the general tendencies toward a "society of increased performance" (Coenen, 2008b), we can detect the contours of a new area of necessary social self-understanding, which will encounter substantial foreseeable normative uncertainty. The topic of enhancement ethics (Schöne-Seifert *et al.*, 2009) that covers the normative aspects of a society that is permeated by the most varied forms of "everyday doping," such as the fairness imperative, issues of distribution and access, risks, possibly endangered rights, and social dislocations, could

— in collaboration with anthropology, the philosophy of technology, and social theory — develop into a promising and efficacious field of philosophical work.

This brings this volume to its conclusion. It started with considerations of ethics in and for nano(bio)technology, i.e., oriented on diverse directions of development held together by the guiding motif of attempting to technically shape matter at the level of atoms and molecules. An important outcome is that ethics in and for nano(bio)technology is at least as diverse. It is not structured, however, according to technical features because the latter involve different types of normative uncertainty. It is precisely this diversity of moral problems that prevents nanoethics from developing into a discipline of its own. If the family of subdisciplines in applied ethics is to be joined by a new offspring as a result of the considerations of nanoethics, then this is most likely to take place in the form of enhancement ethics. Ethical considerations on nano(bio)technology — an irony of development — would thus have played an enabling role in the creation of such a new subdiscipline, just as nanotechnology is frequently seen to be an enabling technology (Chapter 2). The resulting ethical questions have however little to do specifically with nano(bio)technology.

In contrast, demands are increasingly being raised for the results of ethical considerations to be embedded in participatory processes and in decision making. These demands are accompanied by questions about this:

> [T]here is a future for a third generation of deliberative processes in the development of nanotechnology. These processes need to be more specifically oriented and more closely linked to decision-making processes. [...] One of the main challenges in the future is the question of who will take the responsibility for running such processes, and independent institutions may be one answer to this question. (Sto *et al.*, 2010)

In the course of efforts to achieve responsible innovation and responsible development, the object on the one hand will be to conduct ethical reflection on future forms of nanotechnology in a professional manner, while on the other moving vigorously out of the ivory tower of philosophy and participating in an increasingly active way in debates on nanotechnology, both in the natural sciences and engineering and in the social and political arena.

# Bibliography


Abbott, A. (2006) Neuroprosthetics: in search of the sixth sense. *Nature* **442**, 125–127.

Ach, J., Pollmann, A. (eds.) (2006) *No Body Is Perfect: Baumaßnahmen am menschlichen Körper — Bioethische und ästhetische Aufrisse*, Bielefeld.

Ach, J., Siep, L. (eds.) (2006) *Nano-bio-ethics: Ethical and Social Dimensions of Nanobiotechnology*, Berlin.

Allhoff, F. (2007) On the autonomy and justification of nanoethics. *NanoEthics* **1**, 185–210.

Allhoff, F., Lin, P. (2006) What's so special about nanotechnology and nanoethics? *International Journal of Applied Philosophy* **20**, 179–190.

Allhoff, F., Lin, P., Moor, J., Weckert, J. (eds.) (2007) *Nanoethics: The Ethical and Social Implications of Nanotechnology*, New Jersey.

Altmann, J. (2006) *Military Nanotechnology: Potential Application and Preventive Arms Control*, London.

Altmann, J., Gubrud, M. A. (2002) Risks from military uses of nanotechnology: the need for technology assessment and preventive control. In: Roco, M. C., Tomellini, R. (eds.), *Nanotechnology: Revolutionary Opportunities and Social Implications*, Luxembourg, pp. 144–148.

Anders, G. (1964) *Die Antiquiertheit des Menschen*, München.

Attfield, R. (1995) Genetic engineering: can unnatural kinds be wronged? In: Wheale, P., McNally, R. (eds.), *Animal Genetic Engineering: Of Pigs, Oncomice and Men*, London, pp. 201–210.

Bachmann, A. (2006) *Nanobiotechnologie: Eine ethische Auslegeordnung*, Bern.

Bachmann, G. (1998) Innovationsschub aus dem Nanokosmos, Technologieanalyse Nr. 28, Düsseldorf.

BAFU — Bundesamt für Umwelt (2007) *Synthetische nanomaterialien: Risikobeurteilung und Risikomanagement*, Bern.

Bainbridge, W. S. (2004) Progress toward cyberimmortality. In: Immortality Institute (Hg.), *The Scientific Conquest of Death*, Buenos Aires, pp. 107–122 (www.imminst. org/book1/).

Baird, D., Nordmann, A., Schummer, J. (eds.) (2004) *Discovering the Nanoscale*, Amsterdam.

Balaban, T., Buth, G. (2005) Biomimetische lichtsammlung. *FZK — Nachrichten 37*, Nr. 4, S. 204–209; see also other work of these authors.

Ball, P. (2003) *Nanoethics and the Purpose of New Technologies*, London (http://www.whitebottom.com/philipball/docs/Nanoethics.doc).

Ball, P. (2005) Synthetic biology for nanotechnology. *Nanotechnology* **16**, R1–R8.

Banse, G., Grunwald, A., Hronszky, I., Nelson, G. (eds.) (2008) *Assessing Societal Implications of Converging Technological Development*, Berlin.

Barthlott, W. (1990) Scanning electron microscopy of the epidermal surface in plants. *Systematics Association's Special* **41**, 69–94.

Baumgartner, C. (2006) Nanotechnologie in der Medizin als Gegenstand ethischer Reflexion: Problemfelder, Herausforderungen, Implikationen. In: Nordmann, A., Schummer, J., Schwarz, A. (eds.), *Nanotechnologien im Kontext*, Berlin, pp. 325–343.

Bayertz, K. (ed.) (2004) *Die menschliche Natur: Welchen und wie viel Wert hat sie?* Paderborn.

Bechmann, G. (1993) Ethische Grenzen der Technik oder technische Grenzen der Ethik? *Geschichte und Gegenwart* **12**, 213–225.

Beck, U. (1986) *Risikogesellschaft: Auf dem Weg in eine andere Moderne*, Frankfurt. English version: *Risk Society: Towards a New Modernity*, London, 1992.

Benner, S. A., Sismour, A. M. (2005) Synthetic biology. *Nature Reviews Genetics* **6**(7), 533–543.

Bensaude-Vincent, B. (2004) Two cultures of nanotechnology? *HYLE — International Journal for Philosophy of Chemistry* **10**, 65–82.

Bensaude-Vincent, B. (2009) *Philosophy of Chemistry*, Boston Studies in the Philosophy of Science, Vol. 276, Part III, pp. 165–189.

Berlosznik, R. (2007) Meeting of minds: a European participatory project. In: Bora, A., Bröchler, S., Decker, M. (eds.), *Technology Assessment in der Weltgesellschaft*, Berlin.

Beyerlein, F. (2000) Nano-Medizin und Gesundheitsvorsorge im 21, Jahrhundert. *Tausend Jahre leben*, Magazin 2000 plus, Medizin 2.

Bhaskar, S., Tian, F., Stoeger, T., Kreyling, W., de la Fuente, J. M., Grazu, V. (2010) Multifunctional nanocarriers for diagnostics, drug delivery and targeted treatment across blood-brain barrier: perspectives on tracking and neuroimaging. *Part Fibre Toxicology* **2010**.

Bhutkar, A. (2005) Synthetic biology: navigating the challenges ahead. *The Journal of Biolaw & Business* **8**(2), 19–28.

Bijker, W. E., Hughes, T. P., Pinch, T. J. (eds.) (1987) *The Social Construction of Technological Systems: New Directions in the Sociology and History of Technological Systems*, Cambridge (Mass.)/London.

Birnbacher, D. (1988) *Verantwortung für zukünftige Generationen*, Stuttgart.

BMBF — Bundesministerium für Bildung und Forschung (2002) *Standortbestimmung: Nanotechnologie in Deutschland*, Berlin.

Boldt, J., Müller, O. (2008) Newtons of the leaves of grass. *Nature Biotechnology* **26**, 387–389.

Boldt, J., Müller, O., Maio, G. (2009) *Synthetische Biologie: Eine Ethisch-Philosophische Analyse*, Bern.

Böschen, S., Kratzer, N., May, S. (eds.) (2006) *Nebenfolgen: Analyse zur Konstruktion und Transformation moderner Gesellschaften*, Weilerswist.

Bostrom, N., Savalescu, J. (2009) *Human Enhancement*, Oxford.

Brenner, A. (2007) *Leben: Eine philosophische Untersuchung*, Bern.

Brown, C. (2004) BioBricks to help reverse-engineer life. *EE Times*, June 11, 2004.

Brown, N., Rappert, B., Webster, A. (eds.) (2000) *Contested Futures: A Sociology of Prospective Techno-Science*, Burlington.

Brown-Weiss, E. (1989) *In Fairness to Future Generations: International Law, Common Patrimony and Intergenerational Equity*, New York.

Bruggeman, F. J., Westerhoff, H. V. (2006) The nature of system biology. *Trends in Microbiology* **15**, 45–50.

BUND — Bund für Umwelt und Naturschutz in Deutschland (2008) *Aus dem Labor auf den Teller: Die Nutzung der Nanotechnologie im Lebensmittelsektor* (http://www.bund.net/fileadmin/bundnet/publikationen/nanotechnologie/20080311_nanotechnologie_lebensmittel_studie.pdf).

BUND — Bund für Umwelt und Naturschutz in Deutschland (2009) *Nanosilber — der Glanz täuscht* (http://www.bund.net/fileadmin/bundnet/publikationen/nanotechnologie/20091202_nanotechnologie_nanosilber_studie.pdf).

Bunge, M. (2003) *Emergence and Convergence: Qualitative Novelty and the Unity of Knowledge*, Toronto.

Burns, T. R., Ueberhorst, R. (1988) *Creative Democracy: Systematic Conflict Resolution and Policymaking in a World of High Science and Technology*, New York.

Caplan, P. (2000) *Risk Revisited*, London.

Caplan, P. (2009) *Do Synthetic Biologists Play God?* (http://www.synbiosafe.eu/uploads///pdf/09_03_SiscoveryTech_Do_Synthetic_Biologists_Play_God.pdf).

Capurro, R. (2004) Informationsethik. Eine Standortbestimmung. *Wechselwirkung* **25**, 132–139; Towards an information ecology. In: Wormell, I. (ed.) (1990), *Information and Quality*, London, pp. 122–139.

Cauller, L., Penz, A. (2002) Artificial brains and natural intelligence. In: Roco, M. C., Bainbridge, W. S. (eds.), *Converging Technologies for Improving Human Performance*, Virginia, pp. 227–230.

Chakrabarty, A. M. (2003) Crossing species boundaries and making human-nonhuman hybrids: moral and legal ramiacations. *The American Journal of Bioethics* **3**(3), 20–21.

Cheng, J., Fleming, G. (2009) Dynamics of light harvesting in photosynthesis. *Annual Review of Physical Chemistry* **60**, 242–261.

Choksi, A. N., Poonawalla, T., Wilkerson, M. G. (2010) Nanoparticles: a closer look at their dermal effects. *The Journal of Drugs in Dermatology* **9**(5), 475–481.

Chopra, P., Kamma, A. (2006) Engineering life through synthetic biology. *Silico Biology* **6**, 401–410.

Church, G. (2005) Let us go forth and safely multiply. *Nature* **438**, 423.

Clausen, J. (2006) Ethische Aspekte von Gehirn-Computer-Schnittstellen in motorischen Neuroprothesen. *International Review of Information Ethics* **5**, 25–32.

Clausen, J. (2009) Ethische Fragen aktueller Neurowissenschaften: Welche Orientierung gibt die »Natur des Menschen«? In: Hildt, E., Engels, E.-M. (eds.), *Der implantierte Mensch*, Freiburg.

Coenen, C. (2006) Der posthumanistische Technofuturismus in den Debatten über Nanotechnologie und Converging Technologies. In: Nordmann, A., Schummer, J., Schwarz, A. (eds.), *Nanotechnologien im Kontext*, Berlin, pp. 195–222.

Coenen, C. (2008a) *Konvergierende Technologien und Wissenschaften: Der Stand der Debatte und politische Aktivitäten zu »Converging Technologies«*, Berlin (summary in English: www.itas.fzk.de/deu/lit/2008/coen08a_summary.htm).

Coenen, C. (2008b) *Von der Leistungs: zur Leistungssteigerungsgesellschaft? TAB-Brief 33*, Berlin, Büro für Technikfolgen-Abschätzung, pp. 21–27.

Coenen, C. (2010) Deliberating visions: the case of human enhancement in the discourse on nanotechnology and convergence. In: Kaiser, M., Kurath, M., Maasen, S., Rehmann-Sutter, Chr. (eds.), *Governing Future Technologies: Nanotechnology and the Rise of an Assessment Regime*, Springer, pp. 73–87.

Coenen, C., Schuijff, M., Smits, M., Klaassen, P., Hennen, L., Rader, M., Wolbring, G. (2009) *Human Enhancement*, Brussels, European Parliament (http://www.itas.fzk.de/deu/lit/2009/coua09a.pdf).

Coenen, Chr., Hennen, L., Link, H.-J. (2009) The ethics of synthetic biology: contours of an emerging discourse. *Technikfolgenabschätzung: Theorie und Praxis* **18**(2), 82–86.

Coffrin, T., MacDonald, C. (2004) Ethical and social issues in nanotechnology. Annotated bibliography (http://www.ethicsweb.ca/nanotechnology/bibliography.html).

COGEM (2006) *Synthetische Biologie: Een Onderzoeksveld Met Voortschrijdende Gevolgen* (http://www.cogem.net/pdfdb/advies/CGM060228-03.pdf).

Collingridge, D. (1980) *The Social Control of Technology*, London.

Colorado Public Commission (1992) Statement of adoption in the matter of the rules for electric utilities of the Colorado Public Utilities Commission. Code of Colorado Regulation-723-3 concerning electric and magnetic fields, Denver CPUC.

Colvin, V. (2003) *Responsible Nanotechnology: Looking Beyond the Good News*. Centre for Biological and Environmental Nanotechnology at Rice University (http://www.eurekalert.org/).

Comstock, G. (2000) *Vexing Nature? On the Ethical Case against Agricultural Biotechnology*, Boston, Dordrecht, London.

Court, E., Salamance-Buentello, F., Singer, P. A., Daar, A. S. (2007) Nanotechnology and the developing world. In: ten Have, H. (ed.), *Nanotechnologies, Ethics, and Politics*, UNESCO Publishing, pp. 155–180.

Dabrock, P. (2009) Playing God? Synthetic biology as a theological and ethical challenge. *Systems and Synthetic Biology* **3**, 47–54.

Danish Council of Bioethics (2008) *Man or Mouse? Ethical Aspects of Chimaera Research*, The Danish Council of Bioethics (http://www.etiskraad.dk/sw16983.asp).

de Vriend, H. (2006) *Constructing Life: Early Social Reflections on the Emerging Field of Synthetic Biology*, Rathenau Institute, The Hague.

Deblonde, M., van Oudheusden, M., Evers, J., Goorden, L. (2011) Co-creating nano-imaginaries. Results of a Delphi. *Bulletin of Science, Technology and Society* **31** (in press).

Decker, M. (2006) Eine Definition von Nanotechnologie: Erster Schritt für ein interdisziplinäres Nanotechnology Assessment. In: Nordmann, A., Schummer, J., Schwarz, A. (eds.), *Nanotechnologien im Kontext*, Berlin, pp. 33–48.

Decker, M., Fiedeler, U., Fleischer, T. (2004) Ich sehe was, was Du nicht siehst ... zur Definition von Nanotechnologie. *Technikfolgenabschätzung: Theorie und Praxis* **13**, 10–16.

Decker, M., Ladikas, M. (eds.) (2004) *Bridges between Science, Society and Policy: Technology Assessment — Methods and Impacts*, Berlin.

DEEPEN (2009) *Reconfiguring Responsibility: Deepening Debate on Nanotechnology* (www.geography.dur.ac.uk/projects/deepen).

DFG — Deutsche Forschungsgemeinschaft (2004) *Tierversuche in der Forschung: Lemmens Verlag & Mediengesellschaft*, Bonn.

Drexler, K. E. (1986) *Engines of Creation: The Coming Era of Nanotechnology*, Oxford.

Dupuy, J.-P. (2005) The philosophical foundations of nanoethics. Arguments for a method. Lecture at the Nanoethics Conference, University of South Carolina, March 2–5, 2005.

Dupuy, J.-P. (2007) Complexity and uncertainty: a prudential approach to nanotechnology. In: Allhoff, F., Lin, P., Moor, J., Weckert, J. (eds.), *Nanoethics: The Ethical and Social Implications of Nanotechnology*, New Jersey, pp. 119–132.

Dupuy, J.-P., Grinbaum, A. (2004) Living with uncertainty: toward the ongoing normative assessment of nanotechnology, *Techné* **8**(2), 4–25; reprinted in: Schummer, I., Baird, D. (eds.) (2006), *Nanotechnology Challenges: Implications for Philosophy, Ethics and Society*, Singapore, pp. 287–314.

Durbin, P. (ed.) (1987) *Technology and Responsibility*, Dordrecht.

Ebbesen, M., Andersen, S., Besenbacher, F. (2006) Ethics in nanotechnology: starting from scratch? *Bulletin of Science, Technology & Society* **26**, 451–462.

Edwards, S. (2006) *The Nanotech Pioneers: Where Are They Taking Us?* Weinheim.

EGE — European Group on Ethics in Science and New Technologies (2005) *Ethical aspects of ICT implants in the human body. Opinion No. 20*, Brussels.

Eigler, D. M., Schweizer, E. K. (1990) Positioning single atoms with a scanning tunneling Microscope. *Nature* **344**, 524–526.

Engelhardt, T. (1976) Ideology and etiology. *The Journal of Medicine and Philosophy* **1**, 256–268.

Engelhardt, T. (1982) The roles of values in the discovery of illnesses, diseases, and disorders. In: Beauchamp, T., Walters, L. (eds.), *Contemporary Issues in Bioethics*, Belmont, pp. 73–75.

Engels, E.-M. (2008) Was und wo ist ein »naturalistischer Fehlschluss«? Zur Definition und Identifikation eines Schreckgespenstes der Ethik. In: Brand, C., Engels, E.-M., Ferrari, A., Kovács, L. (eds.), *Wie funktioniert Bioethik?* Paderborn, pp. 125–141.

ENRHES — (2010) Engineered Nanoparticles: Review of Health and Environmental Safety. Project Report (http://nmi.jrc.ec.europa.eu/documents/pdf/ENRHES%20pdf).

EPTA — European Parliamentary Technology Assessment Network (2006) *ICT and Privacy in Europe* (www.eptanetwork.org).

ESF — European Science Foundation (2005) *Nanomedicine: An ESF Forward Look* (www.esf.org/publications/forward-looks.html).

ETC — The Et-cetera Group (2003) From genomes to atoms. *The Big Down: Atomtech -Technologies Converging at the Nano-scale* (www.etcgroup.org).

ETC — The Et-cetera Group (2004) Down on the farm. *The Impact of Nano-Scale Technologies on Food and Agriculture* (www.etcgroup.org).

ETC — The Et-cetera Group (2007) Extreme genetic engineering. *An Introduction to Synthetic Biology* (www.etcgroup.org).

ETPN — European Technology Platform Nanomedicine (2005) *Vision Paper and Basis for a Strategic Research Agenda for NanoMedicine* (http://www.etp-nanomedicine.eu/public/press-documents/publications/etpn-publications).

European Commission (2004) Towards a European strategy on nanotechnology. *COM* **338**, Brussels.

European Commission (2008) Recommendations toward a code of conduct for nanotechnology. *COM* **424**, Brussels.

Fadeel, B., Garcia-Bennett, A. E. (2010) Better safe than sorry: understanding the toxicological properties of inorganic nanoparticles manufactured for biomedical applications. *Advanced Drug Delivery Reviews* **62**(3), 362–74.

FAO — Food and Agriculture Organization (2006) *Livestock's Long Shadow: Environmental Issues and Options* (http://www.fao.org/docrep/010/a0701e/a0701e00.htm).

Farah, M. J., Illes, J., Cook-Deegan, R., Gardner, H., Kandel, E., King, P., Parens, E., Sahakian, B., Wolpe, P. R. (2004) Neurocognitive enhancement: what can we do and what should we do? *Nature Reviews: Neuroscience* **5**, 421–425.

Felt, U., Fochler, M., Müller, A., Strassnig, M. (2009) Unruly ethics: on the difficulties of a bottom-up approach to ethics in the field of genomics. *Public Understanding of Science* **18**(3), 354–371.

Ferrari, A. (2008) *Genmaus & Co: Gentechnisch veränderte Tiere in der Biomedizin*. Harald Fischer Verlag, Stuttgart.

Ferrari, A. (2010) The control nano-freak: multifaceted strategies for taming nature. In: Kjolberg, K., Wickson, F. (2009), *Nano Meets Macro: Social Perspectives on Nano-Scaled Sciences & Technologies*, Singapore.

Ferrari, A., Coenen, C., Grunwald, A., Sauter, A. (2010) *Animal Enhancement: Neue technische Möglichkeiten und ethische Fragen*, Bern.

Feynman, R. P. (1959) Lecture given at 29.12.1959 at the annual meeting of the American Physical Society at Pasadena (http://www.zyvex.com/nanotech/feynman.html).

Fiedeler, U. (2008) *The Role of Nanotechnology in Chemical Substitution*, European Parliament, Brussels.

Fiedeler, U., Coenen, C., Davies, S. R., Ferrari, A. (eds.) (2010) *Understanding Nanotechnology: Philosophy, Policy and Publics*, Heidelberg.

Fleischer, T. (2003) Technikgestaltung für mehr Nachhaltigkeit: Nanotechnologie. In: Coenen, R., Grunwald, A. (eds.), Nachhaltigkeitsprobleme in Deutschland. *Analyse und Lösungsstrategien*, Berlin, pp. 356–373.

Fleischer, T., Grunwald, A. (2008) Making nanotechnology developments sustainable. A role for technology assessment? *Journal of Cleaner Production* **16**, 889–898.

Fleischer, T., Hocke, P., Kastenholz, H., Krug, H., Quendt, C., Spangenberg, A. (2010) Evidenzbewertung von Gesundheitsrisiken synthetischer Nanopartikel. In: Aichholzer, G., Bora, A., Bröchler, S., Decker, M., Latzer, M. (eds.), *Technology Governance: Der Beitrag der Technikfolgenabschätzung*, Berlin.

Foladori, G. (2008) Converging technologies and the poor: the case of nanomedicine and nanobiotechnology. In: Banse, G., Grunwald, A., Hronszky, I., Nelson, G. (eds.), *Assessing Societal Implications of Converging Technological Development*, Berlin, pp. 193–216.

Foladori, G., Invernizzi, N. (2007) Nanotechnology and developing countries: will nanotechnology overcome poverty or widen disparities? *Nanotechnology Law and Business* **2**, Paper 11.

Frankena, W. (1939) The naturalistic fallacy. *Mind* **48**, 464–477.

Freitas, R. A. Jr. (1999) *Nanomedicine, Vol. I: Basic Capabilities*, Georgetown, TX.

Freitas, R. A. Jr. (2003) *Nanomedicine, Vol. IIA: Biocompatibility*, Georgetown, TX.

Friends of the Earth (2006) *Nanomaterials, sunscreens, and cosmetics: Small Ingredients* (http://www.foe.org/camps/comm/nanotech/nanocosmetics.pdf).

Fuentes, R., *et al.* (2009) Spinal cord stimulation restores locomotion in animal models of Parkinson's disease. *Science* **323**(5921), 1578–1582.

Funtowitz, S., Ravetz, J. (1993) The emergence of post-normal science. In: von Schomberg, R. (eds.), *Science, Politics and Morality*, London.

Gannon, F. (2003) Nano-nonsense (editorial). *EMBO Reports* **4**(11), 1007.

Gee, D., Greenberg, M. (2002) Asbestos: from "magic" to malevolent mineral. In: Harremoes, P., *et al.* (eds.), *The Precautionary Principle in the 20th Century: Late Lessons from Early Warnings*, London, pp. 49–63.

Gehlen, A. (1986) *Der Mensch: Seine Natur und seine Stellung in der Welt*, Wiesbaden.

Geiser, M., Kreyling, W. G. (2010) Deposition and biokinetics of inhaled nanoparticles. *Particle and Fibre Toxicology* **7**, 2.

Gerlinger, K., Petermann, T., Sauter, A. (2008) *Gene Doping: Scientific Basis – Gateways – Monitoring*. Technology Assessment Report 3, Office of Technology Assessment at the German Bundestag, Berlin (http://www.bundestag.de/htdocs_e/bundestag/committees/a18/translations/gene_doping.pdf).

Gethmann, C. F. (1979) *Protologik: Untersuchungen zur formalen Pragmatik von Begründungsdiskursen*, Frankfurt.

Gethmann, C. F., Mittelstraß, J. (1992) Umweltstandards. *GAIA* **1**(1992), 16–25.

Giddens, A. (1984) *The Constitution of Society: Outline of the Theory of Structuration*, Cambridge.

Goodman, N. (1954) *Fact, Fiction, and Forecast.*

Goodsell, D. S. (2004) *Bionanotechnology: Lessons from Nature*, New York.

Gottweis, H. (2008) Participation and the new governance of life. *BioSocieties* **3**, 265–286.

Grin, J., Grunwald, A. (eds.) (2000) *Vision Assessment: Shaping Technology in 21st Century Society*, Berlin.

Grobe, A. (2007) Europa setzt auf Dialoge: Neue Wege der (Risiko-) kommunikation für Nanotechnologien. In: Gazsó, A., Greßler, S., Schiemer, F. (eds.), *Nano. Chancen und Risiken aktueller Technologien*, Wien, pp. 199–214.

Grolle, J. (2010) Konkurrenz für Gott. *DER SPIEGEL* 1/2010, 110–119.

Grunwald, A. (1999) Verantwortungsbegriff und Verantwortungsethik. In: Grunwald, A. (ed.), *Rationale Technikfolgenbeurteilung*, Berlin, pp. 172–195.

Grunwald, A. (2003) Methodical reconstruction of ethical advises. In: Bechmann, G., Hronszky, I. (eds.), *Expertise and Its Interfaces*, Berlin, pp. 103–124.

Grunwald, A. (2004) Strategic knowledge for sustainable development: the need for reflexivity and learning at the interface between science and society. *International Journal of Foresight and Innovation Policy* **1**(1/2), 150–167.

Grunwald, A. (2005) Nanotechnology: a new field of ethical inquiry? *Science and Engineering Ethics* **11**, 187–201.

Grunwald, A. (2006a) Nanotechnologie als Chiffre der Zukunft. In: Nordmann, A., Schummer, J., Schwarz, A. (eds.), *Nanotechnologien im Kontext*, Berlin, pp. 49–80.

Grunwald, A. (2006b) Technikfolgenabschätzung als Nachhaltigke-itsbewertung. Konzeptionelle Herausforderungen und methodische Probleme. In: Kopfmüller, J. (Hg.), *Ein Konzept auf dem Prüfstand: Das integrative Nachhaltigkeitskonzept in der Forschungspraxis*, Berlin, pp. 39–61.

Grunwald, A. (2007a) Converging technologies: visions, increased contingencies of the conditio humana, and search for orientation. *Futures* **39**, 380–392.

Grunwald, A. (2007b) Working towards sustainable development in the face of uncertainty and incomplete knowledge. *Journal of Environmental Policy & Planning* **9**, 245–262.

Grunwald, A. (2008a) *Technik und Politikberatung: Philosophische Perspektiven*, Frankfurt.

Grunwald, A. (2008b) Ethics of nanotechnology state of the art and challenges ahead. In: Schmid, G., *et al.* (eds.), *Nanotechnology. Vol. 1: Principles and Fundamentals*, Weinheim, pp. 245–286.

Grunwald, A. (2008c) Nanoparticles: risk management and the precautionary principle. In: Jotterand, F. (ed.), *Emerging Conceptual, Ethical and Policy Issues in Bionanotechnology*, Berlin, pp. 85–102.

Grunwald, A. (2008d) Ethical guidance for dealing with unclear risk. In: Wiedemann, P., Schütz, H. (eds.), *The Role of Evidence in Risk Characterization: Making Sense of Conflicting Data*, Weinheim, pp. 185–202.

Grunwald, A. (2009a) Technology assessment: concepts and methods. In: Meijers, A. (ed.), *Philosophy of Technology and Engineering Sciences*, Vol. 9, Amsterdam, pp. 1103–1146.

Grunwald, A. (2009b) Vision assessment supporting the governance of knowledge — the case of futuristic nanotechnology. In: Bechmann, G., Gorokhov, V., Stehr, N. (eds.), *The Social Integration of Science: Institutional and Epistemological Aspects of the Transformation of Knowledge in Modern Society*, Berlin, pp. 147–170.

Grunwald, A. (2010a) From speculative nanoethics to explorative philosophy of nanotechnology. *NanoEthics* **4**(2), 91–101.

Grunwald, A. (2010b) Virtualisierung von Kommunikation und Handeln im Pervasive Computing — Schritte zur Technisierung des Menschen? In: Bölker, M., Gutmann, M., Hesse, W. (eds.), *Information und Menschenbild*, Berlin, pp. 79–101.

Grunwald, A. (2011a) Chances and risks of nanotechnology. In: Sattler, K. D. (ed.), *Handbook of Nanophysics: Nanomedicine and Nanorobotics*, Boca Raton, FL, pp. 13-1–13-16.

Grunwald, A. (2011b) Ten years of research on nanotechnology and society: outcomes and achievements. *Proceedings of the 2nd S.NET Conference*, Darmstadt 2010 (in press).

Grunwald, A., Hocke-Bergler, P. (2010) The risk debate on nanoparticles: contribution to a normalisation of the science/society relationship? In: Kaiser, M., Kurath, M., Maasen, S., Rehmann-Sutter, Ch. (eds.), *Governing Future Technologies: Nanotechnology and the Rise of an Assessment Regime*, Dordrecht, pp. 157–177.

Grunwald, A., Julliard, Y. (2007) Nanotechnology: steps towards understanding human beings as technology? *NanoEthics* **1**, 77–87.

Grunwald, A., Kopfmüller, J. (2006) *Nachhaltigkeit*, Frankfurt/New York.

Guston, D. (2010) *Encyclopedia of Nanoscience and Society*, London.

Gutmann, M. (2004) *Erfahren von Erfahrungen: dialektische Studien zur Grundlegung einer philosophischen Anthropologie*, Bielefeld.

Habermas, J. (1970) *Toward a Rational Society*, Beacon Press, First publication. In: Habermas, J. (ed.) (1968) *Technik und Wissenschaft als »Ideologie«*, Frankfurt.

Habermas, J. (1973) Wahrheitstheorien. In: Fahrenbach, H. (ed.) *Wirklichkeit und Reflexion. Walther Schulz zum sechzigsten Geburtstag*, Pfullingen, pp. 211–265.

Habermas, J. (1988) *Theorie des kommunikativen Handelns*, Frankfurt, English version: *The Theory of Communicative Action.*

Habermas, J. (2001) *Die Zukunft der menschlichen Natur*, Frankfurt, English version: *The Future of the Human Nature*, Polity Press.

Hampp, N., Noll, F. (2003) Nanobionics II — from molecules to applications. *Physik Journal* **2**(2), 56.

Hansson, S. O. (2006) Great uncertainty about small things. In: Schummer, J., Baird, D. (eds.) *Nanotechnology Challenges — Implications for Philosophy, Ethics and Society*, Singapore, pp. 315–325.

Hansson, S. O. (2010) Risk and safety in technology. In: Meijers, A. (ed.) *Philosophy of Technology and Engineering Sciences*, Vol. 9, Amsterdam, pp. 1069–1102.

Harremoes, P., Gee, D., MacGarvin, M., Stirling, A., Keys, J., Wynne, B., Guedes Vaz, S. (eds.) (2002) *The Precautionary Principle in the 20th Century: Late Lessons from Early Warnings*, London.

Harris, J. (1975) The survival lottery. *Philosophy* **50**, 81–87.

Hatsopoulos, N. G., Donoghue, J. P. (2009) The science of neural interface systems. *Annual Review of Neurosciences* **32**, 249–266.

Haum, R., Petschow, U., Steinfeldt, M., von Gleich, A. (2004) Nanotechnology and regulation within the framework of the precautionary principle. *IÖW 173/04*, Berlin.

Heckl, W. (2004) Molecular self-assembly and nanomanipulation — two key technologies in nanoscience and templating. *Advanced Engineering Materials* **6**, 843–847.

Hennen, L., Grünwald, R., Revermann, C., Sauter, A. (2007) *Hirnforschung. Büro für Technikfolgen-Abschätzung beim Deutschen Bundestag*, Berlin (www.tab-beim-bundestag.de).

Hermerén, G. (2003) Ethical implications of nanotechnology. Lecture given at EuroNanoForum, Trieste, 9–12 December 2003.

Hermerén, G. (2004) Nano ethics primer. *Conference "Mapping out nano risks,"* Brussels, March 1, 2004.

Hildt, E. (2009) Fremdes im Gehirn. In: Engels, E., Hildt, E. (eds.) *Der implantierte Mensch*, Tübingen.

Hildt, E., Engels, E.-M. (Hg.) (2009) *Der implantierte Mensch*, Freiburg.

Höffe, O., *et al.* (Hg.) (2002) *Gentechnik und Menschenwürde: An den Grenzen von Ethik und Recht*, Köln.

Holm, S. (2005) Does nanotechnology require a new "nanoethics"? Paper published by Cardiff *Centre for Ethics, Law & Society*, August 2005.

Holtug, N. (1996) Is welfare all that matters in our moral obligations to animals? *Acta Agriculturae Scandinavica Sect. A, Animal Science Supplement* **27**, 16–21.

Hook, C. (2004) The techno sapiens are coming. *Christianity Today Magazine*, January 2004.

Hornberg-Schwetzel, S. (2008) Therapie und Enhancement. Der Versuch einer wertvollen Unterscheidung, *Jahrbuch für Wissenschaft und Ethik* 2008, Bonn, Institut für Wissenschaft und Ethik, pp. 207–221.

Hubig, C. (2007) *Die Kunst des Möglichen II: Ethik der Technik als provisorische Moral*, Bielefeld.

Ilulissat Statement (2008) Synthesizing the future: a vision for the convergence of synthetic biology and nanotechnology. *Kavli Futures Symposium "The merging of bio and nano: towards cyborg cells,"* 11–15 June 2007, Ilulissat, Greenland.

Invernizzi, N. (2008) Nanotechnology for developing countries. Asking the wrong question. In: Banse, G., Grunwald, A., Hronszky, I., Nelson, G. (eds.) *Assessing Societal Implications of Converging Technological Development*, Berlin, pp. 229–239.

IPCS (2004) *Risk Assessment Terminology — Part 1 and Part 2*, Geneva, Switzerland, World Health Organization.

IRGC — International Risk Governance Council (2009) *Risk Governance of Synthetic Biology*, Genf.

Irrgang, B. (2005) *Posthumanes Menschsein?: Künstliche Intelligenz, Cyberspace, Roboter, Cyborgs und Designer-Menschen Anthropologie des künstlichen Menschen im 21. Jahrhundert*, Stuttgart.

ITA — Institute of Technology Assessment at the Austrian Academy of Sciences (2010), *Nano-dossiers* (http://nanotrust.ac.at/nano.ita.en/dossiers.html).

Janich, P., Rüchardt, C. (eds.) (1996) *Natürlich, technisch, chemisch: Verhältnisse zur Natur am Beispiel der Chemie*, Berlin.

Janich. P. (2006) Wissenschaftstheorie der Nanotechnologie. In: Nordmann, A., Schummer, J., Schwarz, A. (eds.) *Nanotechnologien im Kontext*, Berlin, pp. 1–32.

Jasanoff, S. (1990) *The Fifth Branch: Science Advisers as Policymakers*, Cambridge, MA.

Jasanoff, S. (2009) The essential parallel between science and democracy. *Seed Magazine*, http://seedmagazine.com/content/article/the_essential_parallel_between_science_and_democracy.

JCP (2008) Sustainable nanotechnology development. *Journal of Cleaner Production* **16**. Special issue.

Jömann, N., Ach, J. S. (2006) Ethical implications of nanobiotechnology — State-of-the-art survey of ethical issues related to nanobiotechnology. In: Ach, J., Siep, L. (eds.) Nano-bio-ethics. *Ethical and Social Dimensions of Nanobiotechnology*, Berlin, pp. 13–62.

Jonas, H. (1979/1984) The imperative of responsibility (1984) Jonas, H. (1979) *Das Prinzip Verantwortung. Versuch einer Ethik für die technologische Zivilisation*, Frankfurt.

Jonas, H. (1987) Lasst uns einen Menschen klonieren. Von der Eugenik zur Gentechnologie. In: Jonas, H. (ed.) *Medizin und Ethik. Zur Praxis des Prinzips Verantwortung*, Frankfurt, pp. 162–203.

Jones, R. A. L. (2004) Soft machines. *Nanotechnology and Life*, Oxford.

Joseph, T., Morrison, M. (2006) *Nanotechnology in Agriculture and Food.* A Nanoforum Report, http://files.nanobio-raise.org/nfnaf.pdf.

Jotterand, F. (ed.) (2008a) *Emerging Conceptual, Ethical and Policy Issues in Bionanotechnology*, Berlin.

Jotterand, F. (2008b) Beyond therapy and enhancement: the alteration of human nature. *Nanoethics* **2**, 15–23.

Joy, B. (2000) Why the future does not need us. *Wired Magazine*, pp. 238–263.

Kaiser, M., Kurath, M., Maasen, S., Rehmann-Sutter, C. (eds.) (2010) *Governing Future Technologies: Nanotechnology and the Rise of an Assessment Regime*, Dordrecht.

Kamlah, W. (1973) *Philosophische Anthropologie: Sprachkritische Grundlegung und Ethik*, Mannheim.

Karafyllis, N. C. (2006) Biofakte. Grundlagen, Probleme und Perspektiven. *Erwägen Wissen Ethik* (*EWE*) **17**(4), 547–558.

Karafyllis, N. C. (2008) Ethical and epistemological problems of hybridizing living beings: biofacts and body shopping. In: Wenchao, Li, Poser, H. (eds.) Ethical considerations on today's science and technology. *A German-Chinese Approach*, Münster, pp. 185–198.

Karpowicz, P., *et al.* (2005) Developing human-nonhuman chimeras in human stem cell research: ethical issues and boundaries. *Kennedy Institute of Ethics Journal* **15**, 107–134.

Kastenhofer, K., Lansu, A., van Dam-Mieras, R. and Sotoudeh, M. (2010) The contribution of university curricula to engineering education for sustainable development. *GAIA* **19**(1), 44–51; see also other contributions in that volume.

Keeney, R. L. (1984) Ethics, decision analysis, and public theory. *Risk Analysis* **4**(1984), 117–129.

Keiper, A. (2007) Nanoethics as a discipline? The new Atlantis. *A Journal of Technology & Science*, 55–67.

Khushf, G. (2004a) The ethics of nanotechnology — visions and values for a new generation of science and engineering. In: National Academy of Engineering (ed.) *Emerging Technologies and Ethical Issues in Engineering*, Washington, pp. 29–55.

Khushf, G. (2004b) Systems theory and the ethics of human enhancement: a framework for NBIC convergence, *Annals of the New York Academy of Sciences* 1013, pp. 124–149.

Khushf, G. (2007) The ethics of NBIC convergence. *Journal of Medicine and Philosophy* **32/3**, 185–196.

Kobayawa, K., Kobayakawa, R., Matsumoto, H., *et al.* (2007) Innate versus learned odour processing in the mouse olfactory bulb. *Nature* **450**(7169), 503–508.

Koch, C. (2004) *The Quest for Consciousness. A Neurobiological Approach*, Englewood Colorado.

Kraft, J. C., *et al.* (2009) *In vivo* dopamine release and uptake impairments in rats treated with 3-nitropropionic acid. *Neuroscience* **161**(3), 940–949.

Kralj, M., Pavelic, K. (2003) Medicine on a small scale. How molecular medicine can benefit from self-assembled and nanostructured materials. *EMBO reports* **4**, 1008–1012.

Krug, H. F., Fleischer, T. (2007) Nanotechnologie — eine Bestandsaufnahme. *Umwelt — Medizin — Gesellschaft* **20**, 44–50.

Kurath, M. (2010) Negotiating nano: from assessing risks to disciplinary transformations. In: Kaiser, M., Kurath, M., Maasen, S., Rehmann-Sutter, Chr. (eds.), *Governing Future Technologies: Nanotechnology and the Rise of an Assessment Regime*, Dordrecht.

Kurath, M., Maasen, S. (2006) Toxicology as a nanoscience? Disciplinary identities reconsidered. *Particle and Fibre Toxicology* **3**, 6.

Latour, B. (1987) *Science in Action: How to Follow Scientists and Engineers through Society*, Harvard.

Latour, B. (1995) *Wir Sind Nie Modern Gewesen*, Berlin.

Lau, C. (2002) Nanotechnology and the Department of Defense. In: Roco, M. C., Bainbridge, W. S. (eds.), *Converging Technologies for Improving Human Performance*, Virginia, pp. 308–309.

Laurent, L., Petit, J.-C. (2006) Nanosciences and their convergence with other technologies: new golden age or apocalypse? In: Schummer, J., Baird, D. (eds.), *Nanotechnology Challenges: Implications for Philosophy, Ethics and Society*, Singapore, pp. 249–286.

LBNL — Lawrence Berkeley National Laboratory (2006) *Homepage* (http://www.lbl.gov/pbd/synthbio/default.htm).

Lenk, H. (1992) *Zwischen Wissenschaft und Ethik*, Frankfurt.

Lewenstein, V. B. (2006) What counts as a "social and ethical issue" in nanotechnology? In: Schummer, J., Baird, D. (eds.), *Nanotechnology Challenges: Implications for Philosophy, Ethics and Society*, Singapore, pp. 201–216.

Lin, P., Allhoff, F. (2007) Nanoscience and nanoethics: defining the disciplines. In: Allhoff, F., Lin, P., Moor, J., Weckert, J. (eds.), *Nanoethics: The Ethical and Social Implications of Nanotechnology*, New Jersey, pp. 3–16.

Link, H.-J. (2010) In-depth analysis of outstanding philosophical issues. In: *Synth-Ethics: Report on Work Package 1* (http://www.synthethics.eu/documents/REPORT WP1 synthethics - ethics+public discourse.pdf).

Lorenzen, P. (1987) *Lehrbuch der konstruktiven Wissenschaftstheorie*, Mannheim.

Lösch, A. (2008) The defuturization of the media: dynamics in the visual constitution of nanotechnology. In: Maasen, S., Rehmann-Sutter, C. (eds.), *Deliberating Future Technologies: Identity, Ethics, and Governance*.

Luhmann, N. (1997) *Theorie der Gesellschaft*, Frankfurt.

MacDonald, C. (2004) Nanotech is novel; the ethical issues are not. *The Scientist* **18**, 3.

Macnaghten, P., Davies, S., Kearnes, N. (2010) Narrative and public engagement: some findings from the DEEPEN project. In: von Schomberg, R., Davies, S. (eds.) *Understanding Public Debate on Nanotechnologies*, European Commission, Brussels, EUR 24169, pp. 13–30.

Marchant, D., Sylvester, G., Abbott, K. (2008) *Risk Management Principles of Nanotechnology* (http://www.law.upenn.edu/academics/institutes/regulation/papers/MarchantRiskManagementPrinciples.pdf).

Marcinkowski, F., Kohring, M., Friedemann, A., Donk, A. (2008) Risikowahrnehmung beim Thema Nanotechnologie: Analyse der Medienberichterstattung für den Analysezeitraum 1/2006 – 6/2007, Berlin.

Margolis, J. (1981) The concept of disease. In: Caplan, A. (ed.), *Concepts of Health and Disease: Interdisciplinary Perspectives*, Reading, pp. 561–577.

Margolis, J. (1981) The concept of disease. In: Caplan, A. (ed.), *Concepts of Health and Disease: Interdisciplinary Perspectives*, Reading, pp. 561–577.

Marquis, B. J., Love, S. A., Braun, K. L., Haynes, C. L. (2009) Analytical methods to assess nanoparticle toxicity. *Analyst* **134**(3), 425–439.

Maurer, S., Lucas, K., Terrel, S. (2006) *From Understanding to Action: Community-Based Options for Improving Safety and Security in Synthetic Biology*, Berkeley.

Maynard, A. D., Aitken, R. J., Butz, T., Colvin, V., Donaldson, K., Oberdorster, G., *et al.* (2006) Safe handling of nanotechnology. *Nature* **444**(7117), 267–269.

Mehta, M. D. (2002) On nano-panopticism: a sociological perspective. *Canadian Chemical News* **54**(10), 31–33.

Merkel, R., Galert, T., *et al.* (2007) *Intervening in the Brain: Changing Psyche and Society*, Berlin.

Meyer, R., Bütschi, D., Gram, S., Haugen, J. M., Sauter, A., Steyaert, St., Torgersen, H. (2009): *Genetically Modified Plants and Foods: Challenges and Future Issues in Europe*, Final Report, EPTA, Berlin (www.eptanetwork.org).

Miller, S., Selgelid, M. (2006) *Ethics and the Dual-Use Dilemma in the Life Sciences*, Berlin.

Mitcham, C. (1994) *Thinking through Technology: The Path between Engineering and Philosophy*, Chicago.

Mitcham, C., Briggle, A. (2010) The interaction of ethics and technology in historical perspective. In: Meijers, A. (ed.), *Philosophy of Technology and Engineering Sciences*, Vol. 9, Amsterdam , pp. 1147–1192.

Mittelstraß, J. (1991) Auf dem Wege zu einer Reparaturethik? In: Wils, J.-P., Mieth, D. (eds.), *Ethik ohne Chance? Erkundungen im technologischen Zeitalter*, Tübingen, pp. 89–108.

Mnyusiwalla, A., Daar, A. S., Singer, P. A. (2003) Mind the gap: science and ethics in nanotechnology. *Nanotechnology* **14**, R9–R13.

Monteiro-Riviere, N. A., Tran, C. L. (eds.) (2007) *Nanotoxicology: Characterization, Dosing and Health Effects*, New York.

Moor, J., Weckert, J. (2004) Nanoethics: assessing the nanoscale from an ethical point of view. In: Baird, D., Nordmann, A., Schummer, J. (eds.), *Discovering the Nanoscale*, Amsterdam, pp. 301–310.

Moore, D. (2007) Nanotechnology and the military. In: Allhoff, F., Lin, P., Moor, J., Weckert, J. (eds.), *Nanoethics: The Ethical and Social Implications of Nanotechnology*, New Jersey, pp. 267–275.

Moritz, C. T., Perlmutter, S. I., Fetz, E. E. (2008) Direct control of paralysed muscles by cortical neurons. *Nature* **456**(7222), 639–642.

Moshammer, H., Wallner, P. (2007) Gesundheitsrisiken durch Nanopartikel? In: Gazsó, A., Greßler, S., Schiemer, F. (eds.), *Nano. Chancen und Risiken aktueller Technologien*, Wien, pp. 165–180.

Müller, S. (2006) Minimal-invasive und nanoskalige Therapien von Gehirnerkrankungen: Eine medizinethische Diskussion. In: Nordmann, A., Schummer, J., Schwarz, A. (eds.) *Nanotechnologien im Kontext*, Berlin, pp. 346–370.

Munich Re (2002) *Nanotechnology: What Is in Store for Us?* (http://www.munichre.com/publications/302-03534_en.pdf).

Myhr, A. I., Dalmo, R. A. (2007) Nanotechnology and risk: what are the issues? In: Allhoff, F., Lin, P., Moor, J., Weckert, J. (eds.) *Nanoethics: The Ethical and Social Implications of Nanotechnology*, New Jersey, pp. 149–159.

Nachtigall, W. (2002) *Bionik: Grundlagen und Beispiele für Ingenieure und Naturwissenschaftler*, Berlin.

Nair, I., Morgan, G., Florig, H. K. (1989) *Biologic Effects of Power Frequency Electric and Magnetic Fields*. Office of Technology Assessment Report OTA-BP-E-53, Washington, DC.

Nanoforum (2004) *Nanotechnology: Benefits, Risks, Ethical, Legal, and Social Aspects of NANOTECHNOLOGY* (www.nanoforum.org).

Neuberger, M. (2007) Umweltepidemiologie und Toxikologie von Nanopartikeln (Ultrafeinstaub) und Feinstaub. In: Gazsó, A., Greßler, S., Schiemer, F. (eds.) *Nano — Chancen und Risiken aktueller Technologien*, Wien, pp. 181–197.

Neumann, S. (2008) Cosmetic surgery: customer service or professional misconduct. *Canadian Veterinary Journal* **49**(5), 501–504.

Nida-Rümelin, J. (ed.) (1996a) *Angewandte Ethik: Die Bereichsethiken und ihre theoretische Fundierung*, Stuttgart.

Nida-Rümelin, J. (1996b) Risikoethik. In: Nida-Rümelin, J. (ed.) *Angewandte Ethik: Die Bereichsethiken und ihre theoretische Fundierung*, Stuttgart, pp. 863–887.

NNI (1999) National Nanotechnology Initiative, Washington.

Nordenfelt, L. (1993) *Quality of Life, Health, and Happiness*, Avebury.

Nordmann, A. (2004) *Converging Technologies: Shaping the Future of European Societies*, High Level Expert Group "Foresighting the New Technology Wave," Brussels.

Nordmann, A. (2007a) If and then: a critique of speculative nanoethics. *NanoEthics* **1**, 31–46.

Nordmann, A. (2007b) Entflechtung — Ansätze zum ethisch-gesellschaftlichen Umgang mit der Nanotechnologie. In: Gazsó, A., Greßler, S., Schiemer, F. (eds.) *Nano: Chancen und Risiken aktueller Technologien*, Wien, pp. 215–229.

Nordmann, A. (2008) Philosophy of "nanotechnoscience." In: Schmid, G. (ed.) *Nanotechnology: Principles and Fundamentals*, Weinheim, pp. 217–244.

Nordmann, A. (2010) A Forensic of wishing: technology assessment in the age of technoscience. *Poiesis & Praxis* **7**, 5–15.

Nordmann, A., Rip, A. (2009) Mind the gap revisited. *Nature Nanotechnology* **4**, 273–274.

Nordmann, A., Schummer, J., Schwarz, A. (eds.) (2006) *Nanotechnologien im Kontext: Philosophische, ethische und gesellschaftliche Perspektiven*, Berlin.

Nowotny, H., Scott, P., Gibbons, M. (2001) *Re-Thinking Science: Knowledge and the Public in an Age of Uncertainty*, London.

Nuffield Council of Bioethics (2005) *The Ethics of Research Involving Animals*, Nuffield Council on Bioethics, London (http://www.nuffieldbioethics.org/fileLibrary/pdf/RIA_Report_FINAL-opt.pdf).

Oberdorster, G. (2010) Safety assessment for nanotechnology and nanomedicine: concepts of nanotoxicology. *Journal of Internal Medicine*, **267**(1), 89–105.

Oberdörster, G., Oberdörster, E., Oberdörster, J. (2005) Nanotechnology: an emerging discipline evolving from studies of ultrafine particles. *Environmental Health Perspectives* **113**(7), 823–839.

Oertel, D., Grunwald, A. (2006) *Potenziale und Anwendungsperspektiven der Bionik*, Büro für Technikfolgen-Abschätzung beim Deutschen Bundestag (TAB), Berlin.

Oppenheim, P., Putnam, H. (1958) *The Unity of Science as a Working Hypothesis*, Minnesota.

Pacurari, M., Castranova, V., Vallyathan, V. (2010) Single- and multi-wall carbon nanotubes versus asbestos: are the carbon nanotubes a new health risk to humans? *Journal of Toxicology and Environmental Health, Part A* **73**(5), 378–395.

Parens, E. (ed.) (1998) *Enhancing Human Traits: Ethical and Social Implications*, Washington, DC.

Parens, E., Johnston, J., Moses, J. (2009) *Ethical Issues in Synthetic Biology: An Overview of the Debates*. Woodrow Wilson International Center for Scholars (http://www.synbioproject.org/library/publications/archive/synbio3/).

Paschen, H., Coenen, C., Fleischer, T., Grünwald, R., Oertel, D., Revermann, C. (2004) *Nanotechnologie: Forschung und Anwendungen*, Berlin.

Pedrono, E., Durukan, A., Strbian, D., *et al.* (2010) An optimized mouse model for transient ischemic attack. *Journal of Neuropathological Experimental Neurology* **69**(2), 188–195.

Pereira, A. G., von Schomberg, R., Funtowicz, S. (2007) Foresight knowledge assessment. *International Journal of Foresight and Innovation Policy* **4**, 65–79.

Perelmann, C. (1967) *Justice*, New York.

Petersen, A., *et al.* (2009) Opening the black box: scientists' views on the role of the news media in the nanotechnology debate. *Public Understanding of Science* **18**, 512–530.

Peterson, M. (2007) The precautionary principle should not be used as a basic for decision-making. *EMBO Reports* **8**, 10–14.

Phoenix, C., Treder, M. (2003) Applying the precautionary principle to nanotechnology (http://www.crnano.org/Precautionary.pdf).

Picht, G. (1971) *Prognose Utopie Planung*, Stuttgart.

Presidential Commission (2010) *Recommendations on Synthetic Biology*, Washington.

Psarros, N. (2006) Diskussionsbeitrag. *Erwägen Wissen Ethik* (*EWE*) **17**, 594–596.

Qiu, L. (2006) Mighty mouse. *Nature* **444**, 814–816.

Quante, M. (1994) Natur, Natürlichkeit und der naturalistische Fehlschluß. Zur begrenzten Brauchbarkeit eines klassischen philosophischen Arguments in der biomedizinischen Ethik. *Zeitschrift für Medizinische Ethik* **40**, 289–305.

Radder, H. (2010) Why technologies are inherently normative. In: Meijers, A. (ed.), *Philosophy of Technology and Engineering Sciences*, Vol. 9, Amsterdam, pp. 887–922.

Renn, O., Roco, M. C. (2006) The risk governance of nanotechnology. *Journal of Nanoparticle Research* **8**(2), 153–191.

Renn, O., Schweizer, P. J., Dreyer, M., Klinke, A. (2008) *Risiko: Über den gesellschaftlichen Umgang mit Unsicherheit*, München.

Rescher, N. (1983) *Risk: A Philosophical Introduction to the Theory of Risk Evaluation and Management*, Lanham.

Rescher, N. (1988) *Rationality: A Philosophical Inquiry into the Nature and the Rationale of Reason*, Oxford.

Reuter, K. (2007) Tierzucht für den ökologischen Landbau: Probleme, offene Fragen, Lösungsansätze. *Bochum* (http://orgprints.org/15131/1/reuter-etal-2007-Tagungsband_Tierzucht_Kassel.pdf).

Reuzel, R. P. B. (2001) *Health Technology Assessment and Interactive Evaluation: Different Perspectives*, Nijmegen.

Rip, A., Swierstra, T. (2007) Nano-ethics as NEST-ethics: patterns of moral argumentation about new and emerging science and technology. *NanoEthics* **1**, 3–20.

Rip, A., Misa, T., Schot, J. (eds.) (1995) *Managing Technology in Society*, London.

Rippe, K.-P. (2002) Vorsorge als umweltethisches Problem. *Im Auftrag der Eidgenössischen Ethikkommission für den außerhumanen Bereich* (www.ethikdiskurs.ch/Umweltethik/Vorsorgeprinzip.pdf).

Roache, R. (2008) Ethics, speculation, and values. *Nanoethics* **2**, 317–327.

Robert, J. S., Baylis, F. (2003) Crossing species boundaries. *The American Journal of Bioethics* **3**(3), 1–13.

Robischon, M. (2007) Von Fischen und genen. *Amazonas* **14** (November/Dezember 2007), Seite 64–68; (http://www.amazonas-magazin.de/index.php?id=1142).

Roco, M. C. (2007) Foreword: ethical choices in nanotechnology development. In: Allhoff, F., Lin, P., Moor, J., Weckert, J. (eds.) *Nanoethics: The Ethical and Social Implications of Nanotechnology*, New Jersey, p. 5f.

Roco, M. C., Bainbridge, W. S. (eds.) (2001) *Societal Implications of Nanoscience and Nanotechnology*, Boston.

Roco, M. C., Bainbridge, W. S. (eds.) (2002) *Converging Technologies for Improving Human Performance*, Arlington.

Rohbeck, J. (1993) *Technologische Urteilskraft: Zu einer Ethik technischen Handelns*, Frankfurt.

Rolston, H. (1988) *Environmental Ethics*, Philadelphia.

Ropohl, G. (1996) *Ethik und Technikbewertung*, Frankfurt.

Rorty, R. (1998) *Truth and Progress: Philosophical Papers*, Cambridge.

Royal Society (2004) *Nanoscience and Nanotechnologies: Opportunities and Uncertainties*, London.

Sandel, M. (2004) *The Case Against Perfection*, Boston.

Sarewitz, D., Karas, T. H. (2006) *Policy Implications of Technologies for Cognitive Enhancement*, Arizona State University, Consortium for Science, Policy & Outcomes.

Sattler, K. D. (ed.) (2011) *Handbook of Nanophysics: Nanomedicine and Nanorobotics*, Boca Raton, FL.

Schepelmann, P., Ritthoff, M., Jeswani, H., Azapagic, A., Suomalainen, K. (2009) Options for deepening and broadening LCA. *CALCAS — Co-ordination Action for Innovation in Life-Cycle Analysis for Sustainability*, Brussels.

Schmid, G. (ed.) (2008) *Nanotechnology: Principles and Fundamentals*, Weinheim.

Schmid, G., Decker, M. (eds.) (2003) *Small Dimensions and Material Properties: A Definition of Nanotechnology*, Bad Neuenahr-Ahrweiler.

Schmid, G., Ernst, H., Grünwald, W., Grunwald, A., *et al.* (2006) *Nanotechnology: Perspectives and Assessment*, Berlin.

Schmidt, K. (2008) *Tierethische Probleme der Gentechnik: Zur moralischen Bewertung der Reduktion wesentlicher tierischer Eigenschaften*, Mentis, Paderborn.

Schmidt, J. (2002) Vom Leben zur Technik? Kultur- und wissenschaftsphilosophische Aspekte der Natur-Nachahmungsthese in der Bionik. *Dialektik* **2**, 129–143.

Schmidt, J. C. (2008) Unbestimmtheitssignaturen der Nanotechnologie. In: Köchy, K., Norwig, M., Hofmeister, G. (eds.), *Nanobiotechnologien. Philosophische, anthropologische und ethische Fragen*, Freiburg, pp. 47–66.

Schöne-Seifert, B., Ach, J. S., Talbot, D., Opolka, U. (eds.) (2009) *Neuro Enhancement: Ethik vor neuen Herausforderungen*, Paderborn.

Schummer, J. (2007) Impact of nanotechnologies on developing countries. In: Allhoff, F., Lin, P., Moor, J., Weckert, J. (eds.), *Nanoethics: The Ethical and Social Implications of Nanotechnology*, New Jersey, pp. 291–307.

Schummer, J., Baird, D. (eds.) (2006) *Nanotechnology Challenges: Implications for Philosophy, Ethics and Society*, Singapore.

Scott, N. R. (2005) Nanotechnology and animal health. *Revue Scientifique et Technique (International Office of Epizootics)* **24**(1), 425–432.

Scott, S. (2006) Converting thoughts into action. *Nature* **442**, 141–142.

Seaton, A., Tran, L., Aitken, R., Donaldson, K. (2010) Nanoparticles, human health hazard and regulation. *Journal of the Royal Society Interface* **7**(1), 119–129.

Seeman, N. C., Belcher, A. M. (2002) Emulating biology: building nanostructures from the bottom up. *PNAS* **99**(2), 6451–6455.

Selgelid, M. (2007) Ethics and drug resistance. *Bioethics* **21**, 218–221.

Selin, C. (2006) Time matters: harmony and dissonance in nanotechnology networks. *Time & Society* **15**(1), 121–139.

Selin, C. (2007) Expectations and the emergence of nanotechnology. *Science, Technology and Human Values* **32**(2), 196–220.

Sethe, S. (2007) Nanotechnology and life extension. In: Allhoff, F., Lin, P., Moor, J., Weckert, J. (eds.), *Nanoethics: The Ethical and Social Implications of Nanotechnology*, New Jersey, pp. 353–365.

Sharp, P. A. (2005) 1918 flu and responsible science. *Science* **310**(5745), 17.

Shrader-Frechette, K. S. (1991) *Risk and rationality: Philosophical Foundations for Populist Reforms*, Berkeley.

Shriver, A. (2009) Knocking out pain in livestock: can technology succeed where morality has stalled? *Neuroethics* **2**, 115–124.

Shumyatsky, G. P., Malleret, G., Shin, R. M., *et al.* (2005) Stathmin, a gene enriched in the amygdala, controls both learned and innate fear. *Cell* **123**, 697–709.

Siep, L. (2006) Die biotechnische Neuerfindung des Menschen. In: Abel, G. (ed.) Kreativität, *Akten des XX. Deutschen Kongresses für Philosophie*, Hamburg, pp. 306–323.

Simko, M., Mattsson, M.-O. (2010) Risks vom accidental exposures to engineered nanoparticles and neurological health effects: a critical review. *Particle and Fibre Toxicology* **7**, 42–54.

Siune, K., Markus, E., Calloni, M., Felt, U., Gorski, A., Grunwald, Rip, A., de Semir, V., Wyatt, S. (2009) *Challenging Futures of Science in Society*, MASIS Expert Group, European Commission, Brussels.

Smalley, R. E. (2001) Of chemistry, love and nanobots. *Scientific American* **285**, 76–77.

Stehr, N. (2004) *The Governance of Knowledge*, London.

Steinfeldt, M., Gleich, A. von, Petschow, U., Haum, R. (2007) *Nanotechnologies, Hazards and Resource Efficiency*, Heidelberg.

Stephens, L. F. (2005) News narratives about Nano S&T in major U.S. and Non-U.S. newspapers. *Science Communication* **27**, 175–199.

Stern, P. C., Fineberg, H. V. (eds.) (1996) *Understanding Risk: Informing Decisions in a Democratic Society*, Washington DC.

Stieglitz, T. (2006) Neuro-technical interfaces to the central nervous system. *Poiesis & Praxis* **4**, 95–110.

Stirling, A. (2008) "Opening up" and "closing down": power, participation, and pluralism in the social appraisal of technology. *Science Technology Human Values 2008* **33**, 262–294.

Stø, E., Scholl, G., Jègou, F., Strandbakken, P. (2010) The future of deliberative processes on nanotechnology. In: von Schomberg, R. (ed.) *Understanding Public Debate on Nanotechnologies. Options for Framing Public Policy*, Brussels, pp. 53–80.

Sunstein, C. (2005) *The Laws of Fear: Beyond the Precautionary Principle*, Cambridge.

Swiss Re (2004) *Nanotechnologie: Kleine Teile — große Zukunft?* Risk Perception, Zürich.

Synbiology (2005) *SYNBIOLOGY* (SYNBIOLOGY_Literature_And_Statistical_Review.pdf in www2.spi.pt/synbiology/documents/).

Synth-Ethics (2010) Homepage of the EU-funded project "Ethical and regulatory issues raised by synthetic biology" (http://synthethics.eu/).

Talwar, S. K., Xu, S., Hawley, E. S., *et al.* (2002) Behavioural neuroscience: rat navigation guided by remote control. *Nature* **417**, 37–38.

Taniguchi, N. (1074) On the basic concept of "nano-technology". *Proceedings of the International Conference on Production Engineering, Tokyo*, Part II., Japan Society of Precision Engineering.

ten Have, H. (Hg.) (2007) *Nanotechnologies, Ethics and Politics*, Paris.

Truin, M., van Kleef, M., Verboeket, Y., *et al.* (2009) The effect of spinal cord stimulation in mice with chronic neuropathic pain after partial ligation of the sciatic nerve. *Pain* **145**(3), 312–318.

Tucker, J. B., Zilinskas, R. A. (2006) The promise and perils of synthetic biology. *The New Atlantis* **12**, 25–45.

UK Govt., H. M. Government (2005) *Response to the Royal Society and Royal Academy of Engineering Report* (http://www.ost.gov.uk/policy/issues/nanotech_final.pdf).

van de Poel, I. (2010) Values in engineering design. In: Meijers, A. (ed.) *Philosophy of Technology and Engineering Sciences*, Vol. 9, Amsterdam, pp. 973–1006.

van den Hoven, J. (2007) Nanotechnology and privacy: instructive case of RFID. In: Allhoff, F., Lin, P., Moor, J., Weckert, J. (eds.) *Nanoethics: The Ethical and Social Implications of Nanotechnology*, New Jersey, pp. 253–266.

van Gorp, A. (2005) *Ethical Issues in Engineering Design, Safety and Sustainability*. Simon Stevin Series in the Philosophy of Technology, Delft/Eindhoven.

van Merkerk, R. (2007) *Intervening in Emerging Technologies: A CTA of Lab-on-a-Chip Technology*, Utrecht.

VDI — Verein Deutscher Ingenieure (1991), *Richtlinie 3780 Technikbewertung, Begriffe und Grundlagen*, Düsseldorf.

VDI — Verein Deutscher Ingenieure (2000) *Ethische Ingenieurverantwortung: Handlungsspielräume und Perspektiven der Kodifizierung*. VDI-Report 31, Düsseldorf.

VDI-TZ (VDI-Technologiezentrum) (2002) *Nanobiotechnologie I: Grundlagen und technische Anwendungen molekularer, funktionaler Biosysteme*. Zukünftige Technologien 38, Düsseldorf.

Velliste, M., Perel, S., Spalding, M. C., *et al.* (2008) Cortical control of a prosthetic arm for self- feeding. *Nature* **453**, 1098–1101.

von Gleich, A., Pade, C., Petschow, U., Pissarskoi, E. (2007) *Bionik: Aktuelle Trends und zukünftige Potentiale*, Bremen.

von Schomberg, R. (2005) The precautionary principle and its normative challenges. In: Fisher, E., Jones, J., von Schomberg, R. (eds.) *The Precautionary Principle and Public Policy Decision Making*, Cheltenham, UK/Northampton, MA, pp. 141–165.

von Schomberg, R. (2007) *From the Ethics of Technology Towards an Ethics of Knowledge Policy & Knowledge Assessment*. A Working Document from the European Commission Services, EUR 2429, Brussels.

von Schomberg, R., Davies, S. (eds.) (2010) *Understanding Public Debate on Nanotechnologies*, European Commission, EUR 24169, Brussels.

Wagner, P. (2005) Nanobiotechnology. In: Greco, R., Prinz, F. B., Lane, R. (eds.) *Nanoscale Technology in Biological Systems*, Boca Raton, FL, pp. 39–55.

Wardak, A. (2003) *Nanotechnology and Regulation: A Case Study Using the Toxic Substance Control Act (TSCA)*, Woodrow Wilson International Center, Foresight and Governance Project, Paper 2003-6.

Warneken, F., Tomasello, M. (2006) Altruistic helping in human infants and young chimpanzees, *Science* **311**(5765), 1301–1303.

WCED — World Commission on Environment and Development (1987), *Our Common Future*, Oxford.

Weckert, J. (2007) Editorial. *Nanoethics* **1**, 1–2.

Weckert, J., Moor, J. (2007) The precautionary principle in nanotechnology. In: Allhoff, F., Lin, P., Moor, J., Weckert, J. (eds.) *Nanoethics: The Ethical and Social Implications of Nanotechnology*, New Jersey, pp. 133–146.

Whitehouse, D. (2003) GM fish glows in the bowl, *BBC News*, 27 June 2003, (http://news.bbc.co.uk/2/hi/science/nature/3026104.stm).

Wiedemann, P., Schütz, H. (eds.) (2008) *The Role of Evidence in Risk Characterization: Making Sense of Conflicting Data*, Weinheim.

Williams, E., Frankel, M. S. (2006) *Good, Better, Best: The Human Quest for Enhancement.* Summary Report of an Invitational Workshop Convened by the Scientific Freedom, Responsibility and Law Program, American Association for the Advancement of Science.

Wilson, E. O. (1999) *Consilience: The Unity of Knowledge,* New York.

Winner, L. (1982) *Autonomous Technology: Technics-Out-of-Control as a Theme in Political Thought,* Cambridge.

Woese, C. R. (2004) A new biology for a new century. *Microbiology and Molecular Biology Reviews* **68**(2), 173–186.

Wolbring, G. (2006) *The Triangle of Enhancement Medicine, Disabled People, and the Concept of Health: A New Challenge for HTA, Health Research, and Health Policy,* Alberta Heritage Foundation for Medical Research (AHFMR), Edmonton. HTA Initiative #23.

Wolbring, G. (2008a) Oscar Pistorius and the future nature of Olympic, Paralympic and other sports. *SCRIPTed* **5**, 139–160.

Wolbring, G. (2008b) Why NBIC? Why human performance enhancement? *The European Journal of Social Science Research* **21**, 25–40.

World Transhumanist Association (2006) For the ethical use of technology to extend human capabilities (www.transhumanism.org/index.php/WTA/index/).

Wynne, B. (1991) Knowledges in context. *Science, Technology and Human Values* **16**(4), 1–19.

Xia, T., Li, N., Nel, A. E. (2010) Potential health impact of nanoparticles. *The Annual Review of Public Health* **30**, 137–150.

Yoshinaka, Y., Clausen, C., Hansen, A. (2003) The social shaping of technology: a new space for politics? In: Grunwald, A. (ed.) *Technikgestaltung: Zwischen Wunsch oder Wirklichkeit,* Berlin, pp. 117–131.

Zuin, S., Pojana, G., Marcomini, A. (2007) Effect-oriented physicochemical characterization of nanomaterials. In: Monteiro-Riviere, N. A., Tran, C. L. (eds.) *Nanotoxicology: Characterization, Dosing and Health Effects,* New York.

# Index


abilities
  acquired 279
  cognitive 237, 241
  night-vision 271
abiotic nanoparticles 184, 204, 220
abortions, late 277
accident 70, 197, 260, 263, 270, 289
  Bhopal 70
  Chernobyl 70
  Fukushima 70
active packaging 150
additives 113, 150–151
  chemical 202
adenosine triphosphate synthase 213
advice, normative 75, 190
aerosols 135, 149
AFM *see* atomic force microscopy
agglomeration 136, 141, 155
aging 123, 255, 261–262, 273, 300
agricultural research 228
air conditioners 153
air vehicles 24
alarmist approaches 45
Alzheimer's disease 69, 121, 230, 285
American National Nanotechnology Initiative 90
amino acids 196
AML animals 242, 245
AMLs *see* animal microencephalic lumps
animal enhancement 11, 131, 227–228, 232, 240, 243–244
animal ethics 132, 239–240, 242–243, 248, 335
animal experiments 62, 122, 130–131, 135, 154, 209, 231–232, 238–240, 243, 245, 247–248
animal genome 243
animal-machine hybrids 247
animal medications 246
animal microencephalic lumps (AMLs) 240–241, 335
animal models 230–231, 238–239, 241
animal protection 132, 245
animal protection organization (PETA) 245
animal welfare 238, 241–242, 247
animals 11, 69, 130–132, 145, 154, 163, 209, 227–38, 240–249, 335
  cloning 229
  transgenic 130
anthropocentrism 244
anthropogenic combustion processes 148
anthropological questions 12, 90, 128
anthropology 4, 96, 100, 102, 117, 132, 217, 245, 275, 284, 289, 304, 310, 321, 341–342
antiaging 261
antinano movement 331, 333
antiwrinkle applications 261
aporias 164, 307
appellative 80
applied ethics 4, 6, 8, 11–12, 49, 52–53, 89, 95–97, 99, 101–102, 118–120, 303–304, 310–311, 318–321, 340–342
approval 104
Aristotelian approach 171
Aristotelian ethics 169
arms races 46
arteriosclerosis 154
artificial accommodation systems 259
artificial antennas 193
artificial brains 43
artificial cells 196, 201
artificial eye 271
artificial hand 285

artificial hippocampus 285
artificial implants 27, 121
artificial life 11, 113, 145, 191, 194, 201, 216–218, 334
  creation 11, 192, 194, 196, 198, 200, 202, 204, 206, 208, 210, 212, 214
artificial limbs 259–260, 265
artificial problems 92, 309
artificial replacements 265
Aryan ideals 257
asbestos fibers 45, 153
asphyxiation 155
atomic force microscopy (AFM) 16, 20
atomic power plant 72
atomic reductionism 16, 36–37, 253–254, 293, 321
atomic structures 40
atomic weapons 56
Atomtech 112
autonomous systems 132

Bacon 36
Baconian approach 209, 323
bacteria 19, 27, 42, 148, 156, 193, 198, 208
  engineered 197
  real 42
bacteriochlorophylls 193
bacterium 113, 192, 208
  artificial 113
beings 218–219, 241–43, 293
  nonhuman 269
beliefs 64, 257
  fundamental 322
  popular 202
  spiritual 210
Big Down, the 111
Bio2Nano 192
biocompatible materials 26, 120, 197
biodegradable plastics 197
biodesigner-hackers 199
bioethics 5–6, 96, 99–100, 102, 120, 145, 208, 238, 243–244, 263, 340–341
biofuels 198
bioinformatics 33, 229
biological principle 29
biological systems 156, 191, 193, 197, 199, 221
biological weapons 127, 133, 199, 202–203
biologists 110, 209–210, 223
biology 33–35, 98, 102, 145, 193–195, 199, 213, 216–219, 321
  classical 194
  natural 195
  open source 113
biomedical challenges 111
biomedical sciences 117
biomedicine 127, 130, 198, 230
biomembranes 127, 198
biomineralization 198
bionanotechnology 5, 26, 117, 191, 201
bionic approach 214
bionic equipment 261
bionic idea 216
bionic robotics 246
bionics 110, 214–216
biosafety 113, 127, 200–201, 212
biosecurity 113, 127, 200, 212
biosensors 127, 197–198, 229
biosphere 42, 134, 152–153, 183
biotechnology 12, 19, 34, 38, 43–44, 46, 112, 123, 126, 128, 193, 197, 251–252, 259, 293
bioterrorists 222
biotic systems 194, 214
bioweapons/bioterrorism 113
blood–brain barrier 155
botox medications 261
boundary objects 32
bovine spongiforme encephalopathy (BSE) 83
brain, human 128, 255, 259–260
brain chip 128, 259, 285
brain–computer stimulations 288
brain drain/depopulation economics 283

brain–machine interfaces 231
brain research 110, 228, 251, 253
BSE *see* bovine spongiforme encephalopathy

cancer 121, 153
lung 45, 153
stomach 154
cancer diseases 153
cancer therapy 121
canon 117–118, 143, 273
capacity, cognitive 243–244, 246, 248, 252
carbon nanotubes 22, 25, 44, 153–154, 199
cardiovascular problems 121
Cartesian interpretation 246
CAS *see* Chemical Abstracts Service
catalysts 24–26, 153, 192, 323
catastrophic accidents 106, 346–347
cathode ray tubes 22
cellular metabolism 126, 192
chain
cause–effect 83, 230
if–then 66, 306
Charybdis 318
Chemical Abstracts Service (CAS) 110
chimeras 43, 238, 243–244
civilization 256–257, 264, 291
technical 43, 291–292
cloning 62, 229
animals 229
human 335
reproductive 56, 62, 95, 281
cloning technology 335
$CO_2$ emissions 246
code of conduct 175
cognitive enhancements 259–260, 273
COMEST *see* World Commission on the Ethics of Scientific Knowledge and Technology
communication technologies 23, 25, 38, 123–124, 253, 259
comparative risk evaluations 92
complex carbohydrates 193
composite beings 296–297
composite creature, man–animal 243–244
composite creatures 243–244
conceptual framework 5, 7, 9
conditions of coherence 57–58
conference of Asilomar 220
consequentialism 159, 283
consequentialist approach
utilitarian 160
utilitarian-style 163
consumer behavior 77–78
consumer freedom 174
consumer protection 110, 135, 177
convergence 18, 90, 97–99, 128, 228, 253–255, 322
converging technologies 3–4, 18, 34, 43, 69, 129–130, 145, 228–229, 243–244, 246–247, 251–253, 257–259, 261, 290–291, 312–313
corporate commitments 319
corporate guidelines 55
cosmetics 23, 47, 112, 134, 147–148, 152, 173, 179, 240, 256
cyborgs 42, 44, 119, 128, 294, 296–297, 309, 314, 322

data protection 125–126, 339, 341
deaths, cancer 153
deficient being 256
deforestation 246
deontological advice 159, 163
deontological approach 163
deontological arguments 172
deontological ethics 163
deterministic thinking 36
deterministic work 37
devices

electronic 25
nano-enabled 304
optoelectronic 25
undetectable 98
diagnostic instruments, nanotechnology-based 26
diagnostics
medical 120
prenatal 277
procedures 78, 197
digital data storage units 213
dignity 56–57, 131, 208, 211, 242, 244, 248, 287
acquired 118
ethical 286
diseases 26, 121, 156, 230, 234, 236–237
human 230
infectious 121
prevention 120
disinfectants 138, 156
DNA 19, 195, 215, 219, 322
artificial 192
DNA-based artificial organism 200
DNA molecules 199
DNA sequences 196
doping 12, 128–129, 229, 256–257, 263, 265–269, 283, 298, 301, 320, 341
gene 268
doping medication 229
dot-like structures 21

ecological damage 45
ecology 136, 185
ecotoxicology 136
EHS *see* environmental, health, and safety
electrical fences 213
electromagnetic fields 169
electronic circuits (EMFs) 196
electrotechnology 196
ELSI *see* ethical, legal, and social implications
ELSI studies 109, 137
embryos 56, 62, 69, 320, 333
human 69
EMFs *see* electromagnetic fields
empirical research 47, 84, 93, 136, 142
energy technologies 38
enhanced animals 231, 235, 244
enhancement 12, 101, 128, 130, 139, 228–237, 240, 244–245, 252, 257–258, 263–274, 276–277, 283–285, 296–298, 341
ethics of 129
goals of 236, 248
neurocognitive 139, 276–277
technological 10
enhancement ethics 293, 341–342
environment-safety-health 93
enhancement society 300–301
enhancement technologies 93, 110, 117, 129, 240, 273–278, 281–282, 300–301
functioning 271
environmental chemistry 136, 185
environmental damage 200
environmental degradation 166
environmental, health, and safety (EHS) 115, 134, 306
environmental medicine 84
environmental quality 157
epidemiological research 149, 155
epistemology 4, 12, 37, 84, 117, 304, 320–321, 324
ESH *see* environment-safety-health 93
ethical, legal, and social implications 1, 35, 89, 108–109, 192
ethical advice 63–64, 75–76, 88, 101, 307, 337
ethical analysis 5, 52, 69, 128, 160, 173–174, 272, 275, 305, 325, 336
ethical approaches 159–160, 170
ethical argumentation 182, 272, 275, 282, 330
ethical arguments 4, 62, 234, 276, 286, 330, 339

ethical challenges 6, 11, 70, 95, 104–105, 127, 135, 204, 217, 227, 298, 338
ethical consultation 9, 75, 281
ethical counseling 53, 63
ethical criteria 57, 142
ethical debates 13, 71, 107, 145, 288
ethical evaluations 7, 219, 237, 242
ethical expertise 63–67, 337
ethical exploration 4
ethical guidance 161–162, 165, 170–171
ethical inquiry 74, 84–86, 160, 328, 335
ethical issues 84, 116, 203, 205, 207, 209, 211
ethical judgment 231, 284, 288, 303, 336, 339
ethical parallel research 337–338
ethical principles 123, 220
ethical problems 63, 74, 163, 174, 209, 284, 286, 311, 334
ethical questions 5, 7, 10, 19, 32–33, 43, 95–96, 99–100, 102, 117–119, 121–123, 125–131, 280, 297, 340
ethical reasoning 134, 326
ethical reflection 9–10, 50–52, 56–58, 67–68, 74–77, 82, 86–92, 94–95, 99–104, 129, 143–145, 297–300, 318–319, 330–331, 335–340
ethical standards 134, 287
ethical studies 13, 94
ethicists 66, 137, 210, 305, 337–338, 341
ethics 6–7, 9, 49–54, 62–68, 70, 74–76, 80–90, 93–103, 115–118, 307, 317–318, 327–330, 335–336, 338–339, 341–342
  pathocentric 131, 240
ethics last models 308
European Commission 47, 80, 110, 177–178
European Technology Platform NanoMedicine 123
evildoers 222
experimental animals 130, 228, 230, 238, 241
explorative nanophilosophy 303–304, 306, 308, 310, 312, 314, 316, 318, 320, 322, 324, 326
extropian movement 43

Federal Ethics Committee for Nonhuman Biotechnology 130
fibroses 154
flavored substances 151
forest fires 148
forests 17, 246
Frankenstein's monster 210
free oxygen radicals 154
fullerenes 21–22, 135, 148, 151, 156
  antibacterial properties 156
fulminant momentum 327
functional biomolecules 192, 198
functionality 133, 149, 285
futuristic visions 43, 324–325

garbage incineration plant 72
gene encoding stathmin 230
gene expression 196
gene pharming 239
gene regulations 196
gene sequences 196, 204
gene therapy 202
genes 202, 229
  artificial 195
  manipulation 112
genetic diagnosis 99
genetic diagnostics 56
genetic engineering 6, 11, 73, 125, 181, 209, 220, 224, 227, 238–241, 244–245, 247, 252, 293, 331–332

genetic interventions 131, 209, 277, 281–282
genetic modification 86, 126
genetic pool 200
genetic procedures 228
genetic sequences 196
genetic techniques 196
genetic technology 74, 95, 127, 200, 228
genetic transfer 200
genetically modified organisms (GMO) 69
genetically modified plants 233
genomes 86, 112–113, 203, 229
genus 290, 292
German government 177
German law 236
German Parliament 46, 177
GMO *see* genetically modified organisms
graphite 20–22
gray goo 46, 200
green goo 200
Greenpeace 221
grey goo 42, 118
ground water quality 157

handicapped people 69, 71
hard disks 22
HCFCs *see* hydrochlorofluorocarbons
healing 12, 123–124, 128, 263–269, 271–272, 276, 285, 287, 341
health 2, 10, 44–45, 71, 93, 111–112, 115–116, 134–135, 142, 149, 155, 157, 185–186, 197, 199–200
health problems 148, 277
health standards 135
healthcare 1, 121, 252
heritable code 219
hermeneutic analysis 219
hermeneutic diagnosis 213
hermeneutic problems 128
hermeneutic questions 191, 216, 247, 251
hermeneutical processes 54
hermeneutics 7, 11, 96, 190, 213, 227, 245, 304, 320, 323
heuristics 205, 320
heuristics of fears 161, 164, 205–206
Holstein Friesian 237
*Homo faber* 36
*Homo homini lupus* 257
*Homo sapiens* 280
hubris 37, 201, 212
human–animal relationship 132, 244–245
human breeding 257
human dignity 123, 243, 286
  radical 109, 265
  technical 261, 265
human essentialism 273
human genome projects 91
human hubris 127, 204, 210, 212
human–machine interface 119, 252
human mental disorders 231
human rights 46, 63, 70, 111, 129, 134, 163–164, 222, 275, 294, 297
humanity, subsuming 290
Huntington's disease 232
hydrochlorofluorocarbons (HCFCs) 153, 156, 181
hyperthermia 120

ice crystal 17
ICT *see* information and communication technology
ICT systems 128
illnesses 69, 120–121, 125, 231, 237–238, 255, 263, 270, 285, 287, 320
implants 119–121, 123, 145, 230, 258–259, 288
  cochlear 289
  neural 44

neuroelectric 33, 288
retinal 121, 270, 289
infectious agent 202
information ethics 6, 99, 101–102, 126, 340–341
information technology 12, 25, 38, 43, 251, 253, 259
information and communication technology (ICT) 123
innovations 3, 23, 60–62, 70, 72, 91, 100, 105, 107, 121, 140, 207, 214, 219, 335
clinical 116
technical 60, 70, 72, 101
technological 42, 60, 167
instrumentalization 282, 286, 290, 315
human 70
integrated circuit 199
intensive animal farming 242
interdisciplinarity 7, 75, 80, 91, 103, 224, 253, 337
interface
brain–computer 231, 288
brain–machine 27
broad-bandwidth 255
computer 289
intermeshing 126
interventions
human 199, 217, 238
technical 73, 123, 209, 215, 218
technological 145
island of coherence 59, 66–67

jurisprudence 147
justice 63, 102, 119, 137, 139, 144, 221, 223, 273, 341

lab on a chip 125
lab-on-a-chip technology 26
labeling 175, 184
labeling requirement 173–175, 189, 319, 330
labor market 274
laissez-faire 273
laws 16, 32, 63, 66, 76, 81–82, 117–118, 158, 176, 178, 188
classical physical 29
environmental protection 57
laws of nature 306–307
legislation 75, 184, 223, 238
life cycle approach 231
life cycle assessment 142
life expectancy 273
human 278, 294
lightweight design, wear-resistance 23
limb reconstruction 259
limbs 260, 265
natural human 271
liquid film 20
lithography 20–21
living beings
artificial 217
human 243
natural 216
synthetic 220
living machines 208
living objects 203
modified 201, 203
living systems, natural 214
lotus effect 23, 150
lung damage 155

machine language 11, 213–215
machines
animated 246
anthropomorphic 246
complex 292
cybernetic 292
molecular 193
mad cow phenomenon 83
magnetic fields 120, 169
man-made radioactivity 148
man-made situations 87
mass destruction 112, 118
mass media 41, 44, 77, 251
mastitis 237

matter
  inanimate 112
  self-organizing 28
MCDA *see* multi-criteria decision-making approach
means–end schema 53
meat 241–242
  in vitro 245–246
media 41, 78, 192, 286, 299, 330
media coverage 42, 210
medical care 26, 138
medical ethics 53, 96, 99, 101–102, 120, 123, 264, 269, 272, 287, 340–341
medications 27, 230, 258
medicine
  human 185, 236
  intensive 56
  reproductive 56, 95
  veterinary 236, 247
medicine ethics 100
membranes 213, 215
  biological cell 21
memory 230
  long-term 285
  lost 285
  reduced 230
mere possibility argument (MPA) 79, 306–307, 323
mesothelioma 45, 153
metaknowledge 337
metaphysics 117, 243
mice
  transgenic 230, 241
  zombie 241
microimplants 121
micromachines 196
micrometer sphere 28
microprocess engineering 19
microscope
  atomic force 20
  scanning tunneling 19
microscopy
  atomic force 20
  scanning electron 117
  scanning probe 37
  scanning tunnel 16
microsystems 254
  technology 1, 16
microtechnology 35
microtubules 230
mitochondria 126, 192
MNT *see* molecular nanotechnology
models
  labeling 174
  mouse 230
  rat 231
modified cells 202
molecular nanotechnology (MNT) 46
MPA *see* mere possibility argument
multi-criteria decision-making approach (MCDA) 160
Münchhausen trilemma 64
muscle mass 245
mutagenic changes 154

nano anthropology 322
nano developments 176
nano epistemology 321
nano futures 4, 117, 316–317, 324–325
nano hype 41
nano-objects 178–179
nano-particles 23, 152
nano scientists 334
nano speculations 310
nano-technology 327, 329
Nano2Bio 192
Nano2Life project 337
nanobiology 216
nanobionics 26, 214–217
nanobiotechnology 2–3, 6, 8, 11, 26–27, 73, 111–112, 126–127, 145, 191–197, 205, 213–216, 220, 222–225, 322
  developments 213
nanobiotics 215
nanobots 42–43
nanocapsules 21, 148, 151
nanocommunity 33, 35
nanoelectronic modules 22

nanoelectronics nanoscale pattern 20, 25, 124, 229, 340
nanoengraving 20
nanoethics 3–4, 6–10, 89–91, 93–104, 107–109, 114, 118, 204, 303–310, 318, 321, 327–329, 335, 337, 339–342
  terminology 96, 103–104
nanoethics community 118
nanofabrication 111
nanofood 113, 150, 174
nanogrooves 22
nanoinformatics 121, 259
nanomachines 17, 93, 150, 192, 205, 322
nanomanufacturing 197
nanomaterials 2, 21, 44, 78, 85, 92, 109–111, 113, 135, 141, 147, 151, 171, 173–179, 181
  classification 110
  synthetic 23, 44
nanomedicine 2, 26–27, 78, 116, 120–123, 138, 261, 340–341
nanometer 19, 21–22, 28
nanopanopticism 124
nanoparticle contamination 45
nanoparticle dosage systems 27
nanoparticle dust 155
nanoparticle film 150
nanoparticle risks 157
nanoparticles
  amorphous 21
  artificial 87, 137
  bioactive 26
  carbon 27, 155
  carcinogenic properties 149
  catalyst 24
  engineered 110, 177
  harmful effects 82–83
  man-made 148
  metal oxides 152
  natural 148, 154, 156
  nomenclature 110, 185
  silicon dioxide 151
  silver 27
  silver oxide 220
  synthetic 10, 44–45, 113, 134, 141, 144, 147–152, 154–156, 158, 160, 162, 170–176, 178–184, 186–188, 204–205
  $TiO_2$ *see* titanium dioxide
  titanium dioxide ($TiO_2$) 21, 23, 27, 112, 151–152, 155, 220
  toxicity 2, 136
nanoreactor 21
nanorobots 42–43, 109, 133, 216, 262, 292
  self-duplicating 292
  self-replicating 42, 200, 336
nanoscale additives 25
nanoscale objects 8, 205
nanoscale particles 148
nanoscale powder 132
nanoscale salt crystals 148
nanoscale structures 193
nanoscaliness 28
nanosciences 4, 7, 9, 23, 34, 80, 92, 102, 178, 305, 309, 321, 340
  definition 114
nanoscience and nanotechnology 80–81
nanoscientists 37, 78, 331
nanosilver ions 113
nanostructured absorber layers 24
nanostructured photovoltaic devices 24
nanostructures 20, 28, 120
  biological 214–215
  functional 215
nanotechnological developments 6, 337
nanotechnological methods 27, 322
nanotechnologists 40, 78, 150, 171
nanotechnology (NT) 1–13, 15–47, 67–70, 78–81, 89–98, 100–105, 107–111, 113–126, 132–145, 176–178, 251–254, 306–313, 315–319, 321–325, 327–342
  application of 15, 119, 122, 134
  applications 4, 120, 145, 333
  benefits 114, 118, 141
  emerging technologies 8, 35, 47, 111, 254, 328, 339

ethics 5–6, 8, 13, 49, 78, 89, 91, 93–95, 107, 117, 339
explorative philosophy 12, 317, 319, 321, 323, 325
guiding motif 16, 332–333
history of 15, 17, 116, 205
implications of 97–98
its nature 117
metaphysical foundations 116
research 140
scientists 32
nanotechnology-based medicine 138
nanotechnology hermeneutics 323
nanotechnology products 45, 216
nanotechnology R&D 28
nanotechnology transport 40
nanotoxicology 329
nanotransporter 21
nanotubes 21–22, 135, 193
nanovisionaries 39
nanowires 21–22
National Nanotechnology Initiative (NNI) 17, 43, 111, 114, 191, 323
National Science Foundation (NSF) 252
natural artificiality 279
natural radiation 87
natural radioactive radiation 148
natural radioactivity 87
naturalistic fallacy 65, 279
naturalization 292
naturalness 12, 275, 278–280, 284
nature at work 213
Nazi Germany 257
Nazi regime 257
NBIC convergence 38, 244, 252
NBIC technologies 298
NBIC tools 252
needs science 52
neuro-morphic engineering 252
neurobionics 121
neuroelectric brain–computer interfaces 284
neuroelectric interfaces 251, 260, 263, 284–288
neuroenhancement 258–259
neuroethics 289
neuroimplants 145, 198, 259, 286, 288–289, 299
nanoelectronic 121
neuronal signals 260
neurophysiology 123, 228, 259, 293, 322
neuroprostheses 296
neurosciences 12, 128, 341
neurotechnologies 289
new arms race 133
NF I *see* normative frameworks 61
NF II 61
NGOs *see* nongovernmental organizations
N&N *see* nanoscience and nanotechnology
N&N research 81
NNI *see* National Nanotechnology Initiative
non-zero weight 206
nongovernmental organizations (NGOs) 111–112, 132
noninstrumentalization 123
normalizations 183, 330
normative choices 167
normative commitments 168
normative components 264, 287
normative conditions 59, 236–237, 268, 338
normative consequences 283
normative content 50, 324–325
normative context 79, 263, 269, 272, 276, 287, 296
normative conventions 272
normative criteria 87, 233, 263–264
normative evaluative criterion 264
normative factors 189
normative footing 7
normative frameworks 55–63, 66–67, 85, 104–105, 108, 127, 204, 207, 220, 237, 239, 241, 243, 264, 287–288
normative fundamentals 68
normative questions 162, 190

normative reflection 147, 167, 171, 188, 190
normative uncertainties 7, 9–10, 12, 32, 52–53, 70–71, 74–79, 94–95, 99–104, 144–145, 267–268, 272–273, 286–288, 295–296, 341–342
normativity 58–59, 75
NSF *see* National Science Foundation
NT *see* nanotechnology
nuclear fission 6
nuclear power 331
nucleosomes 213
numismatics 31

objectifiability 314
objects
  natural 79, 297
  nonhuman 40
  nonliving 207
obligations, moral 244
olfactory systems 230
opportunity costs 168, 308, 310–311
optical circuits 25
optical zoom 259
organisms
  artificial 113, 197
  dangerous 197
  natural 269
  synthetic 113, 198
organs
  destroyed 287
  natural 270
  porcine 230
oxidative stress 154
ozone holes 70, 153
ozone layer 74, 141, 153, 174

Pandora's box 47
panic attacks 231
Parkinson's disease 232
particles
  carbon 148
  dust 149
  nanosized 110, 177
  titanium dioxide 135, 148, 152
  ultra-small 44
  ultrafine 148–149, 177
pathogens 200, 202
  virulent 113
patient care 116
performance
  enhanced 277
  human 252, 266–267
  visual 270
pharmaceuticals 81, 86, 158, 178, 209, 257, 268
  performance-enhancing 274
pharmaceutics 122
philosophical epistemology 321
philosophical ethics 5, 103, 275, 281, 298
philosophy
  moral 320
  natural 132
  political 50, 129, 211
  technical 341
  technological 96
  theoretical 100
photonic crystals 25
photosynthesis 193
photovoltaics 127, 198
physics
  classical 16
  mesoscopic 18
  quantum 16
  semiconductor 25
  solid-state 18, 33
physiological processes 291
plants
  fabrication 157
  incineration 153, 174
  nuclear power 157
pluralism 50, 58
  moral 50
pluralistic society 94
policy
  government risk 175

privacy protection 126
social 300
polis 80
positions
biocentristic 208
bioethical 207
deontological 163–164, 170
pathocentric 209
transhumanistic 272
post-normal science 182
post-traumatic stress syndrome 231
power plants 157
pragmatic coherence 57
precautionary principle 40, 78, 87, 92, 111, 114, 117, 119, 127, 164–168, 170–173, 175–176, 178–181, 183–185, 205–207
productivity 301
protective layer 22
proteins 11, 126, 192, 215, 322
artificial 194
proton pump 193
prudent avoidance principle 168
pseudoproblems 296
psychic changes 278
public debate 47, 52, 66, 76–78, 279
public involvement 103
pulmonary penetration 135, 153
Pyrrhic victory 292

quantum computing 25
quantum dots 24–25
quantum mechanical effect 19, 25, 28–29, 33
quantum mechanics 16
quasi-ontological predicate 79
quasidisciplinary institutionalization 97

radiation
artificial 87
electromagnetic 266
radioactive 87
ultraviolet 21
radicalization 291
radioactive markers 154
radionuclides 22
rat neurons 247
reductionism 16, 18, 219, 254, 291
reductionist approach 213
reductionist–materialistic view 219
reductionist understanding 238
reflective research 92
reflective science 3, 88, 309
regulations
animal welfare 241
legal 55, 333
nanotechnology 46
nanoparticle 45
reinsurance companies 44
risks, health 118, 137, 154, 305, 333
robot arms 231

scanning probe techniques 16, 20
scanning tunnel microscopy (STM) 16
science, technology, and society (STS) 32, 96, 119, 311, 318, 328
scientific view, reductionistic 216
self-cleaning surfaces hydrophobic 23
self-cleaning surfaces, lipophobic properties 23
smart treatments 121
stem cell research 56, 76, 95, 99, 333
stem cells 69, 320
STM *see* scanning tunnel microscopy 16
storage 126, 135, 202
storage capacity 132, 266
storage function 259
storage layer 22

storage technology 124
strategies, wait-and-see 45, 82, 166, 172, 188, 309
STS *see* science, technology, and society
STS studies 35, 331
studies
  empirical 155
  epidemiological 155
  nanoethical 305
subdisciplines 5, 33, 35, 96–99, 102, 104, 342
sun creams 4, 27, 134–135, 152, 155
sunscreen creams 152
sunscreen lotions 21, 23, 148
sunscreens 112, 147
superhuman powers 256
superhumans 267, 276
  enhanced 274
superplasticity 23
surface electrodes 285
surface water 141
surveillance 124–125, 133
sustainability assessments 142
sustainable development 1, 40, 72–73, 135, 140–143, 252, 257, 327
sweat-repelling mattresses 138
Swiss Federal Ethics Committee on the Non-Human Field 227
Synth-Ethics project 210
synthetic biologists 113, 194, 210, 212, 217–218
synthetic biology 3, 6, 8, 11, 18, 26, 73, 112–113, 126–127, 137, 145, 184, 191–205, 207–213, 217–225
synthetic chemicals 149, 183, 204
synthetic life 113
systems
  hybrid 198
  self-organizing 199

TAB *see* Technology Assessment at the German Bundestag
technical artifacts 290, 293
technical enhancements
  animals 232
  consequences 293
  enhancement society 300
  ethical aspects 12, 131, 272
  humans 11, 101, 145, 251, 271
  meaning 232, 234,
  neuroenhancement 258
  occupational groups 277, 281
  specific concerns 278, 283, 294
  informed consent 282
  responsible actions 295, 296, 298
technicalization 71, 213–214, 216, 249, 270, 284, 290, 292
technological Baconism 39
technological determinism 140
technology
  computer 232
  environmental 127, 198
  gene 207
  light-based 25
  military 76, 133, 144
  nanoscale 118
  neuro-prosthetic 285
  nuclear 44, 181
  nuclear power 46
  physical 34
  propulsion 1
  prosthetic 285
  software 218
  surveillance 101, 125
Technology Assessment at the German Bundestag (TAB) 109
technomorphic 215
technomorphic modeling 216
technomorphic terms 213, 245
technophobes 210, 260
technosciences 73, 244
teleology 161
terrorist groups 199
theologians 210–211
thermal shoes 138
threats
  apocalyptic 94
  biological 255

global 181
tissues 230, 242–243, 245
artificial 245
human 154
tumor 121
title 15, 17, 80, 111, 210, 252
top-down approaches 34, 49
top-down reasoning 59
totalitarian regimes 257–258
totalitarian traits 258
Toxic Substance Control Act (TSCA) 82, 158, 176
toxicity 110, 154–155, 177, 333
toxicological research 238
toxicological test procedure 154
toxicologists 92, 110
toxicology 84, 136, 147, 158, 185
transdisciplinarity 223, 253
transgressions 59, 63
transhumanism 43, 252, 291, 295
transhumanists 38, 44, 254, 292, 330
transistors 22, 196
transplantation 243
transsubjectivity 63–64, 315
traumatic injuries 120, 231
treated surfaces 4, 149
troposphere 153
TSCA *see* Toxic Substances Control Act
TSCA Chemical Substance Inventory 176
TSCA regulations 176
tumors 120, 154
tunneling flow 20

ultraviolet light 152
UN declaration of human rights 163
UNCED *see* United Nations Conference on Environment and Development
uncertainties
epistemological 74, 83
moral 338
scientific 167–168, 176
situation of 86, 167, 179
unclear risk 57, 81–88, 136, 144, 157–165, 167–171, 173, 182–184, 188, 203, 206, 220, 224
consequentialist approach 159–160, 163
harmful effects 82, 84
philosophical approaches 159
situation of 82–83, 160–161, 168
UNESCO 3, 115–116
unified science 34
United Nations Conference on Environment and Development (UNCED) 73, 166
unity of science 253
universalism 52
US National Nanotechnology Initiative 91
utilitarian perspectives 235, 237
utilitarian principle 160
utilitarian version 163
utopia, scientific 12
utopias
positive 41–42
technical 258

value judgments 87, 264, 315
value neutrality 68, 204
Viagra 268
viral vector 202
virus 113, 202–203, 220
influenza 203
modified 200
mousepox 203
polio 203
virus imitations 194
viruses 19, 21, 127, 194, 202, 205, 208, 216
modified 133
reprogrammed 202
vision assessment 308, 325–326
visions
apocalyptic 94
dystopian 294

visual acuity 276
visual nerve 259
visual prosthesis 259
volcanic eruptions 148

walls 149, 261
  multiple 21
  single 21
war 119, 133, 256, 265
waste 42, 72, 110, 140, 177, 333
  hazardous 72
  problematic 193
  radioactive 57, 72, 174, 332
water pollution 70, 246
water technologies 137, 141
weapons 112, 118, 132–133
  efficient 133
  improved 132
  nanotechnological 133
  nuclear 69
white noise 188
WHO *see* World Health Organization
world
  biotic 216
  macroscopic 16, 321
  natural 200, 215
  real 96
World Commission on the Ethics of Scientific Knowledge and Technology (COMEST) 115, 225
World Health Organization (WHO) 200

yuck factor 245